The Making of
Technological Man

The Making of Technological Man

The Social Origins
of French Engineering
Education

John Hubbel Weiss

The MIT Press
Cambridge, Massachusetts
London, England

Publication of this volume has been
supported by a grant from the Hull Fund
of Cornell University

This book was set in VIP Trump
by Graphic Composition, Inc.
and printed and bound by
The Murray Printing Company
in the United States of America.

Library of Congress Cataloging in
Publication Data

Weiss, John Hubbel.
    The making of technological man.

    Bibliography: p.
    Includes index.
    1. Engineering—Study and teaching—
France—History.   2. Paris.   École
centrale des arts et manufactures—
History.   I. Title.
T121.W44   620'.0071144
81–17223
ISBN 0–262–23112–3   AACR2

For Heather

Contents

List of Tables

Foreword

There was a time when historians concerned themselves only with "important" people—kings and generals—and important subjects—reigns and wars. That was a long time ago, and history has since learned to concern itself with anything and everything touching the condition and fate of our species and the world in which we live. One result of this diversification of interest has been the fragmentation of history into many subdisciplines: political, economic, intellectual, social, psychological (psychohistory); and with this fragmentation it has become ever harder to write general history—history *tout court*.

Of these subdisciplines, the most amorphous and unmanageable has been social history, which was long defined negatively as the residuum of everything not included in the other branches. What, after all, does not touch on society, its membership, institutions, ways, and values? Much social history, therefore, has been a grab bag of anecdote and detail, sometimes illustrative, sometimes amusing, sometimes instructive.

The modern social historian—I mean by that, scholars since World War II—understandably recoils from this kind of refuse collection. He has been much influenced by the more theoretical concerns of ancillary disciplines: sociology in the first place, but also demography, anthropology, and economics among others; and much of his effort has been to order this heterogeneous, uneven, adventitious array of relevant evidence so as to infer regularities, test hypotheses, establish links between social institutions and attitudes and other aspects of historical change. In the course of this effort, some branches of history have so matured as to constitute separate subdisciplines of their own: historical demography is the best example, with its ingenious methods (such techniques as family reconstruction), often abundant numerical data, and extensive recourse to statistical analysis. The signs of demarcation and autonomy are unmistakable: separate courses, specialized periodicals, new professional associations and research centers, even perhaps a separate degree program or (and this comes last) teaching department.

The history of the professions has not come this far, but it has come a long way. A growing stream of articles and books has proclaimed the importance of the subject; the Davis Center of Princeton University recently devoted two years to the topic; courses are beginning to find their way into university catalogs; and I cannot count the conference sessions that have met to hear and discuss papers in this field.

Clearly we have the beginnings here of a new subdiscipline, a new window on the story of social change. Its significance, hence its intellectual interest and appeal, lies in its ability to integrate a wide variety of aspects of life, culture, and work around the theme of occupational role. In this respect, the history of the professions is related to, indeed a subbranch of, the larger history of work and workers.

The importance of such a theme is obvious. For one thing, the story of work and work roles touches the lives of all but the very young and the very old; indeed, throughout history the vast majority of people have spent the vast majority of their waking hours at work. For another, the theme of work roles ramifies in all directions and, in so doing, illuminates just about all major aspects of historical change.

It was Adam Smith who enunciated the famous dictum, "That the division of labour is limited by the extent of the market." He drew his examples from manual occupations, but the rule applies to the so-called professions as well. He did not say anything in this connection about technology as a *source* of role differentiation (though he did discuss technological change as a *consequence*), for he was strangely oblivious of, or indifferent to, the Industrial Revolution that was gathering around him. Yet changing techniques do almost invariably generate new work roles, especially in conjunction with expanding markets and growing volume of production.

The Industrial Revolution combined both forces and redefined as never before both the content of work and division of labor. It generated an endless series of new specializations, among them that multifarious one that we know as engineering, which proliferated in the course of time into its own family of subspecialties: mechanical, chemical, electrical, civil, aeronautical, naval, and so on. Here was a new profession that embodied as much as anything the marriage of science and technique that marked the new industrial age. The engineers were the custodians of the knowledge—more and more esoteric and closed to noninitiates—that was the basis of the new productivity. As such they were the priests and doctors (representatives of much older professions) of the new age.

Whoever says profession says credentials: It is not for just anyone to be a priest or doctor or engineer. When the content of the relevant knowledge and functions is still loosely defined, entry is fairly easy. But the elaboration of that special knowledge that constitutes the *raison d'être* of any profession militates against usurpation, and besides, the accredited professionals work hard to limit competition to those chosen few who meet what they define as acceptable standards. Their income and status are both at stake; there are no stronger motives. Hence we are not surprised to find that in all countries caught up in the Industrial Revolution, there soon appears a corps of specialists denominated as engineers, charged with mediating between an increasingly complex body of scientific knowledge and its applications in industry and transport; that these engineers form themselves into ranks in professional societies that bestow on their members identifying credentials—those sacred initials after the name that announce to the world one's recognized competence; and that they work toward gaining from the state exclusive rights to practice.

The underlying tendency is the same, but the modalities vary from one jurisdiction to another. In England, the first industrial nation, the profession of engineering grew like enterprise itself—by private initiative and small increments. There were no formal engineering schools, no sense at first that there even was such a profession. Indeed the very use of the word in Britain reflects other concerns: It was applied at first (and is still so applied in certain contexts) to those skilled workers who built and operated steam engines. Only with time and practice did the word also acquire its "professional" connotations. As Musson and Robinson have shown us, would-be engineers, that is, people who wanted to do those things that we define as engineers' work, did have access to instruction in applied science, through both occasional university courses and private, often itinerant, lecture series. But above all, they learned from their predecessors, on the job, at the bench, by word of mouth and example. There was, to be sure, a growing body of technical literature for those who could and would read it; but experience was the principal avenue of learning and training, even for the builders of roads and canals.

For all later industrializers, the process was necessarily different. For one thing, they were in a hurry to copy Britain and catch up with her new productivity, which constituted a political as well as economic challenge. For another, the process of industrial revolution had so changed technique and the content of the knowledge required to employ it that the older ways of apprenticeship were no longer effective. How could an older specialist, however gifted, teach something he did not know himself? So the European nations and the United States had to rely far more than Britain on formal training, offered in courses or schools established for the purpose. The United States came closest to the British pattern, partly because it could draw so well on British immigrants to teach the new technology; though in one area, that of civil engineering (construction of roads and bridges), it depended largely on the education provided by the military academy at West Point.

On the European continent, it was France that provided the model. There was already a strong tradition of formal training under the Old Regime: A strong, centralizing monarchy needed trained technicians to design and build roads and canals and to exploit that underground wealth that by French law was the property of the state. (The state could concede mineral rights to private operators, but even then it retained an interest in the mode and effectiveness of exploitation.) So it was that Ponts-et-Chaussées was established in 1747; Mines in 1783.

This tradition was only reinforced by the Revolution, which, as Tocqueville pointed out, was in a number of critical areas the heir and intensifier of older trends. The Revolution saw a mobilization of talent, among other things, for the training of cadres in a national campaign of defense and aggrandizement. The keystone of this educational effort was the Ecole Polytechnique, founded in 1794 to train officers for the French army, but much more than a mere military academy. The faculty included the finest mathematicians and scientists in the country, men of international renown and enduring fame, and the curriculum from the beginning went far beyond imparting military techniques. Rather, the emphasis was on basic math and science, so that the graduates could in principle turn to any application to which career and necessity called them.

In the century that followed, the Polytechnique became a model for engineering schools through Europe. In France it remained the school of preference for the most brilliant young men the society produced. (The only exceptions were those very few who wanted to pursue pure science; these went to Normale.) It provided hundreds of gifted, well-educated men for government service, not only in the army, but in all branches touching industry, mining, transport, and communication. And it provided increasing numbers to private enterprise; for the graduates of Polytechnique were the best of each year's cohort and much sought after by a sector ready and able to pay much higher salaries than the state.

Yet Polytechnique could not do everything. The training it provided fitted its students more for the older forms of engineering (construction and mining) than for such newer branches as chemical manufacture. And the heavy emphasis on theory in combination with generally recognized presumptions of social status unfitted the graduates for those activities long associated with "unprofessional" practitioners—much of what came to be known as mechanical engineering, for example. There was a demand, then, for trained personnel that older institutions were not able to satisfy, and into this gap there moved a number of private schools, some local, some confined to a single branch of production. The most important of these, in scope of curriculum, quality of training, role of graduates, and influence on other edu-

cational institutions, was the Ecole Centrale des Arts et Manufactures, founded in Paris in 1829.

In view of the significance of Centrale, it is astonishing that it
did not long ago attract the attention of historians. John Weiss's
study is the first major effort to study the character of the school,
the content and nature of its curriculum, its place in the larger
system of French technical education, its role in the formation
and definition of an engineering profession, its contribution to
the competition of humanistic and scientific cultures. His book
is a case study of social history as it should be done: open-ended,
intensely curious, alert to relations and interactions, sensitive to
the illuminating detail while aware of the large picture. It exem-
plifies, in short, the maturation of this new branch of knowledge.
As such, the book will be of interest to many different kinds of
historians, but also to sociologists, educationists, and, not least,
engineers, and will take its place from the start as one of the
major elements in the large and growing corpus of literature on
this subject—required reading for both teachers and students.

David S. Landes

Acknowledgments

The PhD dissertation upon which this book is based was written under the supervision of David S. Landes. I was thus a fortunate apprentice: his learning, his advice, and his constant, patient support were as generously given as they were indispensable.

Institutions also sponsored the research: the Harvard Graduate Prize Fellowship Fund, which allowed me to conduct a year of research in France; the James Fund of Harvard University, which supported my conversion of part of the results of that research to machine-readable form; and the Western Societies Program of Cornell University, which supported the purchase of microfilm and xerographed material in France. Finally, I owe a special debt to the Harvard Program on Technology and Society and its director, Emmanuel G. Mesthene, who gave me the chance to cast my net broadly in an attempt to understand the relation between technology and social history.

During and after my research in France I had the pleasure of discussing my work with a group of most hospitable and helpful French scholars: Louis Bergeron, Roger Chartier, Paul Gerbod, the late Roger Lévy, Georges Ribeill, André Thépot, Léon Velluz, and Olivier Zunz. I was also given assistance by three exceptional archivists: Mme Pierre Gauja, director of the Archives of the Académie des Sciences; Jean Lasson, who at the time of my first visit had just volunteered for the task of putting in order the documents retained by the Ecole Centrale des Arts et Manufactures; and Nicole Magnoux, currently head of the documentation service at the Ecole Centrale.

At various times over the past several years, American and Canadian scholars have also given me help in my efforts to tie together threads from the history of education, science, technology, and social structure. Everett Mendelsohn, Craig Zwerling, Barbara Buck, and Peter Buck shared with me their knowledge of the history of science and at the same time allowed me to benefit from the special imaginativeness that seems to be part of the endowment of that breed of historian. The detailed criticisms of Peter Buck and Cecil O. Smith, Jr., readers of the entire manuscript, helped me avoid many errors and misplacements of emphasis. I also profited considerably from Rod Day's knowledge of French trade schools, Harry Paul's insights concerning nineteenth-century science, Fritz Ringer's understanding of European educational systems, Margot Stein's experience with French occupational hierarchies, and Edwin Layton's expertise in the history of French and American engineering. I would like to give special thanks to Gordon M. Jones, a historian whose field of interest lies far from my own but whose generosity of spirit prompted him to take time from a vacation to assist me in Parisian archives. Since 1978 Donna Evleth has been my principal assistant on other research projects, but she also deserves thanks for help in the research for the final revisions of this book. Finally, the patient work of Bill and Kay Gilcher prevailed against a host of typographical errors infesting the final manuscript.

One's indebtedness to his family is usually too general to seem directly relevant to a particular study. Such, however, was not my experience. In the final stages of the research I received indispensable help from Gerard A. Weiss, who checked sources in Paris at a time when I could not get there, Emily H. Weiss, who dug a crucial document out of the New York Public Library on very short notice, and Emily W. McClung, whose gift of time was used for many purposes.

The muse of history is a jealous mistress. My wife has not only tolerated that long rivalry but given up much to promote the liaison. The affair continues, but it also brings fruit. It is thus as an expression of the promise of things to come, as well as in remembrance of things past, that this book is dedicated to Heather.

The Making of
Technological Man

1

The Origins

Engineers and Society

As Blake wished to see a world in a grain of sand, so every student of history hopes that his own work will contribute to an understanding of some of the larger questions about human experience over time. To this end the present study hopes to illuminate aspects of the interaction between technological changes and changes in social structure during the course of industrialization. Because engineers' training and professional activities bring them into close and frequent contact with the most powerful types of industrial technology, the investigation of their education seems a logical way to begin to examine this broader problem.

This study deals with engineers' education as it took place in schools, both technical and nontechnical. One might object that engineers were given neither their individual social identities nor their collective status as a profession solely by their schooling: family loyalties, hereditary status, political groupings, bureaucratic ties, and, of course, personal achievements all performed similar functions. Yet education has the special advantage of providing clearly perceivable links between the history of engineers and the history of the larger social structure of which they form a part. By its very nature as an institutionalized collective introduction, education tends to make explicit—and to express in written documents—its claim to bestow upon the student his identity as a generally educated man or as a member of a particular profession. At the same time, education both unifies and stratifies, binding together those who attend the same schools while separating them from those who do not.

As they impart a particular body of information, then, schools carry out two other processes: molding and labeling. Their official pronouncements and their daily institutional routines are designed to make changes in the outlook and behavior of their students that endure long after graduation. This is the sense in which the French refer to an education as a *formation*, a moral and intellectual molding. I shall offer evidence concerning the nature of the *formation* that schools intended to carry out and shall indicate ways one may infer that such intentions were realized, but only a full collective biography of all the students—if it overcame the formidable barrier posed by engineers' characteristic laconism about their social and political philosophies—could claim to offer a complete composite portrait of the Technological Man who was "made" in French secondary and technical schools in the years before 1848. Investigating the labeling function of schools, on the other hand, is a more circumscribed task because it deals with identity not as a matter of profound personal psychology, into which a *formation* may or may not have intruded, but as a question of perceptions expressed in a public discourse of more limited vocabulary and more widespread conventions. Occupational titles, diplomas, and professional ranks may reveal little about the inner moral or technical capacities of the individuals so labeled, but they are nevertheless the stuff of which national systems of social stratification are built. In the case of French engineers, moreover, the struggle over the design and content of these labels, their connotative values as well as their denotations, became part of a more general conflict between differing visions of what constituted a legitimate high culture, a debate especially intense during the first half of the century.

The particular species of engineer that receives most attention here is the *ingénieur civil*. In France this term came to designate *all* engineers not employed by the State, whatever their specialty, whereas nonmilitary engineers who worked for the State carried special titles such as *ingénieur des ponts et chaussées* (Engineer of the Bridges and Road Corps) or *ingénieur des mines* (Engineer of the Mining Corps). The term *ingénieur civil* was not in use before 1815, and men performing functions that might have qualified them for the title were all but nonexistent before 1800.[1] The study of the French "civil engineer" would thus seem to present the student of social stratification with the relatively rare opportunity to investigate the emergence of a new professional and social type whose appearance accompanied the onset of the French Industrial Revolution.

This study argues, however, that the *ingénieur civil* cannot be seen merely as product and agent of nineteenth-century industrialization. Nor can his place in French society fully be assessed, in the manner of a contemporary American stratification survey, by assigning him a place on a unilinear prestige scale according to some "combined measure of income and education," at a location peopled by other managers or technical executives, "high-level manpower for modernization."[2] Even today, when nearly all engineers in all countries receive their professional training in universities or technological institutes, this broad similarity in education has not produced a uniform social type. The title *ingénieur* still carries in France connotations noticeably different from those of its counterparts in other countries.[3] Like other professions, the *ingénieur civil*, considered as a category in a system of social stratification—a complex lattice of class, status-honor, and power—can only be understood as the product of *particular* stratifying processes, cultural traditions, and ideological conflicts.[4] By the same token, as Fritz Ringer has argued, schools cannot be seen as passive conduits that merely insert information into sons from selected social strata and then certify that this curriculum has been properly assimilated. Especially in the early phases of industrialization, schools were not merely "adjuncts of more decisive [economic] processes or institutions."[5] Their educational policies were based upon a wide range of considerations, only some of which stemmed, in a not very precise way, from what their guiding spirits saw as the imperatives of industrialization.

This study gives most attention to a single Parisian school, the Ecole Centrale des Arts et Manufactures, during the first twenty years of its existence. Founded in 1829, when Paris was "the engineering capital of the world," and its scientific center as well, the Ecole Centrale soon became the largest single source of *ingénieurs civils*, and it remains so today.[6] Graduates of Centrale, the *Centraux*, filled most of the important positions in the Société des Ingénieurs Civils, the profession's major association. As this is written, in fact, a descendant of a prominent member of the school's first graduating class, Baron Jules Pétiet, is president of the Société. The school's impact, moreover, extended well beyond the borders of France. Not only did *Centraux* direct major projects in Europe, Africa, and Latin America; the example of the Ecole Centrale influenced the structure and content of engineering education at a number of other institutions such as the Ecole des Arts, Manufactures, et Mines in Belgium and the Massachusetts Institute of Technology.[7]

To understand how the Ecole Centrale during its first generation carried out its training, its molding, and its labeling of *ingénieurs civils* and captains of industry, this study places the school at the intersecting point of five perspectives:

1. The present chapter discusses aspects of the history of Napoleonic and Restoration France that influenced the conception and birth of the school: economic changes, institutional models, and interpersonal networks.

2. When students arrived at Centrale, they already carried labels acquired during their preparatory years in institutions of secondary education (*enseignement secondaire*). Their status as products of *enseignement secondaire*, with its unique ideology, curriculum, and status connotations, became one of the crucial and problematic elements in their social identity. Because of *secondaire*'s dominant position, even those engineers who were not its products were likely to define themselves in relation to it. In fact, engineers were more likely than most other groups to be affected by this heritage since important aspects of the conflict within it were results of attempts to create educational institutions congenial to the fostering of a technological culture. The social history of *enseignement secondaire* and the cultural stalemate within that system thus form the central focus of chapter 2.

3. Entering students were molded and labeled not only by secondary education but also by their social backgrounds. Discovering what social groups contributed what proportions of students—an exercise the French call the analysis of "social recruitment"—thus offers one of the best ways to understand how processes of stratification shaped the engineering profession. Recently published studies of the clientele of other engineering schools, moreover, now offer to this analysis the precision and nuance lent by a comparative perspective. Interpreting the Ecole Centrale's pattern of social recruitment thus will be the task of chapter 3.

4. The study next turns to a discussion of engineering education at the school. My principal concern in chapters 4–6 is with the labeling process that accompanied instruction, especially the school's development of the concept of "industrial science" (*la science industrielle*) and its construction of a portrait of the engineer who learned such a science as a man who was simultaneously scientist, generalist, and expert in finding practical solutions to technological problems. In looking at that same instruction, specialists in economic history would probably ask how well the Ecole Centrale trained its engineers: what it taught them and how they were taught. Chapters 4–6 do investigate the school's textbooks, teaching techniques, and curricular structure. But whether this constituted fruitful technical instruction is only a secondary concern of these chapters. To give a persuasive answer to such a question, after all, may not be possible. Graduates of Centrale, of course, went on to achieve great things. But would they not have achieved just as much if they had been trained in some other school? As will be noted, engineers trained in "lesser" schools or merely by apprenticeship made outstanding contributions to French industrialization, a pattern that only the most recent scholarship has begun to elucidate. Could not Centrale's education have stifled many engineers' talents by penning them up in a school for three years and exhausting their energies on an exaggerated range of minutely prescribed exercises? Did the *Centraux* not learn their most important skills on the job, as is the case with most other occupations? What Randall Collins has called the Technocracy Myth has conditioned the history of education ever since Condorcet; its influence is not entirely absent from these pages. Yet Collins's careful sifting of the evidence suggests that claims that the technical quality and duration of formal schooling constitute the principal determinants of job performance spring not so much from methodologically convincing tests of such propositions as from the needs of educators to justify their activities.[8] In any case, for the purposes of the present study the claim that the quality of instruction was the decisive influence upon the later achievements of *Centraux* need only be awarded the Scottish verdict of Not Proven.

5. Chapter 7 shifts the emphasis from labeling to molding. Like almost any full-time school, the Ecole Centrale encompassed within its institutional boundaries a small society with its own

structure of interpersonal relations, norms, and daily routines. Since these things shaped students' outlooks and disciplined their behavior, they command our attention. Studying this aspect of Centrale's history, moreover, brings out most clearly the changes that occurred during the school's first twenty years. In moving from turbulence and experiment in the early years to a relatively quiet routine that obtained until the Revolution of 1848 broke out (and was reestablished soon after that upheaval), the constructors of the Centrale mold bought precision at the expense of flexibility. In some respects the pattern set in the years before 1848 endures to the present.

The concluding chapter of the study examines the assessment that the Ecole's guiding spirits made of their achievement and of the role of their school in French society. They were not given to probing, self-conscious exercises in social philosophy, despite the contemporary vogue for such philosophizing; they did not see themselves as builders of a systematic ideology. Yet their social vision was no less powerful for being embodied not in newspaper columns or weighty treatises but in their daily activities, decisions, and discussions. Technological Man, the *ingénieur civil* from the Ecole Centrale, sprang from a specific set of legitimations of interests, from an ideology, no less than did other social and professional types. The unmasking approach permitted by the easy wisdom of hindsight, however, should not obscure the distinct probability that in many instances the ideological labels became psychological molds. However confused, incoherent, and altogether unlikely were some of the claims made for the supremely versatile, autonomous, resourceful, technically sublime, work-dominated, nonpolitical Centrale graduate and his comprehensive, all-conquering *science industrielle*, there were undoubtedly some French engineers who believed it all, and tried to act upon that faith.

Backwardness and the Founding of Schools

The Napoleonic period is perhaps not the most useful for an attempt to understand the general patterns of interaction between industrialization, taken as a broadly based process of technological change and economic reorganization, and the development of institutions of technical education. A number of factors do more to obscure the relation than illuminate it. The most important discontinuities, of course, arose from the exigencies of imperialism and war. The loss of communication, colonies, and overseas markets could not be compensated for by the "forced markets" created in areas of the Continent under French domination. "Cut off from the world outside, France also lost contact with technical progress."[9] At the same time, short-term but severe economic crises, such as that of 1810–1811, had largely political origins; the connection with a lack of technically trained men seems rather remote. Ministers such as the chemist Chaptal found that ambitious plans begun in the short intervals of peace were all but forgotten when war came.[10] Despite the flood of decrees supposedly instituting long-term policies, Chaptal often found himself in a situation like that of the early years of the Revolution, when "everything had to be improvised."[11] Although he and men like him did manage to establish some institutions that were of lasting value to French economic growth, such as the Society for the Encouragement of National Industry, these victories were not enough to prevent the country from entering the reign of Louis XVIII much farther behind Britain than it had been before the Revolution.[12] The larger question, then, is the extent to which the so-called imperatives of industrialization, especially French attempts to overcome "backwardness," determined the nature of the configuration of technical schools that appeared in France during the nineteenth century.

Both indirect and direct evidence indicate that the demand for persons with the technical skills necessary to create a modern industrial economy increased substantially during years after the collapse of the Empire. Despite the Bourbon regime's later reputation as a less-than-enthusaistic midwife with French industry, the period was far from stagnant. Whatever the effect of the high tariffs on everything from coal to manufactured products imposed after Waterloo, the government did take certain positive steps; the engineers of the Ponts et Chaussées (Bridges and Roads Administration) added 930 kilometers of canals to the existing network during the fifteen years of Bourbon rule, four and one half times as many as Napoleon. They also made important improvements to such existing waterways as the Saint-Quentin canal and the Oise river.[13] A number of economic sectors in which machine technology and engineering played a substantial and highly visible role witnessed dramatic growth. The number of spindles employed in the cotton industry of Haut-Rhin increased nearly tenfold, from 48,000 to 466,000, in the years from 1812 to 1828.[14] Clapham reports that between the same years "the output of pig iron rather more doubled," as did that of coal.[15] By 1823 twenty French ironmasters were using the English puddling and rolling process, unknown in France before 1817.[16] Other new products and processes also began to appear in numbers large enough to be generally noticed. Although use of the steam engine remained far below the British level, the *rate* of increase in installation, psychologically the most important aspect of an innovation's impact, was fairly rapid during the Restoration years: the number in use rose from 65 in 1820 to 625 in 1830.[17] By 1826 fifty steamboats operated on French waterways, the first having appeared only after Napoleon's fall.[18] Gas lighting was introduced in 1815, and soon the Royal Palace itself was illuminated.[19] Dynamic regions such as the Nord, with its textile boom towns of Roubaix and Tourcoing, and Alsace, where "the power loom was adopted perhaps more quickly than anywhere in Europe, not excluding Lancashire," came to the attention of the entire country through the writings of men like the engineer and politician Baron Charles Dupin.[20] In short, French industrialization had begun in earnest: Tihomir Markovitch concludes in his quantitative history of the French economy that the period of the Restoration marks "the greatest growth in industrial production" of the entire period from 1796 to 1913.[21]

The direct evidence of demand for technological skills tells the same story. The British showed their respect for such demands by banning the emigration of artisans and mechanics, who nevertheless crossed the Channel in large numbers.[22] In 1825, when the British repeal of the ban finally acknowledged its futility, there were at least 2,000 skilled mechanics on the Continent, most of them working from job to job in France or Belgium.[23]

The real problem lies in understanding the supply side of the equation. It may be granted that the nature of an institution of technical education, such as the Ecole Centrale des Arts et Manufactures, affects a country's pattern of social mobility and

stratification through its choice of students, educational ideology, teaching competence, certifying system, and power to
place its graduates. The question is whether that *particular* type
of institution was called for by the imperatives of technological
development.

In retrospect, the Ecole Centrale does seem to have filled a need
for the education of a corps of engineers, entrepreneurs, and managers who performed a vital role in French industrial growth. Not
too surprisingly, the histories of that institution and references to
the school in other literature support this claim unanimously.[24]
The school's graduates played important roles in industrial
growth; therefore without the school that growth would have
been far slower and would have taken a different form. To accept
this argument without investigation, however, is to subscribe to
a kind of institutional determinism that assumes that industrialization requires a hierarchical, carefully differentiated set of institutions in which the curriculum at each level—and, presumably,
the size of the student population—corresponds to a particular
range of skills needed by the economy, from bare literacy—or il-
literacy—for operatives to the most advanced mathematical and
scientific knowledge for the managers and research personnel.[25]

The British needed no such institutional configuration. During
the eighteenth and nineteenth centuries, British engineers, entrepreneurs, managers, and scientists emerged from a striking variety of educational backgrounds. Brian Simon points out that
John Metcalf, James Brindley, Thomas Telford, John Rennie, and
George Stephenson all began their careers as ordinary mechanics
and learned as they went along, "picking up experience and ideas
in the workshop and on the new construction sites of developing
capitalist industry."[26] Sidney Pollard presents an even longer list
of prominent engineers and managers who had the village elementary school or Scottish parish school "as the main, or sole,
basis of their formal education."[27] He points out that many Scottish grammar schools not only taught the Latin necessary for admission to Scottish universities but also "achieved standards in
mathematics and other subjects which fitted their alumni for the
highest technical and commercial posts."[28] Each study of the educational institutions which produced the officers and sergeants
in Britain's industrial army lengthens the honor roll: the village

elementary schools, an assortment of private elementary schools, the famous (and somewhat overrated) Dissenting Academies, the technical academies, institutes for adult education, even proverbial backwaters such as Oxford and Cambridge.[29] Nor does it suffice to list merely schools. At the same time that A. E. Musson and Eric Robinson have all but effaced the previous image of the British innovator as an unlettered, prescientific tinkerer, they have demonstrated the importance of networks of subscription libraries, periodicals, itinerant lecturers, and local "Lit and Phil" societies.[30] French "diffusers of Useful Knowledge" like De Lasteyrie and Dupin would not have been surprised at this.

The experience of Britain thus suggests that the establishment of institutions of higher technical education does not automatically accompany industrial growth. If such were the case, Britain should have set up such schools long before France or Germany, but in fact the closest approximation to an engineering school to appear at this time was to be found in the courses on "engineering and the application of Mechanical Philosophy to the Arts" first given at University College, London, in 1835.[31] On the other hand, the German-speaking countries, though lagging economically behind France as well as Britain, founded schools later designated as technical universities (*Technische Hochschulen*) in Vienna (1816), Karlsruhe (1825), Munich (1827), Dresden (1828), Stuttgart (1829), and Hanover (1831).[32]

The very backwardness of the Continental countries may be the key to the connection between their industrialization and their technical schools. As Alexander Gerschenkron has argued, the relative backwardness of certain countries meant that the speed of their industrial development and the "productive and organizational structures of industry" that emerged from that process necessarily differed from that of advanced countries. The faster growth of a "backward" country cannot be explained, however, solely by the greater efficiency of technology transfer, that is, the possibility of borrowing quickly from the advanced country's existing stock of technological knowledge. Gerschenkron instead gives great importance to the specific social conditions—constraints on the labor force, resource endowment, strength of potential entrepreneurial groups—that existed in each country before the race to industrialize began. Along with imitation of advanced countries came totally new ideological and organizational combinations: "These differences in the speed and character of industrial development were to a considerable extent the result of the application of institutional instruments for which there was little or no counterpart in an established industrial country."[33] In Gerschenkron's discussion the principal examples of such new institutional instruments were the industrial investment banks, "mixed banks," and universal banks that were created on the Continent after the middle of the nineteenth century.[34]

For David Landes the institutions of technical education established on the Continent can be viewed as examples of Gerschenkron's "new institutional instruments." France and Germany created "a veritable hierarchy" of such institutions:

On the highest level, the Ecole Polytechnique (and its graduate affiliates of Mines and Ponts-et-Chaussées), the Berlin Gewerbe-Institut, the Prussian Hauptbergwerks-Institut; a middle range of mechanical training schools, the *écoles des arts et métiers* in France and provincial Gewerbeschulen in Prussia; and at the bottom a heterogeneous group of local courses, sometimes private, sometimes public, in manual arts, design, and the rudiments of calculation.[35]

In short, these schools were evidence of the way the Continental countries reached high rates of growth through "the cultivation of systematic instruction as a cure for technological backwardness."[36]

As an aid to the understanding of economic development, such an application of Gerschenkron's approach is undoubtedly useful. As will be discussed, a sense of backwardness was commonly felt by educational reformers in France. Von Dyck similarly concludes that in the case of Germany "the desire to become independent of England gave an essential impetus to the growing demands for the founding of higher technical schools."[37]

If one is not too strict about when particular countries began to overcome their "economic backwardness," this argument is difficult to refute because technical schools were constantly being founded since the seventeenth century. The Ecole des Ponts et Chaussées (1747, becoming a full-fledged engineering school in

1775), the Ecole des Mines (founded in 1783 and inspired not by British advances but by the establishment of the Freiburg *Bergakademie* in 1765), and the Ecole de Génie at Mézières (1748) all preceded the French campaign to emulate British industrialization.[38] Eighteenth-century French lower-level technical schools taught everything from drawing to baking to "subterranean geometry."[39] Other mining schools were founded in German-speaking lands long before those areas' growth spurt: Brunswick in 1745, Schemnitz in 1763, Clausthal in 1775. The first of the non-classical middle-level *Realschulen* was also founded early in the eighteenth century.[40] Von Dyck even mentions lectures on engineering sciences (*Ingenieurwissenschaften*) at the University of Prague as early as 1717.[41] In the case of the United States, a list of Gerschenkron's "new institutional instruments" for technical education would have to include canal companies, machine shops, and railroad projects. Despite early exceptions such as West Point, the "cultivation of systematic instruction" in full-time schools accelerated only after the middle of the nineteenth century.[42] Moreover, promoters of engineering education during this later period embraced science more as a means to consolidate professionalization than as a way to overcome economic backwardness,[43] a pattern applicable to the French case as well. Even Britain soon found itself feeling backward; the founding of South Kensington—the Department of Science and Art and its affiliated schools—was a reaction to the technological achievements of Continental countries demonstrated at the Crystal Palace Exhibition of 1851.[44]

In any case, this study is concerned with the experience of economic development only to the extent that this experience provides a context for understanding its connection to changes in social stratification. In this connection the specific characteristics of this "veritable hierarchy" of educational institutions become of central importance. The general observation that a nation began to systematize and augment its technical instruction at a particular point in time may help to explain that nation's overall economic development, but such broad statements do not address themselves to the differences in social structure that might have been associated with different configurations of educational institutions. In his initial statement of the problem, for example, Gerschenkron discusses the different types of banks, but he does not discuss the implications of any hierarchy that may have existed. And tracing the impact of credit practices and the flow of money produced and distributed might have proved far simpler than the educational counterpart—tracing the diffusion of knowledge and measuring its effects. Landes, for example, doubts that the establishment of technical schools had an immediately economic payoff in the sense implied by Gerschenkron's argument: "The growing technological independence of the Continent resulted largely from man-to-man transmission of skills on the job. Of less immediate importance, though of greater consequence in the long run, was the formal training of mechanics and engineers in technical schools."[45] The concern of this study, on the other hand, lies with both the "short-run" and "long-run" social consequences of establishing such schools in an environment where "man-to-man" transmission was playing the central role in diffusing technical knowledge.

The Model: The Ecole Polytechnique

Brief reference to a single example, the Ecole Polytechnique, may help to make the point more clear. The principal accounts of the origins of this school in 1794 make no mention of French economic retardation. Most of the founders, men such as Gaspard Monge, Lamblardie, Prieur de la Côte d'Or, and Fourcroy, viewed England more as the traditional enemy and recent counterrevolutionary power than as the initiator of a radically new form of industrial production and economic organization. The immediate reason for the founding of the school was the sad state of recruitment to the various government engineering schools occasioned by the collapse of the educational system during the early years of the Revolution. To men like Lamblardie, moreover, students entering these latter schools needed not only better scientific training; they also needed a uniform preparation that would enable graduates of the Ecole des Mines and graduates of the Ecole des Ponts et Chaussées to cooperate on projects and to exchange assignments without encountering the problems of translation and differing methods then plaguing the engineering services.[46] In addition, the founders thought that inviting the extraordinary group of scientists then assembled in Paris to educate future engineers would tie this group more closely to national service.[47] The demands of almost incessant warfare imposed a third mission upon the schools, the training of military engineers and artillery officers.[48] In short, even though strengthening the engineering capabilities of a central government can be seen as contributing to the economic advance of the nation as a whole, the Polytechnique's relation to French social structure must also be understood from the perspective of the school's juxtaposition of scientific elite and student engineers, its fundamental commitment to forming servants of the State, with all the implications such service has for traditional status rankings, and its role as a source of recruitment for the top ranks of both the military hierarchy and the scientific community.

In offering instruction intended to lay down a base of common knowledge for all engineers, the Polytechnique of the 1790s in fact offered the beginnings of a general scientific education: plane geometry, algebra, the calculus, elementary chemistry, and mechanics. In their second year at the school the students repeated each of these subjects, but "explored them in greater depth."[49] Such "general" and "theoretical" subjects were thus separated from the engineering specialties themselves, which were taught at the "schools of application," as the Ecole des Ponts et Chaussées, the Ecole des Mines, and the Ecole du Génie Militaire now became known.

Especially in the Ecole Polytechnique's first twenty years, brilliant teachers, the cream of the scientific community (Lagrange, Laplace, Berthollet, Fourcroy, Vauquelin, Monge), produced brilliant students: Arago, Biot, Malus, Cauchy, Sadi Carnot, Fresnel, Poinsot, Gay-Lussac. A description of the school's glorious early decades, when "scientists became professors . . . for the first time," has become an inescapable obligation for any history of modern science.[50] Less attention has been paid to the role of the school in determining the social status, patterns of behavior, and collective identification of the profession of engineering. The Polytechnique, however, became the single most important source of the image of the French engineer as scientist, formally educated in his *science* by long, intensive years of study. Never again would the unschooled role models of the previous century, Vauban, builder of forts (for the military engineers), and Perronet, builder of bridges (for the nonmilitary), inspire such emulation from younger members of the profession.[51] As will be discussed in greater detail, the dominant personality at the Polytechnique during this period, the man referred to as "the father of *Polytechniciens*," was Gaspard Monge. A member of the Academy of Sciences since 1780, Monge made major contributions to the various kinds of geometry as well as to chemistry and physics.[52] His greatest intellectual impact on engineers, however, came from his virtually single-handed invention of descriptive geometry, perhaps the most generally useful subject taught at the Polytechnique.[53] At Monge's insistence, moreover, certain other applied subjects were not left for the *écoles d'application* but taught at the Polytechnique itself: "descriptive geometry applied to public works," architecture, and mining techniques.[54]

To the link to science the Polytechnique added the link to service to the State. Unlike their British counterparts, of course, French engineers had always been public functionaries. Service to the absolutist monarchy during the century before the Revolution had even brought ennoblement to some of the more successful Ponts et Chaussées engineers.[55] As Revolution and Empire infused the State with nationalism's beguiling thaumaturgy, *Polytechniciens* found their role as executors of a public mission an ever-growing source of prestige. Such traditions proved both an advantage and a handicap for the collective self-definition of *ingénieurs civils* in the nineteenth century. As *ingénieurs*, they profited from more than a century of recognition as a useful and honored profession; as *ingénieurs civils*, they were deprived of a crucial element in that claim to esteem.

In addition, the creation of a single central training institution made engineering education especially vulnerable to military influence. In 1795 the school was given the additional mission of serving as the preparatory school for artillery officers, and during the next sixty years, 2,525 of the 6,131 Polytechnique graduates entered the field or naval artillery.[56] In 1796 the students themselves were required to wear uniforms, apparently in the hope that their new visibility would deter their raucous uproars in Parisian theaters, which were becoming a regular occurrence.[57] Not until 1804, however, when Napoleon imposed barracks residence and a system of ranks, discipline, and drill modeled on the Ecole Militaire at Fontainebleau, did the militarization become complete.[58] The Polytechnique thus provided a convenient point of leverage where rulers of France could introduce attitudes and loyalties into the training not only of future army officers but also of aspirants to nonmilitary careers in roads, bridges, mines, and even the tobacco monopoly. The nearly constant warfare of the Imperial decade, a crucial gestation period for both the *ingénieur* and the industry of the nineteenth century, provided ample incentive to use that leverage. The nonmilitary corps found themselves engaged in a constant struggle to gain recruits for their *écoles d'application*.[59] Even without a militarized Polytechnique the French engineering profession would have been shaped by the demands of war; educational structures cannot explain everything. But although military attitudes are not necessarily incompatible with industrialization, their imposition radically limits the kind of political and social arrangements that will accompany changes in technology and economic organization. A question to be discussed later in this study, then, is the extent to which the education of the *ingénieur civil* was influenced by the military aspects of the Polytechnique.

The Origins

Neither the various dynamic aspects of the Restoration economy nor the existence of the Polytechnique as a kind of institutional template can fully explain, of course, why the Ecole Centrale was founded when and where it was. One must focus the narrative more closely.

A look at the specific origins of the Ecole Centrale itself reveals the network of institutions that brought together the intellectual and economic elites of Restoration France. The man who became the proprietor of the school, Alphonse Lavallée, was a wealthy landowner from Savigné-l'Evêque, near Mans.[60] After he completed his studies at the *collège* of Mans, he went to Paris to study at the Faculty of Law. At the same time, about 1812, three of the other four founders of the school were students in Paris— Eugène Péclet at the Ecole Normale Supérieure and Théodore Olivier and Philippe Benoît at the Ecole Polytechnique—but none apparently met him at that time.[61] Nor has any of them reported what influenced him as a student in Paris during Napoleon's Russian campaign. Not too far from this time, however, Eugène Péclet became a member of Henri Saint-Simon's intimate circle. Sent to teach physics in the *lycée* at Marseilles, he wrote to Saint-Simon on 1 November 1816 that he missed being part of the coterie so much that when he finally received a letter from Saint-Simon, he actually "jumped around the room like a madman." He vowed that he would always teach in accordance with Saint-Simon's system and that if he were ever elected to the Academy of Sciences, he would deliver his opening lecture on the influence of the sciences on industry.[62] Péclet thus represents the single best case of "Saint-Simonian influence" on the founding of the Ecole Centrale.

Meanwhile, Lavallée went to live in Nantes, where he went into shipping with his brother-in-law and began plans for establishing a model farm. One source claims that at this time "his studies on social problems led him to think that a technical school was needed . . . to form men capable of competently directing great enterprises."[63] He may have already been a subscriber to the Liberal journal *Le Globe*. In any case, he returned to Paris in 1827 and became a stockholder in that journal, which closely followed events in industry and education.

At this point Lavallée began to attend the lectures of Jean-Baptiste Dumas at the Athénée, a scientific lecturing society originally endowed by the king's brother in 1781. Dumas, born in Alais in 1800, had come to Paris on the advice of Alexander von Humboldt after making important chemical and biological discoveries in collaboration with the members of the Geneva scientific community. He was a brilliantly successful lecturer and at the same time passionately interested in the practical applications of science.[64] In October of 1828 he, Olivier, and Péclet began discussions with a colleague of Péclet's at the Ecole Normale, François Binet de Saint-Preuve, who had begun raising funds for an "Ecole Polytechnique Civile."[65] When he read of these discussions in the *Globe*, Lavallée approached Dumas at the Athenée and offered his financial backing.[66] Binet de Saint-Preuve, who had envisioned a school of 50 to 80 students along the lines of the Ecole Normale, withdrew when he heard that

Lavallée wanted to enroll 300 to 400. After Binet de Saint-Preuve withdrew, Olivier enlisted his Polytechnique classmate Philippe Benoît, at that time self-employed. (In fact, he was one of the first French *ingénieurs civils*.[67])

Political circumstances favored the founders. At that time, the establishment of an educational institution required the approval of the Ministry of Public Instruction. Péclet was a political opponent of the Bourbon party of order, as was Paul-Francois Dubois, editor of the *Globe*. Fortunately, however, the founders made their request to the ministry just after the Liberal Martignac government took office. The new minister of public instruction, Vatimesnil, not only approved the new school; he also exempted it for one year from the *rétribution scolaire*, the fee all non-*Université* educational institutions were required to pay to the government, which amounted to 20 percent of each day-student's tuition.[68]

Finding a building with facilities to teach 400 students gave Lavallée some difficulties. According to Daniel Colladon, the first nonfounder to teach at the Ecole Centrale, Lavallée finally chose a site in the Marais, the Hotel de Juigné, because he wanted to isolate his students from their disruptuve and politically volatile counterparts in the Latin Quarter. This Right Bank location presented difficulties for certain faculty members, however. Dumas, who lived at the Jardin des Plantes, and Colladon, who lived at the Place de l'Estrapade before taking up residence at Centrale itself, had to walk considerable distances across "dangerous territory" in order to teach at Centrale.[69]

The school was not meant to be a charitable enterprise. In the "act of association" signed by the five founders, Lavallée promised each of them a minimum salary of 3,000 francs per year, plus 12.5 percent of the profits. He calculated that with a full enrollment of 400 students, even if 90 were given half-scholarships, a full tuition rate of 600 francs would bring an excess of income over expenditures amounting to 63,000 francs.[70]

In any case, the Ecole Centrale was an idea whose time had come: a scheme for such a school had occurred independently to Lavallée and to Binet de Sainte-Preuve. In 1830, Charles Dupin, apparently unaware of the Ecole Centrale's existence, called for the establishment of a similar school.[71]

The 3,000-franc minimum salary also points to two other factors that facilitated the establishment of the Ecole Centrale: the centralization of French intellectual life and the system of part-time posts. Men like Dumas, Péclet, and Olivier might hold three or four such positions: an admissions examiner at the Polytechnique, a lecturer in an evening *cours* at the Athénée, or a substitute (*suppléant*) in some other man's course. Even the professorships at the Faculty of Sciences were not meant to be full-time posts. Only Paris offered such opportunities. The whole system brought increasing complaints about pluralism (*cumul*) as the century wore on. But it also meant that Lavallée could recruit top-rank scientists for his school without being forced to offer them full-time support.

If Lavallée's wealth, Dumas' eloquence in advocating technological development, the centralization of opportunities in Paris, and the possibility of flexible salary arrangements thus help to explain the "supply side" of the Ecole Centrale's successful beginnings, the growing number of candidates for the Polytechnique suggests what helped out on the "demand side." Table 1.1 shows that since the fall of the Empire, the latter school had attracted a constantly growing pool of candidates. In 1829, the year of Centrale's founding, 400 young men, who had spent from one to as many as five years in *mathématiques spéciales* preparatory classes, took part in the competition (*concours*) for Polytechnique entry. Only 100 were admitted (*reçus*). An undeterminable additional number found themselves "admittable" (*admissible*) but not admitted, condemned to dwell in the peculiarly French limbo wherein have dwelt so many souls after the spread of *concours* throughout the educational system.[72] Those judged *admissible* to Polytechnique were exempted from further examinations if they applied to Centrale, but many, preferring State service to careers in private industry, may have stayed on for another grinding year of calculus problems and geometry exercises and another day of judgment before their examiners.[73] For others, however, it was too late: unless he were already serving in the army, no candidate could be admitted after his twentieth birthday.[74] The Ecole Centrale, on the other hand, set, in practice, no upper limit: some

Table 1.1
Polytechnique candidates, 1800–1850

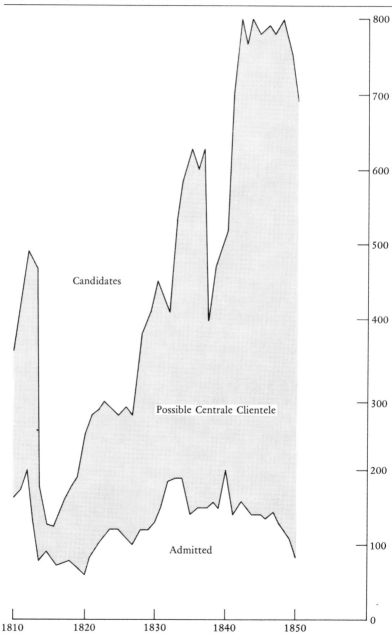

Source: Callot, *Histoire de l'Ecole Polytechnique*, 348.

Frenchmen were admitted at the age of thirty-three.[75] At the same time, the figure of 400 entering the *concours* does not include those who gave up preparing, unable to stomach one more differential equation. Since the Ecole Centrale did not require them to undertake advanced theoretical mathematics, they, too, must be considered potential clients of Lavallée's school.

A second relevant question on the demand side is the extent to which the Polytechnique responded to the growing admissions pressure by attempting to expand its overall size by securing posts for its graduates in areas which the Ecole Centrale wished to mark out for its own. An answer to this question was supplied in 1830 with a work entitled *De la nécessité et des moyens d'ouvrir de nouvelles carrières pour le placement des élèves de l'Ecole polytechnique et de l'utilité de créer de nouvelles chaires dans cet établissement*, written in 1830 by a lawyer named Doré who had served since 1816 as secretary to the director and governors of the Ecole.[76] As a tribute to their part in the July Revolution, all the students in the Polytechnique had been promoted to lieutenant and guaranteed upon graduation a place in the military or in the *services civils*.[77] In Doré's view this latest decision threatened to produce an impossible situation. The *services publics* would become so crowded that they could accept no more graduates. The ruin of the Polytechnique would come "in a few years" if the government did not open new careers to the graduates.[78] Doré then enumerated the government services offering posts to *Polytechniciens*, giving the following estimates of the number of graduates requested in a normal year:

Naval artillery	2
Army artillery	25
Geographical engineering	½ (i.e., one every two years)
Hydrographic engineering	¼
Naval engineering	1
Army engineering	10
Mines	3
Bridges and Roads	20

Powder and Saltpeter	¼
Administration	

62 places[79]

Doré then pointed out that each class at the Ecole had almost twice as many members as this, while at the same time the government had guaranteed places in these services to the two classes attending the school at the time of the *trois glorieuses*. Forced to accept twice as many new lieutenants as they wanted, the various services would be extremely reluctant, even if budgetarily able, to accept the normal number of graduates from among those entering in 1830 and 1831. Each year, moreover, a score of first-year students were judged not admittable to the second year at the Ecole or to the *services*. Despite the fact that according to Doré, they were as well educated when they arrived at the Polytechnique as the graduates of St.-Cyr were when they left their school, these blocked students were not allowed to become officers in any service.[80]

To find new careers for this surplus of students, Doré simply surveyed the principal remaining State military and civilian bureaucracies and argued that their annual intake of new "junior executive" personnel should be drawn from the Polytechnique. Places in the Royal Navy could be taken by the younger *Polytechniciens* since it would be easier to break them in to seafaring life. The corps of *intendants militaires*, the teaching staff in the cavalry and infantry schools, and the cadre of army headquarters would also make excellent places for the highly trained young officers. Even while still in school, *Polytechniciens* could serve as subprefects, pursuing concurrent training in an *école d'application* devoted to administration, as well as courses at the Ecole de Droit and at the Jardin des Plantes. Another *école d'application* attached to the Ministry of Foreign Affairs would complete the training for posts as consuls and embassy secretaries, where the students' scientific knowledge would be especially valuable for assessing the quality of goods involved in foreign trade. Living "in a world of numbers" as they did, *Polytechniciens* should be well qualified to begin a career in public accountancy as *inspecteurs des finances, troisième classe*. By studying during their second year with various government architects, a certain number of students could prepare for careers in that field. The *direction des télégraphes* and the customs service could also absorb a number of graduates. With their special scientific background, *Polytechniciens* should also share more chairs in the Faculties of Science and in the *collèges royaux* with graduates of the Ecole Normale Supérieure (who in 1830 had only just begun to dominate such posts).[81]

At the end of his survey, Doré came to the State-supported manufacturing establishments of Sèvres and Les Gobelins, where "already one can find graduates of the Polytechnique."[82] He also mentioned the possibility that private manufacturers could, by regularly admitting "trained men," make "productive applications of science to the arts." He then referred to the possibility of competition from the Ecole Centrale:

A new school, established in Paris in imitation of the Polytechnique, is attempting to create openings along these lines for its students; but I have no doubt that if the government made an appeal in this connection to all the manufacturers (*fabricans*) of France, their enlightened patriotism and their comprehension of the best interests of their establishments would bring them to prefer the good old Ecole Polytechnique.[83]

In the past Doré's Ecole had received requests from businessmen, but they were often "untimely" and did not permit "advantageous placement." To put some order into this relationship it would "suffice to make a public announcement that those manufacturers who wished to hire students should, at a particular time of year, inform the Ecole's administration of the conditions under which they wish to receive men."[84]

Despite his confidence that the Ecole Polytechnique could find outlets in private manufacturing, Doré apparently did not consider supplying private industry to be a major activity of the school. In his final estimates of the additional places that his scheme would provide, private industry played a small role:

Currently recruiting outlets	62
HQ, cavalry, infantry	50
Corps of Military Intendants	1
Sous-Préfectures	3

Embassy secretaries, consulates	3
Inspection des finances	1
Architects	3
Télégraphes	¼
Instruction publique	3
Manuf. royale et autres, Customs inspectorate	5

	Total	131¼

As the last line in the table indicates, Doré did not even bother to make a separate estimate for the requirements of private industry. As far as one can tell, moreover, his brochure inspired no overtures to the prefectoral corps, the diplomatic corps, the inspectorate of finances, or private manufacturers.[85] The *Polytechniciens* themselves seem to have accepted Doré's view of the range of acceptable careers. Lévy-Leboyer reports that during the first half of the nineteenth century, "no more than eighty *Polytechniciens*," 1.9 percent of the total number, resigned their commissions to accept a position in a business firm.[86] The field was left open to Centrale.

Doré was not precisely correct in claiming that the Ecole Centrale had imitated the Polytechnique. The unpublished history of the school, probably written, at least in part, by Dumas, shows that their model was that of 1794:

We adopted as a model the old Ecole Polytechnique [before it became a militarized boarding school], making the modifications required by the nature of the goal to be attained. Hence we eliminated the teaching of such overly advanced mathematical theories as have not yet lent themselves to practical application.[87]

In describing the type of man they hoped to produce, on the other hand, they took as their model the case of England. Only the formation of an elite of independent (*libres*) civil engineers could supply France with leaders for "that industrial army that the new era would see develop with a frightening but necessary fecundity."[88] The founders believed that "the true cause of the superiority which England acquired in world markets after 1815" was her possession of a corps of civil engineers "carefully supported and renewed, and drawing from the general esteem surrounding them the necessary *élan* and the will to succeed."[89] That England did not in fact possess any such clearly recognized and tightly organized engineering profession during this period—although individuals such as Telford, Watt, and Fairbairn certainly enjoyed esteem—only underlines the importance of such a corps in the founders' thinking.[90] They projected on to Britain the vision they would soon attempt to realize in the Ecole Centrale: the creation of a new breed of men, intellectually armed with a knowledge of *la science industrielle* and morally fortified by the precepts and discipline of three years at their Parisian school.

2

The Matrix:
Secondary Schools
and the Conflict over
Technological Culture

Secondary Education: The
Classicist Heritage

Taking seriously the claim that particular cultural traditions shaped the education and social cachet of the French engineer imposes one task, however, before we can enter the Ecole Centrale: an investigation of the role of secondary education in French society. As was mentioned in chapter 1, during the first half of the nineteenth century *enseignement secondaire* regained its position as the repository of key elements in the high culture of French social and political elites. In addition, this type of education shaped the pattern of entry to state functionary positions and to most of the more important professions. Conflicts over the content, structure, and intellectual fundaments of *enseignement secondaire* thus expressed basic disagreements about how cultural values and social processes should be adapted to political, economic, and technological changes. As products of this tradition-bearing *secondaire* system who nevertheless hoped to direct such changes, engineers found themselves at the center of these disputes.

Those who restored in the nineteenth century the *enseignement secondaire* that had seemed forever shattered by the Revolution drew on a long tradition of which the social and ideological keystones would weigh heavily on their newly created structures. The *collèges* that became the principal dispensers of secondary education under the Old Regime began in the thirteenth century as mere residences for those students attending the University of Paris. The professors in the university's Faculty of Arts gradually began to give their courses in *grammaire* and *philosophie* in the residences themselves. By the middle of the fifteenth century, the *collèges* had become autonomous centers of study.[1] They thus took form as institutions not so much subsequent to elementary education as they were prior to higher education. They grew as stalactites, not stalagmites. Not until the middle of the twentieth century did they merge to any substantial degree with the primary education that had been founded to instruct the masses on the floor of the social cave.

In the Parisian *collèges*, and in the Jesuit *collèges* that rose to rival them at the end of the sixteenth century, the Latin language and works of classical literature formed the core of the curriculum. The Jesuit *Ratio studiorum* of 1599 and the Statutes of the Faculty of Arts published the following year both prescribed that Latin should also be the common medium of communication among students. Penalties were assessed for the use of the vernacular within the *collège*.[2]

The tradition of classical education bequeathed to the nineteenth century by the Old Regime was in fact a complex and changing pattern. Despite the attempts by the *Ratio studiorum* and the *Statuta facultatis artium* to prescribe detailed and definitive rules of daily conduct, teaching methods, and curricular content, the period between the death of Henry IV and the Revolution brought many changes in what actually went on in the schools as well as a good crop of theoretical works on educational reform. For the purposes of the present study, perhaps the most important development was the growth of a close association between *enseignement secondaire* and the formation of the social type known as the *honnête homme*. Three pedagogical changes underlay this development:

1. The explication of complete works by classical authors was superseded by the study of *morceaux choisis*, selections from texts that served as models for moral and aesthetic inspiration. Antiquity became a world of heroes. A carefully chosen set of anecdotes illustrating deeds of virtue and courage, removed from any real historical or literary context, became elements in a "perfectly educative" *montage*. In the words of the Abbé Rollin, whose famous *Traité des études* purported to be a description of education as it was practiced in the *collèges* in 1726: "It was quite ordinary [in these *morceaux*] for great men to die without even the money for funeral expenses . . . , poverty was so much

honored by them."[3] The art of pedagogy consisted of finding men who incarnated particular virtues, building a world that was populated by such men, then inviting the students to live in it, inspired by the plow of Cincinnatus, the magnanimity of Camillus, the temperance of Socrates, the forbearance of Themistocles, the willpower of Demosthenes.[4]

2. By the middle of the seventeenth century, the students in *enseignement secondaire* were being encouraged to imitate these models more actively. The emphasis in the advanced classes shifted from the mastery of increasingly complex rules of grammar and the explication of authors to the composition of poems and speeches, modeled on the *morceaux choisis*, by the students themselves. The classrooms of the *collèges* resounded with adolescent voices intoning eclogues, pastorals, metamorphoses, descriptions, *parallèles*, *plaidoyers*, dialogues, narrations, and *harrangues*.[5] Even the younger students without much grammatical training were encouraged to compose small pastiches called *exercitiola*.[6]

3. The claim of the *collèges* to be the nurseries of *honnêtes hommes* was advanced most effectively by the third innovation, public recitations. They were often combined with theatrical presentations and became events of considerable social importance during the seventeenth and eighteenth centuries.[7] The ballets, plays, and poetry were designed to demonstrate the students' *honnêteté*, the central concept in the educational and social ideology that emerged in the seventeenth century.[8] Having communed spiritually with perfect models of virtue through his study of the *morceaux choisis*, and having absorbed the special discipline imposed by the Jesuit system of competitive emulation among students and constant surveillance by their teachers, the graduate of the *collège* had received the first part of his preparation for the role of *honnête homme*: the education of his judgment.[9] The ballet represented an aspect of the second part of that preparation, the education of his taste (*la formation du goût*.) Dancing, music, and skill in conversation and literary expression were also essential to this second element.[10] The *honnête homme* was a man of refined manners, of *politesse*. He knew how to acquire a mastery of his thoughts, gestures, and attitudes so as not to offend others. A whole literature appeared to help him learn

the rules of sociability, from table manners and handkerchief usage to the proper dignity in gesture and attitude.[11] The Jesuits emphasized that this *politesse* represented the external sign of the internalization of the precepts of their version of Christianized classical humanism. There was an essential continuity between *politesse* and *honnêteté*.

The task of forming the *honnête homme* was thus an essential justification for the role of *enseignement secondaire* in French society. The students in secondary schools were men whose education was Latin, literary, and supposedly nonprofessional. The *honnête homme* as a social ideal had no *métier*; he was especially disdainful of manual occupations.[12] The concept of utility was a source of contempt, a concern that could only detract from *honnêteté* and debase the noble thoughts of the classically educated humanist. The nineteenth-century scholar Abbé Augustin Sicard, who had no little sympathy for the values of classical humanism, described this ideal in his discussion of the resistance of *secondaire* teachers to the attacks on this ideology that became all but universal in the eighteenth century:

The teachers remained faithful to this tradition. For them the most important goal was to develop little by little the appreciation of taste (*le sentiment du goût*) in the souls of their students, to awaken and to sharpen their intellectual faculties by a progressive cultivation (*culture*) always in contact with perfect models, to form in them the *honnête homme* in its seventeenth-century sense, that is, to prepare them to appear in society, to carry on a conversation, to appreciate with competence matters of the intellect, and, finally, to conduct themselves in society with that superiority, that brilliance always bestowed by a strong literary education. . . .

The result was considerable, but it was not material, it was not palpable, it was not immediate. Instead of forming the man of a certain status (*homme de tel état*), the man of a certain profession, they were satisfied with the formation of *the man* in general.[13]

Cut off from the outside world during their school years by the boarding system (*l'internat*), the filtered purity of the *morceaux choisis*, and the concentration upon Latin in a world increasingly pervaded by the use of French, the young men of the *collège* were nevertheless trained in the pleasantry, polish, and social graces that reinforced the claim on social rank first justified by their humanistic learning.[14]

The Napoleonic Mold: 1802–1829

When in 1802 Napoleon's ministers began their reconstitution of the French educational system after the upheavals of the Revolution, the renewed dominance of secondary education by a classical and literary curriculum seemed anything but a foregone conclusion. Of the three councillors of state charged with presenting the government's plan to the *Corps législatif* in 1802, the most prominent was Antoine-François Fourcroy, who had been a first-rank chemist before gaining prominence as a Revolutionary statesman.[15] Fourcroy assured the legislators that the government's attitude was one of constructive eclecticism:

In the *lycées*, what was taught before in the *collèges* will be joined with the subjects of instruction in the *écoles centrales* [the Revolutionary secondary schools]. We shall teach there the study of all levels of ancient and modern literature as well as the study of the mathematical and physical sciences, necessary in the majority of professions. We have suppressed all those matters that were either superannuated or superabundant . . . , the two sins committed by previous institutions. All that which belongs to a liberal education (*une éducation libérale*) will be included in the *lycées*. Nevertheless, they will not be uniform. Diversities in locality, population, resources, habits, dispositions for various types of knowledge, and the varying needs of agriculture and industry will require diversities in the types and numbers of sciences taught. The law nevertheless sets a minimum because none of these schools must be deprived of that character of universality that distinguished the *lycée* as a type of institution.[16]

Fourcroy's conception of the *lycées'* mission thus has more affinities with that of the Revolutionary *écoles centrales* than with that of the *collèges*. The schools were to offer a general education, while at the same time responding to the more utilitarian needs of the local population. Scientific and literary subjects would stand as equals in the curriculum.

When the organization of the *lycées'* curricula was first specified in detail in a decree of 10 December 1802, the sciences did occupy a place very nearly equal to that of literary studies. Article 1 stated that "the *lycées* will teach essentially Latin and mathematics." The Latin classes, however, included instruction in history and geography, and the mathematics classes included instruction in the physical sciences.[17] Even though this decree provided little guidance on the question of responding to the "varying needs of agriculture and industry," it represents a high point for the role of science in the mainstream of secondary education that would not be equalled for half a century.

The beginning of the scientific subjects' loss of parity is usually traced to the year 1808, when Napoleon organized "all public education in the Empire" into the Imperial University, which included the Faculties of Theology, Law, Medicine, Letters, and Sciences; the *lycées* and *collèges communaux;* and the private secular secondary schools (*institutions* and *pensions*).[18] As grand master of the University, Napoleon chose not Fourcroy but the classicist Fontanes. Since few documents concerning the internal debate over such matters among the makers of educational policy have yet come to light, one is left with inferences from the public record. It does seem clear, however, that Fourcroy had opposition as early as 1802. In a speech to the *Corps législatif* on the day following Fourcroy's appearance, his fellow councillor of state, Antoine Roederer, gave a markedly different account of the nature of secondary education:

This education includes (1) one's native language ... ; (2) the Latin language, without which one can understand French only with great difficulty; (3) the Greek language, so necessary for understanding Latin; (4) logic, the art of conducting the mind in the search for truth; (5) rhetoric, poetics, the art of expressing one's thoughts and sentiments in the most vivid and agreeable manner.

These subjects should come before all others. They must thus become the common property of all men, even those who wish to confine themselves to the honorable enjoyment of their leisure in a polished society, such as the French Republic. The first part, that is, the study of French and Latin, may suffice, but is necessary (*peut suffire, mais est nécessaire*) for men whom a certain wealth distinguishes from the working class (*la classe des ouvriers*).

The sciences in the correct sense of the term, the mathematical, physical, moral, and political sciences, cannot, and must not, become the common property of every one. A great geometer cannot become a great legist; a great legist cannot become a great geometer.[19]

At the time Roederer spoke, of course, not only was his fellow councillor, Fourcroy, a chemist, but so was Chaptal, the minister of the interior; the great geometer Monge was serving as a senator, and even the first consul himself was a member of the section of Mechanics of the Institut!

In any case, the new University moved a step closer to Roederer's view. The program announced on 19 September 1809 revived the old names used by the *collèges'* classical curriculum: *grammaire*, the first two years of Latin language, in which neither mathematics nor a physical science was taught, whereas Greek was begun in the second year; *humanités*, two years in which arithmetic, geometry, and elementary algebra were added to the continued study of Greek and Latin; and *rhétorique*, a fifth year in which trigonometry and surveying were to be taught at the same time that one studied "the most beautiful examples of writing from the ancient and modern authors." In the sixth year the students in the largest *lycées* could choose between special mathematics (*mathématiques spéciales*), with additional instruction in the physical sciences, and *philosophie*, the study of a range of authors from Plato to Descartes to Condillac. History and geography had disappeared entirely.[20] The new degrees reinforced the pattern: the *baccalauréat-ès-lettres* was a prerequisite for the *baccalauréat-ès-sciences*.[21] Whereas the former became a requirement for entry into a wide range of professions and state administrative posts, the latter had usefulness as a credential only for medical students. Not until the year 1847 would the number of those awarded the second degree total more than 3 percent of those awarded the first.[22]

In the first major revision of the curriculum under the Restoration, on 4 September 1821, mathematics and science gave up even more ground. The year-long classes of the secondary curriculum received the names that would become customary in the nineteenth century: Sixth (*Sixième*, which one usually began at the age of ten or eleven), Fifth (*Cinquième*), Fourth (*Quatrième*), Third (*Troisième*), Second (*Seconde*), Rhetoric (*Rhétorique*), and two years of Philosophy (*Philosophie*). With the exception of "the elements of natural history," which were to be taught in the *Qua-trième* and *Troisième* (possibly a concession to the naturalist Georges Cuvier, who sat on the Royal Council of Public Instruction), the mathematical and physical sciences did not begin until the years of philosophy.[23] Five years later the University, now headed by Bishop Denis-Luc Frayssinous, decided that the reduction of the sciences had been too extreme; a decree of 16 September 1826 added two two-hour lessons to the weekly programs in *Deuxième* and *Rhétorique*.[24] This pendulum movement continued throughout the nineteenth century; the average lifespan of any particular blend of scientific and literary topics was about ten years.

When one looks beyond the regulations, moreover, one finds abundant evidence that during the years between the founding of the Napoleonic secondary system in 1802 and the opening of the Ecole Centrale des Arts et Manufactures in 1829, the pressures to return French education to the regime and ethos of the pre-Revolutionary *collèges* had a good measure of success. The prospectus of the Collège de Vendôme, where Balzac studied, announced that "this institution prides itself in having constantly been the repository of the old methods and the traditional syllabus."[25] Georges Cuvier, well known for his Protestant conservatism as well as his paleontology, was forced to write articles in the official *Moniteur universel* denying that the sciences were dangerous to religion and public order.[26] At the *collège* in Dinan the slogan was that "every mathematician is an atheist and every naturalist is a materialist."[27] In 1813 the traditional Latin oration was restored to the most important occasion of the academic year, the awards ceremony of the *concours général*, the annual prize competition.[28]

In 1815 the Jesuits returned to France. They soon established eight important secondary schools. Although they did teach some mathematics and science, these subjects were still called "accessory studies" in the revised *Ratio studiorum* of 1832. Their leading historian notes that "the humanities so-called, Latin and Greek language and literature, were still *the* study of the college."[29]

Other students of French education have found a similar empha-

sis throughout all of secondary education. Falcucci and Durkheim both view this period as the low point for scientific
studies.[30] According to Georges Weill, the reaction against Jaco-
binism and against the *écoles centrales* "renewed the culture of
Latin within the bourgeoisie."[31] In his comment on the decree of
1821, Felix Ponteil judges the pedagogical culture of a generation:
"They returned to the formation of the *honnête homme* of the
years before 1789."[32]

The Clientele

If, then, budding *honnêtes hommes*, or young men seeking Roederer's Latin-centered version of *une éducation libérale*, were the
dominant figures in the secondary schools of the Empire and Restoration, the question arises as to just where these young men
came from. What "vertical" sectors—the commercial-industrial
occupations, government functionaries, the liberal professions—
and what "horizontal" strata contributed the most sons to *enseignement secondaire*?

In the absence of any surveys of *secondaire* students' social backgrounds during the first half of the century, the answers must be
drawn largely by inference. Certainly the *lycées* were not the
schools of the "popular classes." For the mathematician and educational official Antoine Cournot their classical curriculum was
the best evidence that the social level of their clientele went no
lower that the bourgeoisie, and later writers have echoed his
opinion.[33] The excluding effects of Latin were even pushed a step
further in the case of the public schools. In a continuation of the
pattern of stalactite growth begun when the *collèges* formed out
of the medieval Faculty of Arts, the *lycées* and *collèges* appended
"elementary classes" in which Latin was begun. These elementary classes illustrated most clearly the way educational institutions provided cultural reinforcement for the social division
between *bourgeoisie* and *peuple*: the assortment of primary
schools (*l'enseignement primaire*) made available to the "popular
classes" taught reading, writing, religion, and arithmetic—but
not Latin.[34] The size of the enrollment in these elementary
classes is difficult to estimate; even when figures were included
in the statistical reports, which was never done systematically,

they referred only to the public schools. Yet the enrollment in private *pensions* was at least as large. The earliest figures, for 1842, show that the forty-six *collèges royaux* (the former *lycées*) enrolled 18,697 students, of which 6,385 were in the *Sixième, Cinquième,* and *Quatrième,* a mean of 2,128 per class. The total enrollment in the *classes élémentaires* (the *Huitième* and *Septième*) was 3,084, a mean of 1,542. Of the ten- and eleven-year-olds who began the regular *secondaire* course in 1842, then, about 72 percent were recruited from the school's own elementary classes.[35]

The financial burden imposed by most secondary education also placed it beyond the reach of the popular classes. Not only was the son prevented from making any contribution to the family income until his late teens; he had to be supported during his years in school. *Enseignement secondaire* could be expensive: the boarding fees in the *lycées* ranged from 600 francs in the smallest provincial school to 900 francs in a large Parisian *lycée,* payable three months in advance. An additional charge of 100 francs was made for "various additional expenses."[36] The distinctive *lycée* uniform, first adopted at the Lycée Louis-le-Grand in Paris, the school that in many ways provided the model for the entire *secondaire* system, cost 500 francs.[37] Nor were the *lycées'* rivals much less expensive: the fees in the Jesuit *collèges* ranged from 450 to 700 francs.[38] This was at a time when the annual salary of a primary school teacher ranged from 200 francs in the numerous small communes to 1,000 francs in large provincial cities, and a Parisian carpenter or blacksmith made less than fifty centimes an hour.[39]

Not all families had to pay these amounts, however. In the first place, not all students were full boarders (*pensionnaires*), at least not in the State system; almost all the schools allowed day students. Many of them boarded at private *institutions* and *pensions* where the fees were usually lower. Parents who lived in cities endowed with secondary schools could avoid boarding fees altogether. The day-student fees could reach as high as 150–200 francs, but the usual range was 60–80 francs.[40] The communally supported *collèges,* moreover, were seldom as expensive as the *lycées,* and they were not as heavily concentrated in the larger cities as were the *lycées*: 337 of them were scattered across

France in 1812; in 1828 the total was 317.[41] In 1842 (the earliest year for which figures are available) 29.7 percent of the *lycée* students and 48.6 percent of the *collège* students lived at home. In that same year, moreover, 25 of the 312 *collèges communaux* gave instruction entirely without charge.[42]

Scholarships were another way that a student could find his way across the financial barrier. In his decree of 1 May 1802, Napoleon had announced that the State would select 6,400 "students of the nation" who would be fully supported by the government during their years in the *lycées*.[43] This goal was never achieved, and the number of scholarship students continued to decline under the Restoration and the July Monarchy.[44] Nor were all expenses met. Approximately 53 percent of these scholarships (*bourses*) paid for half of the students' fees, 33 percent for three fourths, and only 13 percent for the full *pension*.[45] The Jesuit *collèges* also used a sliding scale. At the school in Bordeaux, for example, in 1828, 31 of the 331 students were charged nothing (11 supported by the Jesuits, 20 by the city), whereas another 80 from "honorable and poor" families were exempted from 325 francs of the 600-franc total fee.[46]

In the case of the government's program of financial aid, however, the important point is that need was not the most important criterion for awarding a scholarship. Instead, Napoleon had decreed that 2,400 of the 6,400 "students of the nation" would be chosen from the sons of "soldiers (*militaires*) and civil, judicial, administrative, and municipal functionaries who have well served the Republic."[47] During Napoleon's reign, at least, this type of student predominated among the scholarship holders: Fourcroy reported in 1806 that 3,923 students received scholarships, of whom 35.1 percent were awarded the grants as the sons of soldiers, 9.2 percent as the sons of judges, and 33.2 percent as the sons of "administrators" (*administrateurs*). He added that the remaining 22.5 percent, the sons of "mere private individuals" (*de simples particuliers*), were appointed by a different procedure, but that there were undoubtedly many sons of "soldiers and public functionaries" within this group as well.[48] The scholarship program thus fostered recruitment to the *lycées* largely from this "noneconomic" sector of society.

Although it was not always necessary to have a bourgeois income to send a son to secondary school, then, the possession of wealth certainly made things easier. On the other hand, during this period, *lycées* and *collèges* made little attempt to revise their curricula in a way that might have attracted a larger clientele from the economic sector. The sciences, especially mathematics, and practical or technological courses were rarely introduced. Unlike their more flexible predecessors, the *écoles centrales*, the schools of the new *enseignement secondaire* favored disinterested and "liberal" education over the accounting, drawing, and technical subjects that had once found a place beside them in a more comprehensive view of Useful Knowledge. Not until 26 March 1829 did an ordinance proclaim that private secondary schools—the *institutions* and *pensions*—and some of the smallest *collèges* might, if they wish, offer "the type of instruction that most suits the industrial and manufacturing professions."[49] By that time, however, the image of *enseignement secondaire* had already formed in the public mind. The association with classical education, preparation for a limited range of professions, and the unpleasantnesses of the boarding system had become a mental habit for most Frenchmen who thought about secondary education at all. In 1842, after efforts had begun to broaden the character of secondary education, the minister of education who made the first statistical survey of progress since the Revolution, Abel Villemain, sensed that an opportunity had been lost and that the accommodation between the University and the nineteenth century would not be an easy task. His figures showed that whereas in 1789, 72,747 students, one out of thirty-one of all male children aged eight to eighteen, were receiving some type of secondary education, in 1842 this absolute total had increased only to 89,341, or one in thirty-five of those eligible. He informed the king that this drop was "easily explained by changes in society itself, the smaller place given to the life of leisure and study and the much more general trend toward the industrial and commercial professions."[50]

The increasing importance given to the *baccalauréat-ès-lettres*, moreover, only reinforced the exclusionary effect of classical studies. Families who saw education as an investment good, a way for the recipient to increase his economic capital rather than his cultural capital, wanted something more practical than the classical curriculum, but the *baccalauréat*, by providing the final justification for that curriculum, instead helped to entrench the traditional program.[51] Non-Latin programs, special courses, "attached schools," and other arrangements began slowly to appear, but they had to enter through the back door. Most of the University considered them temporary, embarrassing concessions to the demands of the moment.

Even before the demands of families from the "economic" sectors (the industrial, commercial, and agricultural occupations) began to make themselves felt, the problem of educating engineers challenged the claims of the votaries of classical education and its *baccalauréat*. In administrative pronouncements, in speeches during legislative debates and at prize-day ceremonies, and in their various publications, the classicists had argued that only the *baccalauréat* could certify eligibility for the "most important professions," for "all the high careers of social life."[52] It soon became clear, however, that the great engineering schools—the Ecole Polytechnique and, later, the Ecole Centrale—were recruiting from the very social groups they had hoped to make part of the University's clientele. If the *lycées* and *collèges* were to function fully as the successors to both Old Regime *collèges* and *écoles centrales*, if they were to consolidate their dominance of secondary education, they had to assume the task of preparing candidates for these schools. In the period under consideration, however, such candidates spent their final *lycée* years preparing for the mathematical entry competition. Perhaps one third of them acquired the *baccalauréat*.[53] But were they not men about to enter some of the "most important professions?" As military officers and as engineers, *Polytechniciens* and *Centraux* challenged the University's claim that only its baccalaureate degree could guarantee the quality of the men who served France in high positions.

Counterattack and Cultural Stalemate: 1829–1848

Although the documents permitting a general picture of what went on in secondary schools during the July Monarchy are only slightly less rare than those referring to the period of the Restoration, they all tend to point in the same direction: a few small rafts of candidates for Polytechnique and Centrale in their last two years of the *lycée*, floating in a sea of Latin. Such was the pattern that the novelist Edmond About remembered from his years at the Collège Sainte-Barbe:

The most important part of the curriculum in those days consisted of translating French into Greek or Latin and bantering elegantly in Latin verses. . . . As for the exact sciences, it was good form to ignore them if one were not trying to enter . . . the Polytechnique. . . . Official education tended only to propagate, extend, and perfect the use of Greek and Latin.[54]

His classmates (1848) at the Ecole Normale Supérieure were renowned for their skills in Latin poetry—and their bad spelling in French. Ernest Renan found that in the prestigious Parisian seminary he attended in the late 1830s, "the study of the sciences was all but absent."[55]

When one turns from what students thought they were learning to what *Université* officials thought they were teaching, the picture is one of rapid change, with scientific subjects added in lower classes in 1838, restricted to the last two years of *secondaire* in 1840, then gradually reintroduced into earlier grades. [56] During this time scientific subjects never had a number of class hours comparable to the classical total. Throughout the entire period, moreover, the teachers themselves reported with bewilderment about the curricula prescribed for them. Camille Leroy, a professor at the Grenoble *collège royal*, complained in 1834 that the state of *histoire naturelle* was desperate: "No other course has been abandoned to such vague propositions and so many vicissitudes."[57] In 1842 a group of teachers of *mathématiques élémentaires* (the first of the two years of advanced mathematics preparatory to the Polytechnique) together sent a letter complaining that their situation was "almost hopeless." They had few students, they received no assistance with textbooks or teaching materials, and they were met with indifference, if not hostility, by the administration.[58] It was the *mauvaise volonté* of the students, on the other hand, that most impressed an inspector-general in 1839: They clearly considered scientific courses a waste of time. Lectures on physics and chemistry seemed to be a favorite time for riots (*chahuts*).[59]

At the same time, the content and structure of secondary education became a matter for intense debate in the Chamber and in the press. In a defense of scientific studies in 1837, the *Polytechnicien* François Arago served up a mixture of utility and psychological benefit:

You cannot make beet sugar with beautiful works, nor soda from sea water with alexandrines. . . . As Bacon put it so well, knowledge is power, and the power of science has greatly increased the welfare of our population, not by impoverishing the rich but by enriching the poor. . . . It is not true, moreover, that scientific studies serve only one's material interests. Their burning light dispersed most of the prejudices that had oppressed populations in the past; the sciences have permanently struck down false beliefs. . . . I do not see how scientific studies, in teaching the mind to reason, could ever warp one's judgment.[60]

The great Romantic poet Lamartine, soon to be Arago's partner at the head of the Second Republic but his opponent in this debate, blended conciliation with hyperbole in his reply:

Far from harming each other, far from combating each other, [scientific and literary studies] fortify each other, complete each other. . . . Nevertheless, only classical education assures to the nation a fund of common ideas and the sense of beauty that the Ancients possessed. If man lost a single one of those moral verities of which literary studies are the vehicle, man himself, all of humanity, would perish.[61]

And so it went: no subject could have been better designed to produce an orgy of rhetoric from classically educated politicians.

Paralleling the speeches of Arago and his allies appeared a number of works presenting the case for the inclusion of science and technology in secondary education. The single most important work of this sort was *Essais sur l'enseignement en général et sur celui des mathématiques en particulier*, by the noted geometrician Sylvestre-François Lacroix. The book gained special force because it presented not rhetorical appeals but a concrete vision of an alternative to the reigning *secondaire* system: the *écoles centrales* of the Revolution. Lacroix, who had taught at the Ecole Centrale des Quatre-Nations in Paris, argued that in these schools, the teaching, influenced by the sensationist epistemology of the Ideologues, had combined subjects as diverse as physics, draftsmanship, French literature, legislation, Latin, and "general grammar" into an integrated, coherent scheme that offered education at once moral, general, theoretical, and useful.[62] What had become since the founding of the *Université* two separate ideologies, the language of General Culture and the language of Utility, were really only dialects, or merely accents, of the same tongue.[63]

Even though it was not a textbook in the format of his widely used works on arithmetic, geometry, and algebra, Lacroix's *Essais* was successful enough to merit republication. A second edition appeared in 1816, a third in 1828, a fourth in 1838, and a fifth in 1857. The book became a kind of minor classic. In his pioneering study of the Ideologues (still the most important work on the subject), François Picavet wrote in 1891 that Lacroix's book had become a favorite assignment for the *lycées'* crowning year of *philosophie,* and was "still assigned."[64] Emile Durkheim relied on it heavily in preparing the lectures on the history of French education that he gave at the Ecole Normale Supérieure from 1904 to 1913.[65] As late as the 1930s it served as the principal source for two major scholarly works on the *écoles centrales.*[66]

The book was especially popular with engineers interested in educational reform. In 1819 Charles Dupin, scientist, member of the Naval Engineering Corps, journalist, and statesman, gave Lacroix's book an enthusiastic recommendation:

M. Lacroix, in his *Essais sur l'enseignement,* has demonstrated in great depth the advantages presented by the instruction given in the *écoles centrales.* He has shown how much that instruction was in harmony with the general progress of the Enlightenment and the needs of society for useful knowledge. The *Essais sur l'enseignement,* written always with elegance and frequently with power, is among the small number of works written both to please and to instruct, and which are within the reach of most readers. The work presents observations on the Ecole Normale and the Ecole Polytechnique that carry the precision drawn from the author's personal experience teaching in those two great schools. We give all these details in order to convince all men who wish to form correct ideas on education that they should meditate on the facts and the arguments presented in the *Essais sur l'enseignement.* Never was such study more necessary than today, when public instruction, sapped at its foundations by the imperial regime, crumbles more each day at the highest parts of the edifice.[67]

Nor is it difficult to understand the book's popularity in the first half of the nineteenth century. Lacroix's important scientific positions only reinforced his reputation as the author of widely used textbooks on many subjects.[68] At a time when the role of the church in the educational system was an especially sensitive issue, Lacroix avoided any mention of religion or religious aims. His criticisms of Latin education and the *culture des lettres* were never as harsh as those of the eighteenth-century writers from whom he drew inspiration; his tone was conciliatory. His sharpest attack, that on the boarding system of the *lycées* and *collèges*, nevertheless supported a balance between familial influence and day schools rather than the abandonment of education to private tutors.

Perhaps the book's strongest appeal lay in the fact that it was a practical handbook for both teachers and students of mathematics as well as a discussion of the role of mathematics in culture and history. Thorny questions such as the relationship between Analysis and "synthetic" techniques, the proper way to integrate drawing exercises based upon geometric solid models (polyhedrons, spheres, and so forth) in life-drawing, and the long-debated matter of the proper sequence in which to teach geometry and algebra all receive a balanced treatment full of practical suggestions.[69] And, finally, what nineteen-year-old aspiring engineer about to take a mathematics examination, what instructor whose reputation depended on his students' success in that examination, would have knowingly neglected to ponder the forty-page section entitled "How to Teach Mathematical Subjects and How to Evaluate, in Examinations, the Knowledge (*Savoir*) of Those Who Have Studied Them"?[70] Four copies of the 1828 edition of the *Essais* have survived in the library of the Ecole Centrale des Arts et Manufactures.

The quantity of literature on reform of secondary education under the July Monarchy gave it the status of a major issue of the day.[71] Although it often confuses more than it clarifies, this literature at least helps to suggest some of the reasons for the new interest in secondary education and for the constant vacillation in the decisions made about the proper relation between classical and scientific culture:

1. With the departure of the Bourbons, the influence of the Church upon the University had suffered a sharp setback. In reply, the Church launched a campaign against the University that was repaid in kind by influential professors such as Jules Michelet and Edgar Quinet. The central issues here were the influence of the Jesuits and "freedom of teaching"—that is, the removal of the University's power of approval and supervision over all educational institutions. When the role of classical education was discussed by the parties in the dispute, moreover, the cleavages were not always identical: Jules Michelet was almost as friendly to the classics as he was hostile to the Jesuits. In general, however, few opponents of Latin were found on the side of the Church.

2. Issues concerning education—in fact, issues of all kinds—were quickly finding a wider reading public. This was the golden age of French journals of opinion. As Joseph Moody puts it, "Technological improvements in paper-making and printing made possible what the maturing of representative government made politically necessary."[72] Each of the many small journals became the organ of a particular pressure group, relatively free from the censorship and high surety deposits that had characterized the Bourbon regime.

3. The structure of the University was itself undergoing important changes. It was during the July Monarchy that the Ecole Normale Supérieure finally came into its own, after two previous false starts, as the dominant institution within the University. During the three-year course of study at the school, students selected in nationwide competitions developed personal ties, gained a sense of common tradition, and—in many cases—acquired a privileged position within the teaching corps after they passed the *agrégation*, the examination that entitled them to the rank of *professeur* and a number of special perquisites. The structure and scale of the process were such that a single man at the Ecole Normale or on the Higher Council of Public Instruction could control the appointments to all the posts in his particular specialty. Thus during the greater part of the July Monarchy, most of the new philosophy *professeurs* owed their positions to Victor Cousin, and most of the teachers of the physical sciences were the proteges of Louis Thénard.[73] Each attempt to change the cur-

riculum, then, was likely to meet well-organized and concentrated opposition from those members of the University who felt their interests threatened. By the same token, pressures for change were generated from within the University by the consolidation of such alliances.

4. A number of the leading figures in the political life of the July Monarchy had been holders of high positions within the University: Guizot, Abel Villemain, and Victor Cousin were employed as professors during this time. Educational reforms were thus especially important to each of them, and they knew that judgments of their political performance would rest to an unusual extent on the degree of their success in educational reforms.

In any case, attempts to change the public educational system occurred frequently during the July Monarchy. Investigative commissions proliferated, to the extent that one minister of public instruction, Narcisse Salvandy, became the target of ridicule because he had created so many.[74] In 1846 Salvandy asked the dean of the Paris Faculty of Sciences, Jean-Baptiste Dumas, to investigate the crucial question of science in secondary education.[75] The implications of what Dumas' commission proposed in its report will become more clear after a look at what his contemporary educational thinkers were saying about the "modernization" of French secondary schools.

Two solutions to the problem were envisioned. The first involved the introduction of more science and modern languages into the existing curriculum of the *secondaire* schools. Students would still study mostly Latin, but at some point they would begin to receive lessons in the sciences as well. For reasons already given, this method was highly vulnerable to changes in regime and to an attack from an alliance between the group of teachers whose portion of class hours was slighted and the partisans of their specialty in the Chamber, the press, and the public. The second solution, specific proposals for which were often modeled on the nonclassical German *Realschulen*, was to change the institutional structure. Writers such as Cousin, Saint-Marc Girardin, and Philibert Pompée proposed some form of "intermediate" education that would constitute the University's attempt to meet the "needs of modern society."[76] Each work in effect was a short commentary on French social stratification: classical *collèges* remained at the top of the hierarchy to serve the clientele of the highest status. Their "new" proposals established a type of education that was "intermediate" on a clearly defined three-level vertical scale. For Cousin the French lack of such schools meant that every *père de famille*, "even in the lower part of the bourgeoisie," had no choice but to direct his son to the classical humanities.[77] Since the fall of Napoleon, Cousin's lectures and writings, by propounding a dedication to the Good, the True, and the Beautiful inspired by classical-literary models, had done much to breathe new life into *honnêteté* and to begin its transformation into the *culture générale* that would serve as the integrating concept for secondary education after the middle of the century.[78] This updated classicism was not fit for these lower strata, however. Such students would only obtain "mediocre results" and would "contract relations that would render them unhappy with their position." Moreover, one should not "summon the lower classes indiscreetly to the *collèges*, which is what we shall be doing if we do not found intermediate establishments."[79] The lower bourgeoisie and the more ambitious members of the *classes inférieures* would presumably form together a contented new stratum in the classrooms of the intermediate schools. For Philibert Pompée, the founder of a trade school (*école professionnelle*) at Ivry, the three types of schools would precisely correspond to three social levels already existing in society: the *collèges latins* for the *classes supérieures* (members of the liberal professions), the intermediate schools for the *classe moyenne*, and the primary schools for the *classes inférieures*.[80] Left out of this scheme—or, perhaps, intended for membership in the *classe moyenne*, where they would not have been very comfortable— were the more successful or socially ambitious members of the industrial, commercial, and agricultural sectors. From these groups came many of the families who were sending their sons to be engineers, but as far as Cousin and Pompée were concerned, there was no place for them in the existing system of *enseignement secondaire*. The *Université* was training Technological Man without acknowledging that he existed. As will be shown in more detail, nearly three fourths of the students entering the Ecole Centrale had prepared in an institution of *enseignement secondaire*.

During the first three months of 1847, Dumas and four of his colleagues from the Faculty of Sciences entered this long debate with a tour of public and private secondary schools in Paris. They also talked to a number of "prominent agriculturalists, industrialists, and merchants." On 6 April they submitted their observations to Salvandy.[81]

Their report started with an account of the fall of French scientific education. The experience of the past forty years had come to "shatter the tradition of that vast scientific education set in motion—with a few excesses, perhaps—in the *écoles centrales*, then regularized and continued in such a tempered and profitable manner in the [earliest] *lycées.*"[82] (The influence of Lacroix is evident here: he was, after all, Dumas' colleague and immediate predecessor as dean of the Faculty of Sciences.) Since the appointment of Fontanes, however, scientific education had withered, forced from the curriculum by the march of classical studies.[83] The study of the sciences thus had to "condense itself, to withdraw into a small number of classes, become more concise, more abstract, more dependent upon feats of memory."[84] Scientific education—the term included mathematics as well—survived only in the special classes preparing for the Polytechnique and, in somewhat less advanced classes, for the military academy at Saint-Cyr. Yet it could only do harm for the University to "become accustomed to seeing scientific studies as reserved for exceptional minds or dangerous for the masses." Besides, the scientific education that did take place in the preparatory classes was based on "pure abstraction that only prepared minds accustomed to a disdain for experiment."[85] The *ésprit de système* warped even the best students.

In deciding what changes to recommend, Dumas was faced with a number of problems. He had always considered *mathématiques spéciales* a waste of time. As he put it somewhat understatedly in the report, "Higher mathematics is not especially suitable for the bulk of the nation."[86] On the other hand, the University had no control over the subjects required in the examinations for Polytechnique and Saint-Cyr, not to mention the privately controlled Centrale. If the State system did not try to coordinate its instruction with those requirements, it would lose its *mathématiques spéciales* clientele to the private schools. As the declining enrollments and general lack of enthusiasm for science showed, however, a scientific curriculum dominated by the requirements of the higher schools would not be especially popular.

Dumas clearly desired such popularity: the University was to be an institution embracing as many Frenchmen as possible. He also knew he faced resistance if he tried to cut hours from the classical curriculum; he had seen the uproar when others had tried it. Finally, he had to offer a "general culture" for secondary education. The mathematicians were vulnerable: there was hardly anything "liberal" or "general" about the cramming exercises of the *taupins* (the nickname for students in the advanced mathematics course). But the Latinists were always ready to display their capacity to deliver a "liberal" education.

The proposals themselves were rather complex, but Dumas' role in training engineers at Centrale (and the power he would wield over secondary education during the Second Empire) makes an attempt to find their fundamental principles worthwhile. The Polytechnique-oriented *taupins* were to be isolated from the rest of the curriculum: a small number of Parisian schools would set up "preparatory" courses for them. Everyone would get a good dose of sciences, but it would be "practical science," much like that Dumas had established at the Ecole Centrale: skills in observation and experiment would be nourished as much as those of deduction and clarification. He proposed to divide *all* non-"preparatory" education into three levels:[87]

Age	Designation	Old *Classes*
8–10	elementary school	*élémentaire*
11–13	*collège litteraire*	Sixth to Fourth
14–16	*collège scientifique*	Third to Rhetoric

The bulk of the hours in the *collège scientifique* would be taken up with mathematics, mechanics, physics, chemistry, and *sciences naturelles*. Yet he denied that he wished to "banish the literary element" entirely from the upper *collèges*: students would study history, a modern language, "and perhaps even some simple Latin texts." In the *collèges littéraires*, moreover, elementary science would be taught each year—here he contented himself with the traditional hours-added method of reform.

Dumas emphasized that all these scientific subjects should be taught in a new way, so that they would be "useful" and "suited to the needs of agriculture, industry and commerce."[88] He argued at the same time that even functionaries and lawyers would have to decide "questions affected by the sciences."

On the other hand, for all his emphasis on occupational utility, Dumas at the same time made broader claims for the technological culture with which he hoped to unify and modernize the French educational system. Practical geometry, for example, was "the best preparation for philosophical studies: it gives one a profound introduction to that sense of order and proportion which is the essential character of truth; it forms at an early stage the wisdom to handle syllogism."[89] Addressing himself to an audience that was especially responsive to the ideology of classical learning, Dumas had to be cautious in his language. Nevertheless, he did not hesitate to make his case for the transforming power of technology:

The University knows that the study of *lettres* can form the heart, exercise the mind, and elevate the soul; it now must be convinced that the well-directed study of the sciences has the same effects, and that it adds one more advantage of its own, that of forming men ready to take their place in the movement of progress and to contribute by their personal success to the development of the wealth of France.[90]

Dumas intended his technological education to unify more than just the clientele of *enseignement secondaire*. Although he had been asked only to investigate secondary education, he in fact surveyed the entire system. Not surprisingly, he advocated teaching his technological subjects at every level, from the primary schools to evening courses in the faculties.

But the strongest implications for social stratification lay in his suggestions about the accessibility of the *collèges scientifiques*. They were to become the single dominant type of secondary institution, established in every municipality that could help support one, just as were the communal *collèges*. Most significantly, students from primary schools were to be allowed to transfer to them freely. This was the crucial point. Technical subjects would not provide the cultural barrier that classical studies did. The *collèges scientifiques*, moreover, could send students directly to the Ecole Centrale des Arts et Manufactures. Though he served as minister of agriculture and commerce during the Second Republic, Dumas was hardly a radical democrat. In his plan the now more accessible *collèges scientifiques* would send aspiring Polytechnique engineers and officers, from at least their early classes, to the "preparatory" schools, but he gave no indication that boarding and tuition fees in those schools would be abolished. Still, the logic of a faith in "useful knowledge" could lead not only to a scientific-industrial curriculum but to increased upward mobility.

The Revolution of 1848 sent Dumas' plan to the archives rather than to the drawing board. When the sponsorship of Louis-Napoleon gave him the chance to try again in 1852, he chose "bifurcation," a parallelism of science and classics, rather than a unified program.[91] As his own report suggested, however, that same parallelism, less complete but equally pervasive, had characterized the educational culture of *enseignement secondaire* for at least a generation. Dumas' engineers spent their early years in schools that were locked in a cultural stalemate.

Technological Education outside the University: Promoters and Constituencies

And so during the founding generation of the Ecole Centrale, Ancients and Moderns debated to a deadlock within the University. The former extolled the transfiguring powers of the *baccalauréat-ès-lettres*; the latter tried to limit their *taupins'* Latin to the minimum needed for the brief translation exercise asked of Polytechnique candidates.[92] The classicists had the upper hand in French secondary education during this period, but outside the University things moved in a rather different direction. Because these developments both contributed to tensions within the University and created alternative visions of the Good, the True, and the Beautiful (but eschewing Cousin's rhetoric) that could shape the environment of attitude and ideology surrounding engineers, these matters too, must receive mention in a description of the cultural setting in which the first *ingénieurs civils* were educated.

Dumas himself, not a man of party, was probably least in disagreement with the Liberals, who sought, through constitutional means, especially the actions of a strong, sovereign Parliament, to reconcile their attachment to social stability with the deep political changes brought by the Revolution and the economic changes brought by industrialization. Theirs would be a government of all the elites, of birth, intelligence, and wealth.[93] The legitimacy of the latter's power rested upon the justification for their successful rise offered in the teachings of the most influential proponent of the new science of political economy, Jean-Baptiste Say. The owner of a textile mill at Auchy-les-Hesdins, Say had visited England in the 1780s, discovering a new industrial system aborning and a new Newton to explain the laws of its operation, Adam Smith.[94] In his publications—his *Treatise on Political Economy* first appeared in 1803—and his lectures at the Conservatoire des Arts et Métiers, the great Parisian teaching museum to which many Liberals had ties, Say laid constantly increasing emphasis on the role of the entrepreneur, who enriched himself and society by his skillful combinations of natural resources, labor, and capital.[95]

For the Liberals, moreover, these entrepreneurs were best assisted by a work force endowed with not only diligence but education. The acquisition of instruction also allowed worthy individuals to make the most productive use of their talents. The Liberals thus devoted themselves to providing the common people with both elementary and technological education. In their schools, in their journals, and in the evening courses that they established throughout French cities in parallel with the British mechanics institutes, these men distributed what they unabashedly called Useful Knowledge, an intellectual potpourri that could include agricultural or industrial techniques, contributions to scientific theory, or discussions of legislative progress in foreign countries. By no means all Liberals were opponents of classicism, but the literature of Useful Knowledge contains some of the strongest attacks on "useless Latin" to appear anywhere at that time. One of the most important periodicals diffusing Useful Knowledge, for example, was the *Journal des connaissances usuelles et pratiques,* published by the Liberal Count Charles de Lasteyrie, Lafayette's son-in-law, between 1825 and 1837. An article by the Liberal engineer Charles Dupin appearing in the May 1825 issue spoke disdainfully of the "escapee from a college, ignorant of everything or at most able to stammer a few words of barbarous Latin, who imagines that a mechanical profession is beneath his dignity."[96]

At the beginning of their most active period, the diffusers of Useful Knowledge were joined by the great visionary who would soon become their enemy, Henri Saint-Simon.[97] Although examining the complexities of the movement spawned by Saint-Simon and his followers would require a separate full-length study, certain aspects of their view of the social structure are of particular interest here. This vision stressed the fundamental solidarity of all the truly productive groups in society, the industrials (*industriels*), regardless of their positions in the social and economic hierarchy. The famous Parable printed in Saint-Simon's *L'Organisateur* contrasted the expendability of the noneconomic sectors of officials, politicans, noblemen, and priests with the indispensability of scientists, engineers, entrepreneurs and workers.[98] The solidarity of these productive strata—Saint-Simon usually referred to the industrials as a single "class"—would bring about a totally new society, organized on scientific principles. The new elite of scientists and engineers would manage the society not to exploit men but to administer things for the benefit of all, including *Messieurs les ouvriers.*[99]

Their emphasis upon the common interests of all members of the "productive" sectors explains part of the Saint-Simonians' hostility to the Liberals. The latter were clearly a false party, peopled by men who really belonged with the industrials (bankers, engineers, chemists) allied with old Napoleonic officers, lawyers, university professors, rentiers, and members of the outdated nobility. Most Liberals, moreover, seemed deeply attached to the existing governmental structures, which the Saint-Simonians wished to transform completely. On the other hand, the Saint-Simonians shared with the Liberals a faith in technology and in the value of technological instruction.[100] Only one branch of Useful Knowledge seemed misleading: political economy. The new industrial society would be built not by the operation of impersonal market forces but by consciously organized collective enterprises, especially large-scale projects such as railroads, harbors, and canals. The spirit of solidarity and love with which these projects were to be carried out would create a true social harmony, not the artificial harmony brought by the hidden hand behind individualistic market forces.[101] Harmony had another source as well: religious ritual. To the Liberals' religious heterogeneity, exemplified by the Protestant Guizot, the Voltairean Lafayette, the Catholic Montalembert, the Saint-Simonians opposed a New Christianity; their offshoots, the Positivists, a Religion of Humanity.[102] In the attacks on private property found in their doctrine (but seldom in their practice), in their anticlericalism, in their views on the status of women, in their collectivism, and in their concern for unity with the workers, the Saint-Simonians, in the eyes of the proper bourgeois, seemed a subversive group.

For some of the most prominent Saint-Simonians, however, the social radicalism faded with the passing of youth, and what remained of the master's teachings offered a vision to guide an energetic entrepreneurship worthy of J.-B. Say's highest praise. The Pereire brothers built railroads and pioneered in industrial banking; the Talabot brothers did the same, while also helping Enfantin, former leader of the Saint-Simonian cult, to plan canals in Egypt. Fournel and Duveyrier first alerted the French to the industrial potential of Algeria. Michel Chevalier, who first gained attention for his elaboration of Saint-Simon's plan for the economic development of the Mediterranean basin, became at his post in the Collège de France a teacher of the most orthodox political economy.[103]

Such domesticated Saint-Simonians have been portrayed by historians as the *dei ex machina* of the French Industrial Revolution, demiurges of technology who single-handedly launched the country out of its economic backwater.[104] Such an interpretation promised to simplify considerably the investigation of the relation between Ecole Centrale and *ingénieur civil*: "Saint-Simonianism nurtured in engineering school hotbed produces visionary molders of French industrialization." From the perspective of more recent research, however, the agronomy appears far more complex. In the first place, the role of Saint-Simonians in French economic growth has been shown to have been considerably exaggerated.[105] In the second place, the influence of Saint-Simon upon engineers was neither dominant nor widespread. Michel Bouillé's evidence indicates that less than 4 percent of the students at the Ecole Polytechnique during the period had any contact with Saint-Simonian circles, and by no means all of these accepted the master's ideas.[106] The traces of Saint-Simonianism at the Ecole Centrale are even fainter.

If Liberal political economy and Saint-Simonianism offered alternative ways to organize one's thinking about industrialism, there were also groups within French society that, without necessarily subscribing to either philosophy, shared the enthusiasm for technological education. The more distant the investigation becomes, socially and geographically, from the lecture halls of the Sorbonne and the Rue de Grenelle offices of the University, and from the sectors of State employment and the liberal professions, the more one encounters a lively technological culture. During the Restoration and the July Monarchy, for example, the dynamic and paternalist textile entrepreneurs of the Mulhouse region established several excellent industrial and technological schools in which to give their children an essentially practical education. The local classical *collège* (merely a private *institution* for most of this period) struggled along by teaching mostly the sons of functionaries.[107] Only after midcentury did the attraction of the classical *baccalauréat* begin to create customers for the *collège* among the business elites. In the 1840s, when the government indicated its willingness to establish a third national "school of arts and trades" (*école d'arts et métiers*), chambers of commerce in cities such as Bordeaux, Lille, Nîmes, Toulouse, and Marseilles competed fiercely to acquire the new institution.[108] The Académie of Lyons, bequeathed 2 million francs to establish an "institution for the public good," decided in 1825 that the institution should be a school for popular education in subjects such as mathematics, chemistry, and drawing. The next year the Ecole Martinière was established.[109] There is evidence as well of a growing interest in the acquisition of technical information outside the school framework: the total number of exhibitors in French industrial expositions grew from 110 in 1798 to 3,960 in 1844.[110]

The working classes also demonstrated their faith in technological education, if not in the austerities of political economy often served up in the inaugural lectures of evening courses. They provided most of the students at the evening courses on geometry and applied mechanics that had appeared in fifty-seven cities by 1825.[111] The Association Polytechnique, established in 1830 to offer a wider range of such courses, estimated that during the first twenty years, its enrollments in Paris alone totaled 400,000.[112] Those periodicals published by artisans themselves, such as *L'Atelier*, constantly demanded more technical schools; they applauded the establishment of the Ecole Centrale, which they hoped would become the model for many others.[113] Saint-Simonians or not, these artisans were well aware of the distinctiveness and independent value of the culture of the productive sector. Denis Poulot, a foreman at the Gouin machine shop and graduate of the *école d'arts et métiers* at Chalons, perhaps expressed the parallel best. He called for the creation of a degree to match the *baccalauréat-ès-lettres*: a "baccalaureate of work" (*baccalauréat-ès-travail*).[114]

In his report to Salvandy, Dumas seemed at first to speak for a University on the defensive: State education had to react to competition from private schools. At the same time, it seems clear that he wished something more than a rearrangement of the existing clientele of *enseignement secondaire*. Against the background of the growing consumption of "useful knowledge" within all social levels of the economic sectors, his scheme appears as an attempt to respond to much more widespread demands for an expanded, broad-based, scientifically informed, technologically enriched culture.

The search for a precise view of the social cachet an engineer acquired as a result of his years in *enseignement secondaire* thus produces mixed results. The failure of Dumas and his contemporaries to overcome the cultural stalemates and social rigidities within French education meant that large question marks obscured part of the portrait. To be sure, the considerable financial and cultural barriers to entering the *lycées* and *collèges* suggested that the students could be categorized, broadly speaking, as bourgeois. Spokesmen for various elites disagreed, however, about what, within these schools, constituted the best education, that is, the cultural symbols of highest status. Promoters of the classics and promoters of science did not fully recognize the legitimacy of their opponents' claims. The search for formulas of accommodation continued, but the crucial fact is that during the period under study engineers and their supporters were aware that within the University, at least, if not outside it, theirs was a minority viewpoint. The failure to establish that the secondary education of their inadequately Latinist engineering candidates commanded an esteem equal to that of a full-blown *honnête homme* helps to explain the ambition and breadth of the claims they then made for the educational process whereby these candidates were transformed into finished *ingénieurs* at the Ecole Centrale.

Who Attended and How They Got There

Before examining the substance of Centrale's education and the status-claiming rationales offered for its structure and content, however, let us take a closer look at the young men who were to receive this *formation*. Although the current state of research into the backgrounds of the clientele of the entire *secondaire* system during this period permits only the rather general conclusions presented in the previous chapter, the documents available at the Ecole Centrale allow considerably greater precision about that school's own students. Those upon whom Centrale claimed to work its transforming power were, of course, not just incomplete and suspiciously "scientific" *honnêtes hommes*. In their backgrounds one finds a particular distribution of geographical origins, preparatory institutions, and social strata. The school's records show, moreover, the extent to which the mix within each of these three categories was shaped by recruiting techniques, costs, and financial-aid practices. Finally, a comparison of *Centraux* with other groups whose backgrounds and achievements can be determined helps to clarify the role the school played in shaping the engineering profession and stratifying French society.

Recruiting Techniques

To recruit their new technological elite, the founders of the school used several tactics. In the spring of 1829 they placed their prospectus in the principal nationally circulated newspapers such as the *Journal des débats*, the *Constitutionnel*, and the *Courier français*. In 1830 they also inserted it in twenty-nine local newspapers such as the *Stéphanois* of St. Etienne, the *Echo de Vesoul*, and the *Propagateur du Pas-de-Calais* as well as German, Dutch, Belgian, Swiss, and American papers.[1] The *Globe*, of course, publicized the school heavily.[2] A number of other newspapers printed stories about the founding of the school that summarized the prospectus. Two Parisian bookstores, Béchet *jeune* and Malhet, were made the official distributors of the prospectus to other bookstores and to private parties.[3]

Lavallée also used more selective methods. He addressed a circular letter of 28 March 1829 to persons whose "social position" rendered them "most capable of seconding" his recruiting effort in the *départements*. In a handwritten postscript, he placed at the disposal of each of his correspondents one half-scholarship (*demi-bourse*) for a candidate they judged most worthy.[4]

Next he sought the assistance of the Society for the Encouragement of National Industry. Twenty-eight years after its founding by Chaptal, Monge, and others, the society continued to link government officials, scientists, and entrepreneurs in an effort to supply information and patronage to promoters of technological advance.[5] On 7 October 1829 the chemist Francoeur reported to the society that the school had placed five half-scholarships at their disposal "in order to prove to the public that our organization takes a sincere interest in their establishment." The society duly examined twelve candidates and awarded five half-scholarships. As a (rather conservative) token of their own good will, they decided to support one of these students from their own funds.[6]

Lavallée's third source of help was the Ministry of Public Instruction. At his request, on 25 August 1829 Vatimesnil sent a letter to the *recteurs* of each of the regional academies directing them to establish an inscription list for interested candidates and take charge of collecting the necessary references.[7] He informed them that four half-scholarships were available for the most deserving candidates. Although the letter itself is unclear on the matter, it seems likely that the Ecole Centrale, rather than the ministry, underwrote these scholarships, too. The unpublished history claims that State aid of this kind began only in 1836.[8]

The other founders of the Ecole also used their personal contacts to create a clientele. The notables they had asked to serve on the Council on Improvements (*Conseil de perfectionnement*), such as Héricart de Thury, director of public works for the city of Paris, or Prony, director of the Ecole des Ponts et Chaussées, were encouraged to publicize the school to their colleagues, students, and employees. Olivier and Benoît spoke to *Polytechniciens*, Péclet approached his fellow graduates of the Ecole Normale Supérieure, and Dumas spread the word at the Academy of Sciences. The physicist Daniel Colladon, who began the teaching of steam-engine construction at the school, was charged with sending the prospectus to acquaintances in his native Switzerland.[9] Finally, the founders made recruiting agents of *professeurs* of *mathématiques spéciales* in the provincial *collèges royaux* when they asked their assistance in administering the entrance examination.[10]

Despite the institution of these examinations, the records of the Ecole Centrale indicate that personal networks remained important throughout the first twenty years of the school's existence. Auguste Comte, for example, regularly selected promising students for the Ecole Centrale from among those he examined for admission to the Polytechnique.[11] The economist Michel Chevalier seems to have intervened only in the more difficult cases. When Jean-Victor Duqueyroix, the son of a wigmaker from Limoges, was about to abandon his studies for lack of funds, Chevalier called the attention of the Ecole to the "immense sacrifices" already made by his family, who had already supported him for three extra years of mathematics at the Limoges *collège royal*. The school finally granted Duqueyroix a half-scholarship.[12] Chevalier also supported the request made by Philippe Massabuau, the son of a clothmaker from the Aveyron, to undergo a second entrance examination after he failed the first.[13] To be allowed to take a second examination during the same year was highly irregular, but Dumas occasionally asked for this favor, too, as in the case of Alfred Troupel, whose father had been Dumas' close friend during his years in Alais. Dumas' international reputation also made him the focus for letters of recommendation from men such as the Neuchâtelois geologist Louis Agassiz and the Genevan botanist Pyramus de Candolle.[14] In general, during the school's first twenty years personal interventions served to modify the supposed rigidity of the entrance examinations and the severity of the school's financial burden. After the school had been taken over by the state, few such instances modified the impersonality and objectivity of the *concours*.

Geographical Origins

The student dossiers preserved in the archives of the Ecole Centrale reveal an unusually detailed picture of those whose families responded to the school's recruitment efforts.[15] Not all the dossiers were complete, but 1,093 from the years before 1848 contain usable information. In addition, a sample of 579 dossiers from the end of the nineteenth century, when information about entering students was recorded more systematically, permits certain conclusions about the direction of change over time.

The geographical origins of the students reflect in part the success of the Ecole Centrale's publicity measures. As table 3.1 illustrates, the Ecole of the July Monarchy had a cosmopolitan student body. Slightly more than one fifth of all students were foreigners. By contrast, all but 2.3 percent of the students attending Polytechnique during the period 1805–1883 were Frenchmen. The comparison between the two schools also shows that the Ecole Centrale recruited significantly more heavily from Spain and Latin America. The absolute Centrale total for the period 1829–1847, representing a sample one tenth the size, even exceeds the Polytechnique total.

Within metropolitan France, the pattern of geographical recruitment reflects the dominance of the larger towns as suppliers of future *Centraux*. Paris accounted for 23.6 percent of the 1829–1847 group, a concentration of students from the capital that even exceeds slightly the 20.6 percent reported by Daumard for students entering the Polytechnique from 1815 to 1848.[16] As table 3.2 indicates, towns of 2,000 or less were substantially underrepresented, with 62.1 percent of the population (even by 1851 figures) and only 19.5 percent of the Centrale students. By the end of the century the proportions of entering students corresponded somewhat more closely to the distribution of the population as a whole. Towns of less than 2,000 sent only 19.6 percent of the Centrale students, but they now represented the residences of only 48.9 percent of the population. Paris increased its contingent from 23.6 to 28.8 percent, but by 1901 the district held 9.4 percent of the population, as compared to 4.1 percent in 1851. The effect of nineteenth-century urbanization was to even out the distribution as the smaller towns sent approximately

Table 3.1
Geographical origins of Ecole Centrale and Polytechnique students

Geographical origins	Centrale students, 1829–1847 (N = 1,158)[a]		Polytechnique students, 1805–1883[b] (N = 11,757)	
	Number	Percent	Number	Percent
Metropolitan France and Algeria	891	76.9	11,390	96.9
Born outside France, French parentage	11	0.9		
French colonies	15	1.3	94	0.8
Switzerland	42	3.6	49	0.4
Other German-speaking	21	1.8	44	0.4
Belgium and Luxembourg	29	2.5	45	0.4
Spain and Cuba	53	4.6	16	0.1
Spanish Latin America and Brazil	33	2.8	3	0.03
Italy	9	0.8	62	0.5
United States and Great Britain	15	1.3	30	0.3
Poland	25[c]	2.2		
Other	14	1.2	24	0.2

a. This includes 65 students whose dossiers gave only name and birthplace and who were not entered on the main computer file.
b. Source: Gaston Pinet, *Histoire de l'Ecole polytechnique* (Paris, 1887), 491.
c. Mostly refugees from the Polish Revolution of 1830.

Table 3.2
Distribution of Ecole Centrale students by size of town of birth

Population range	Percentage of ECAM students, 1829–1847 (N = 781)	Percentage of total population, 1851	Percentage of ECAM students, 1910–1917 (N = 403)	Percentage of total population, 1901
Paris (Seine)	23.6	4.0	28.8	9.4
Over 50,000 (other than Paris)	6.5	2.6	7.4	8.1
10,000–50,000	23.6	7.5	18.9	11.6
2,001–10,000	26.9	23.8	25.3	21.9
501–2,000	13.6	48.6	14.1	35.9
0–500	5.9	13.5	5.5	13.0

Sources: *Annuaire statistique de la France: Résumé rétrospectif* (1966), 24–25; L. Duclos, *Dictionnaire général des villes, bourgs, villages, et hameaux de la France* (Paris, 1840): Statistique générale de la France, *Résultats statistiques du dénombrement de 1891* (Paris, 1894); and for towns too small for the 1891 source, Ministère de l'Intérieur, *Dénombrement de la population, 1901* (Paris, 1902).

constant percentages to Centrale during a period when their role in the population as a whole was steadily declining. In absolute terms, however, the larger towns of France, disproportionately endowed with secondary schools and other educational resources, maintained their dominant position as suppliers of future engineers.

When students' birthplaces are examined according to *département*, the number born in each of the metropolitan French *départements* correlates closely with the total number of students enrolled in all secondary schools, both public and private, within each of these administrative units.[17] Certain industrial areas such as the Nord (2.5 percent of the sample in both 1829–1847 and 1910–1917), Pas de Calais (1.3 and 2.7 percent), Meurthe-et-Moselle (1.4 and 1.8 percent) and Meuse (1.3 and 1.1 percent) played prominent roles in sending students, and many of the more backward departments sent only one or two students throughout the period. Yet the level or rate of industrial development cannot alone explain either secondary school enrollment or the number of students sent to Parisian engineering schools.

The detailed research of Pressly and Julia has shown how the demands for child labor and other factors raising the opportunity cost of schooling in rapidly growing areas with textile, mining, and metallurgical industries were reflected in disproportionately low secondary-school enrollments.[18] Certain areas, such as Alsace-Lorraine, joined industrial dynamism with an unusual hunger for education, but during the first half of the nineteenth century, primary and technical education held a higher place on their desired menu than did secondary education; hence the schools of art and trades (*écoles d'arts et métiers*), which required no previous attendance in *enseignement secondaire*, saw these areas much more strongly represented in their student bodies than they were at the Ecole Centrale.[19] Sutter's study of recruitment to the Polytechnique also notes these regional disproportions but considers them of minor significance: "A careful comparison with the [total] population of the region involved shows a remarkable correspondence. Only the East (Western Lorraine, Alsace-Lorraine, Franche-Comté, Burgundy) demonstrates a number of students slightly higher than the expected figure, especially from 1836 to 1900."[20] Daumard's study of *Polytechniciens* in the period 1815–1847 comes to a conclusion equally justified for the case of the *Centraux*: "Regional distribution has less importance here than family origins or the professional milieu to which the parents of students belong."[21] In any case, the high correlation with secondary-school enrollment suggests that even before a State-run national competition was introduced when the Ecole Centrale became a public institution in 1857, the recruitment to that school was spread broadly throughout the *départements*.

Preparing for Entry

On the other hand, dossier notations concerning the last educational institution attended before entering the Ecole Centrale indicate that the broad geographical distribution in student birthplaces gives an incomplete picture of the pattern of recruitment to the school. Throughout the nineteenth century many students who began their instruction in the provinces spent the last one or two years in a large Parisian *lycée* or in one of the private satellite *institutions* that sent their students to a particular *lycée* for most important courses. In their preferences among the half-dozen leading *lycées*, the *Centraux*, not surprisingly, followed the same pattern as their *Polytechnicien* counterparts. The Lycée Saint-Louis's especially powerful scientific classes, which attracted "more than 20 percent" of all nineteenth-century *Polytechniciens*, also drew 8 percent of those who entered Centrale between 1829 and 1847, the largest single group. The Lycée Henri IV, better known for its literary specialties, supplied distinctly fewer *Polytechniciens* and only 2.5 percent of the *Centraux*.[22] As table 3.3 shows, however, the dominance of Parisian institutions was not especially marked during the July Monarchy: 23.6 percent born in Paris, a total of 33.3 percent attending a Parisian school before entering the Ecole Centrale. The smaller secondary schools in the provinces, such as the communal

Table 3.3
Preparatory education of Ecole Centrale students

Type of institution	Percentage attending, 1829–1847 (*N* = 480)	Percentage attending, 1910–1917 (*N* = 439)
I. Paris		
Lycées	22.1	46.2
Private preparatory schools (Ste. Barbe, Duvignau, etc.)	4.4	14.3
Church-affiliated schools	0.6	12.8
Municipal schools	1.9	11.6
Minor *institutions* and other	4.4	0.5
All Paris	33.3	85.4
II. Provincial *lycées* and *collèges*		
Lyons, Versailles, and *première catégorie*	22.7	11.4
Deuxième and *troisième catégories*, communal *collèges*	22.1	1.1
All Provincial	44.8	12.5
III. Private tutors, "*études libres*," "*chez lui*"	10.2	1.6
IV. All others (*arts et métiers, écoles primaires supérieures*)	11.7	0.5

collèges and the *lycées* (*collèges royaux*) designated as "second and third category," were still able to send a relatively large number of students to the Ecole Centrale, especially when one compares this with the pattern later in the century.[23] The division between primary education and *enseignement secondaire* was still fairly sharp, however. During the entire July Monarchy only four students in the sample moved to the Ecole Centrale directly from the primary system, three of them from higher primary schools, a type established after 1833 to give three more years of nonclassical instruction to the most ambitious *primaire* students.[24]

The *baccalauréat* apparently did not have a very strong appeal for students who entered Centrale during the July Monarchy. Although no extra credit (*majoration*) was awarded to *bacheliers* applying for entry, 115 dossiers (10.5 percent) did note that the candidates had the degree. All but fifteen of these, moreover, were awarded during the last eight years of the period. (The precise year of the award was not always specified, but the range of possibility is fairly narrow.) Perhaps, in these cases, the influence of the University's crusade for the degree as the proper culmination of *secondaire* studies was beginning to make itself felt. Or perhaps more families were deciding to cover their bets by ensuring that their sons were qualified to obtain functionaries' posts if they could not gain entry to an engineering school. In any case, like their comrades at Polytechnique, the great majority of the students at the Ecole Centrale, although they could not have avoided studying Latin and literature in their secondary schools, had not obtained the degree that served as the consecration of the claim to "the highest careers of social life."

Costs and Financial Aid

The cost of training a young engineer represented a considerable investment for his family. The Ecole Centrale charged the students for the full cost of their education. In its last year as a private institution, Lavallée reported "net earnings" (*bénéfice net*), which included his salary and interest on capital, of 100,000 francs.[25] Tuition was 800 francs per year. The founders estimated in their promotional literature that a student would need an additional 1,200 francs per year to maintain himself in Paris, but

this may have been optimistic. As early as 1842 Edouard Charton, the author of the best-known guide to choosing a career,
claimed that such a budget would force a young student to live
on a dangerously deficient diet.[26] Three years at even these figures would bring the cost of a diploma to 6,000 francs. According
to the founders, this amounted to little more than the education
of lawyer, a little less than that of doctor.[27] Even two years at the
Polytechnique cost only 3,000 francs according to Charton, but
he advised adding a minimum of two years of special mathematics preparation at 1,800 francs per year.[28] Without considering
scholarship aid, such educational expenses required a distinctly
bourgeois parental income. Charton's guide stated, for example,
that a primary schoolteacher in a large provincial city had an annual income of only 1,200 francs (plus lodging).[29]

Only financial aid from some external source could overcome the
obstacle presented by such charges. Some approximate indications of the extent of such aid can be pieced together from various
sources. The student dossiers from the 1829–1848 period do not
contain any systematic notations identifying the scholarship
holders (*boursiers*). Accompanying correspondence and other occasional notations established the receipt of financial aid in only
twenty-one cases (1.9 percent) during this period. Published histories of the school indicate that the figures were higher, but
their statements about scholarships officially "made available"
do not give information concerning those actually awarded. Dumas made a survey of his own in 1840, which is summarized in
tables 3.4 and 3.5.

Since these figures refer to the number of *boursiers* in each year's
total attendance, they cannot tell us how many individual students were aided: repeated recipients of aid cannot be separated
from those receiving it for the first time. It was also possible for
a single student to receive aid from all three sources—State,
département, and Ecole Centrale—as well as from such organizations as the Société d'Encouragement. In any case, Dumas' survey does make clear that for the vast majority of Centrale
students, the scholarship possibilities cannot have been a very
important influence upon the decision to apply to the school.

The figures do reveal with some precision the changes in the to-

Table 3.4
Financial Aid at the Ecole Centrale, 1829–1840

Source	1829–1830	1830–1831	1831–1832	1832–1833	1833–1834	1834–1835	1835–1836	1836–1837	1837–1838	1838–1839	1839–1840
A. Ecole Centrale itself											
Full tuition				3	10	15	12	12	9	4	1
¾ tuition								3	3	1	2
½ tuition	49	47	29	5	4	6	11	22	15	17	10
¼ tuition	7	0	0	5	13	3	4	4	3	6	8
Per cent who received some aid	(38)	(28)	(18)	(12)	(22)	(17)	(15)	(18)	(11)	(10)	(7)
B. Central government											
Tuition and board								5	8	11	7
Full tuition								3	6	11	10
Part tuition								2	14	21	33
C. Departments											
Tuition and board									2	10	11
Full tuition									3	3	3
Part tuition									0	8	11
Dumas' total of students who appeared on both B and C in a given year								10	30	51	60
Number of total students present at *rentrée* (all years)—Dumas' figures	147	169	159	113	121	142	181	227	267	287	285

Source: Dumas Papers, A. Acad. Sci., Carton 16.

Table 3.5
Total levels of financial aid to the Ecole Centrale, 1829–1840, in thousands of francs[a]

	Ecole Centrale	Central government	Departments	Total
1829–1830	20			20
1830–1831	19			19
1831–1832	11			11
1832–1833	6			6
1833–1834	13			13
1834–1835	15			15
1835–1836	15			15
1836–1837	21	9		30
1837–1838	18	22	5	45
1838–1839	11	32	15	58
1839–1840	5	32	24	61

a. Tuition in the period was 800 francs.

tal level of aid granted during the first eleven years of the school's existence. Even before the crisis caused by the 1832 cholera epidemic (to be discussed), Lavallée and his colleagues had begun to reduce the number of scholarships and the total amount granted, probably because the initial outlays for other expenses required to launch the school proved greater than they had expected. The importance of the contributions from governmental sources also becomes clear. The sums from the Ministry of Public Instruction and from the conseils-généraux of the départements more than doubled both total aid and aid per student while at the same time permitting the school to reduce its own "outlays" (in practice, income foregone) for such purposes to a sum lower than ever before.

The correspondence scattered in individual dossiers allows a glimpse of some of the sacrifices and struggles undergone by those who asked for scholarship aid. The widowed mother of Hector Rigaud enclosed with her request for aid a notarized statement certifying that she was unable to support her son.[30] Jean Zetter's father, a grocer from Solodori (Switzerland), offered as proof of his poverty the fact that all seven of his children had been given to other families to raise.[31] François Delannoy's father, an *employé* in the tax administration (*contributions indirectes*), found himself crippled financially by a retirement pension of 1,200 francs (his salary had been 2,200 francs). Mistakenly thinking that he would receive a scholarship from his *département*, François had not entered the *concours* for the State awards.[32] Gustave Méraux's request for aid described the steps by which the son of a *"simple laboureur journalier"* (day laborer) might come to aspire to enter the Ecole Centrale. He had managed to study at the communal *collège* of his native Salins (Jura) until the end of the *Cinquième*, when his parents could no longer pay the fees. He then served for two years as an apprentice to a local locksmith with the hope of gaining entrance to the *école d'arts et métiers* at Chalons. Illness prevented him from continuing on this path, so he got a job in the office of an architect at Pontarlier. After eighteen months more he secured a post as a proctor-tutor (*maître d'études*) in the Institution Barbet, one of the larger private secondary preparatory schools. Finally, in 1842, twenty years old and completely *"sans fortune,"* he applied for a *bourse* to enable him to attend the Ecole Centrale.[33] Eugène Despeyroux's father, the manager of a tobacco warehouse, earned a salary of 3,000 francs in 1836, to which he could add 250 francs a year as a member of the Legion of Honor. After giving a detailed account of his military activities during the Napoleonic Wars, which he thought entitled him to assistance from the Ecole Centrale, he explained that his military retirement pension was entirely devoted to the support of his octogenarian parents. The real financial blow came, however, when his son drew a bad number in the draft lottery. The cost of hiring a replacement made study for his son at the Ecole Centrale "out of the question" unless he received a scholarship.[34]

Jules Duchamp, who also sought to enter Centrale in 1836, was one of eight children of a "former artist and pensioner of the Académie Royale de Musique." The previous year, during which he had been attending the Collège Charlemagne, his parents had given him only 45 francs a month "to cover all costs." For a "young man living isolated in Paris," this sum had proven insufficient, and he found his "health threatened." Curiously enough, instead of requesting a full scholarship, Duchamp asked only that the tuition be reduced by 300 francs.[35]

Not all of those who sought aid came from the lower strata of society. In 1836 the Société d'Encouragement awarded a half-scholarship to Camille Polonceau, whose father held the exalted title of engineer-in-chief in the Ponts et Chaussées.[36] In 1839 the Vicomte de Bellefonds wrote to request aid, explaining that he had four children and no other income but his pay as a lieutenant-colonel on the inactive list. The vicomte's estimate of the extent of his need even exceeded that of Duchamp the musician's son: he requested a full scholarship.[37]

The cases of Polonceau and Bellefonds suggest, then, that even knowledge of the total number of students who received aid would not permit precise conclusions about the degree to which the granting of scholarships promoted upward social mobility through education. Aid to a viscount would have a converse effect: the prevention of downward mobility. After all, as was the case with *enseignement secondaire*, scholarship in higher technical education had other goals as well, such as rewarding of servants of the State. Shinn's study of the Polytechnique, which finds that 30.6 percent of the students entering between 1830 and 1880 received financial aid, concludes that "it was not precisely a question of assisting families with low incomes, but rather a recompense offered to those students who were already socially privileged." The two most important criteria for awarding scholarships were letters of recommendation from prefects or mayors and distinctions such as military medals or memberhsip in the Legion of Honor.[38] For the Ecole Centrale, in any case, it seems that this form of State assistance to the educationally ambitious had reached its upper limit by the middle of the July Monarchy. In 1839–1840 the total aid from the State and the Ecole Centrale amounted to 39,340 francs for 285 students (Dumas' figures). In 1907–1908, when the Ecole Centrale, now a state institution, contained 720 students, State scholarship aid amounted to only 60,000 francs.[39]

Social Origins

Despite the lack of precise information about the distribution of occupations or incomes within the French population at this time, the student dossiers reveal clearly the role that the Ecole

Centrale des Arts et Manufactures played in stratifying nine-teenth-century France. Lavallée, Dumas, and their colleagues had intended to recruit their future engineers and captains of industry from among the social and economic elite of French society, and they largely achieved their goal. As table 3.6 shows, if one excludes the somewhat question-begging category of *ingénieurs*, two thirds of the students in the period 1830–1847 were the sons of the upper (*haute*) bourgeoisie.[40] Included in this category are 111 *propriétaires* (15.7 percent of the total), a category peculiar to the nineteenth century.[41] Those who used this title tended to be owners of substantial properties, who moved in the same circles as high officials, doctors, lawyers, and other notables.[42] In the case of *Polytechniciens* entering during this period, the research of Terry Shinn concludes that the average net worth of a *propriétaire* family amounted to no less than 210,000 francs.[43] Even if Centrale *propriétaires* averaged only half this sum—and there is no reason to think that they differed in this way—they would still qualify clearly as *hauts bourgeois*. The other subcategories within the upper bourgeoisie include a good number of families from the "noneconomic" sectors: the upper ranks of the liberal professions (physicians, lawyers, notaries), 9.9 percent; and high officials and military officers above the rank of major, 7.4 percent. The "economic" sector of this most privileged stratum sent the largest number of sons, however, including 95 bankers and large merchants (*négociants*) (13.4 percent of the total) and 87 manufacturers and owners of industrial establishments (12.3 percent). Scions of some of the greatest French metallurgical families, such as Ignace Léon Schneider and Léon Muel, appeared as students at the Ecole Centrale, as did Henry Boucart, whose father was an associate of the great textile firm of Schlumberger. The Mulhouse business elite, in fact, had certainly not surrendered all their sons to the liberal professions or the bureaucracy: the dossiers of Charles Dollfus, the son of Jean Dollfus, the leading figure in Mulhouse industry, and Jean-Jacques Heilmann, whose mother was Eugénie Koechlin, from a family as powerful as that of Dollfus, have been preserved in the Ecole Centrale's records. When Charles Dollfus failed his first entrance examinations, his father paid for a year's private tutoring with a faculty member of the Ecole, after which he finally gained admittance.[44] In a sense, Guizot's famous injunction to those who

Table 3.6
Social origins of students at the Ecole Centrale des Arts et
Manufactures, 1830–1847

French students only $(N = 707)$		Percent
A. Rentiers and *propriétaires*		20.4
B. Liberal professions, upper category		9.9
C. High officials and military officers[a]		7.4
D. Engineers	1.3	
E. Large merchants and industrialists[b]		34.6
F. Large merchants, bankers	13.4	
G. Smaller merchants	4.2	
H. Middle functionaries, lower military officers		10.7
I. Liberal professions, lower categories	3.4	
J. Petty functionaries, enlisted men		4.2
K. Primary school teachers	0.3	
L. Artisans and shopkeepers		9.5
M. Shopkeepers	1.4	
N. Artisans[c]	8.1	
O. Laborers, workers		0.3
P. Peasants		1.7
Q. Domestic service		0.3
R. Unclassifiable but known		1.0
		100.0

Grouping for *classe*	
I. Upper bourgeoisie = A + B + C + E − G	68.1
II. Middle bourgeoisie = G + H + M	16.3
III. Employees and lower cadres = I	4.2
IV. "Popular classes" = N + O + P + Q	10.4
	99.0[d]

Source: Student Dossiers. Archives of the Ecole Centrale des Arts et
Manufactures.
a. Major and above.
b. Includes manufacturers, entrepreneurs, and high-level managers.
c. Includes *ouvriers spécialisés*.
d. Unclassifiable but known = 1.0.

wished to wield power in French society, *"Enrichissez-vous par l'épargne et par le travail,"* was a superfluous dictum. They had *already* enriched themselves. An engineering education was a way to make sure that their sons, the inheritors (*héritiers*), could maintain the family fortune in a time of rapid technological change and increasing international competition.

On the other hand, the Ecole Centrale's student body contained a far-from-negligible contingent (10.4 percent) from the "popular classes": carpenters, tailors, locksmiths, stonecutters. As such examples indicate, this group was represented at the school more by artisans and specialized workers (*ouvriers spécialisés*) than by peasants, factory workers, or day laborers. But in view of the fact that as late as 1962, artisans, farmers, agricultural laborers, factory workers, and domestics comprised 69.6 percent of the French population but only 4 percent of *Centraux* parents, the Centrale of the July Monarchy cannot be seen as purely the private preserve of a plutocracy.[45] As Lavallée had suggested with his offer of tuition remittance for deserving but impecunious applicants, a limited amount of individual upward mobility was recognized, and even promoted, as part of the heritage of the Revolution.[46]

Another way to evaluate the position of the Centrale families in the French social structure is to compare the students with other groups for which both the parentage and the social profile of the group itself can be ascertained with reasonable accuracy. Table 3.7 compares the *Centraux* with two other populations of scientifically educated Frenchmen: members of the physical sciences sections of the Académie des Sciences, who by their membership in the Académie and by their other positions held, can be safely placed in the upper bourgeoisie, and teachers of the physical sciences in French secondary schools, a position generally categorized as middle bourgeois. In the distribution of their parents among the four broad social strata, the *Centraux* closely resembled these latter two groups. The members of the Academy of Sciences, with a median birth date fifteen years earlier than the Centrale students, came from almost identical milieux: 64.5 percent of their fathers were of the nobility or upper bourgeoisie compared to 68.1 percent of the Centrale students;[47] 17.1 percent came from the middle bourgeoisie, versus 16.3 percent from Cen-

Table 3.7
Social origins of selected groups[a]

Ecole Centrale 1830–1847	Upper Bourgeoisie	68.1
	Middle Bourgeoisie	16.3
	Employees and Lower Cadres	4.2
	Popular Classes	10.4
Acad. Sci. I		64.5
		17.1
		3.9
		14.5
1846 Teachers		46.6
		35.5
		6.6
		11.1
Ecole Centrale 1881–1917		57.1
		20.5
		14.4
		7.4[b]
Acad. Sci. II		24.0
		36.0
		4.0
		36.0
1881 Teachers		16.1
		30.8
		12.6
		40.6

Sources: Archives Nationales F 17/20,000 series; Dossiers at Archives of Académie des Sciences and ECAM; replies to mailed questionnaires to Services des actes civils.

a. Acad Sci I: ($N = 76$) Members of the Academy of Sciences born 1778–1824; median birth year 1808. 1846 group, *secondaire* teachers: ($N = 45$); median birth year 1813. Ecole Centrale 1830–1847: ECAM students entering those years ($N = 707$); median birth year 1823.
Acad Sci. II: ($N = 50$) Members of Academy of Sciences born 1859–1890; median birth year 1875. 1881 group, *secondaire* science teachers: ($N = 143$); median birth year 1846. Ecole Centrale 1881–1917: ($N = 478$); class entering 1881 + 1910–1917. See note b.
b. Plus 0.5 unclassifiable but known.

trale, 3.9 percent from the "employees and lower cadres" stratum compared to 4.2 percent from Centrale, and 14.5 percent from the "popular classes," versus 10.4 percent from Centrale. The teachers in the secondary schools, closer to the *Centraux* in age, were somewhat less favored by birth: 46.6 percent of the fathers came from the upper bourgeoisie, 35.5 percent from the middle bourgeoisie.

These similarities among the social backgrounds of the three groups tended to diminish by the end of the century. The proportion of *Centraux* from the most privileged category declined to 57.1 percent, but only 24 percent of a somewhat older generation of *Académiciens*—their median birth year was 1875 compared to 1891 for the *Centraux*—came from this upper group. Only 16.1 percent of the considerably older population of *secondaire* science teachers had fathers in the *haute bourgeoisie*. The proportions in the fourth category tell the same story. The later group of Centrale students contained only 7.4 percent sons of the popular classes, whereas the *secondaire* teachers, with 40.6 percent of fathers in that category, and the fifty members of the Academy of Sciences, with 36 percent, were considerably more "democratic" groups. Forty-seven sons of engineers appear in the later Centrale group, 9.8 percent of the total, which represents a minimum for the number of engineering school graduates in the sample since many such men in managerial positions probably listed themselves under some other title, such as *industriel*. During the July Monarchy, then, the foundations began to be laid for the engineering dynasties to which the schools' yearbooks and official histories always referred so proudly.[48] In general, the comparisons suggest that a school like the Ecole Centrale continued throughout the century to act as a device for the transformation of a social elite into a technological elite.

A comparison of the social origins of Centrale students with those of *Polytechniciens* reveals that the two schools had differing appeals for the various sectors of that social elite. To be sure, the Polytechnique of the July Monarchy, with its 1,000-franc tuition fee and its 600-franc charge for room, board, and uniform, proved to be even less "popular" in its social recruitment than Centrale: 79.5 percent of its students' fathers fell into the categories defined by Shinn as *rentiers* and *propriétaires*, liberal professions, high-ranking officials and military officers, and large merchants and industrialists (which roughly corresponds to the category of upper bourgeoisie used above), whereas 72.3 percent of the Centrale fathers can be placed in these groups.[49] Only 4.2 percent of the *Polytechniciens'* fathers were artisans, shopkeepers, workers, or peasants, as compared to 11.8 percent for the *Centraux.* But it is the differences among the sectors that are most striking. As table 3.8 shows, large merchants and industrialists were nearly three times more strongly represented at Centrale, whereas the liberal professions and high-ranking servants of the State showed a clear preference for Polytechnique. *Rentiers* and *propriétaires*, prominent in the clientele of both schools, exceeded their Centrale proportion at Polytechnique by a margin of 80 percent. By contrast, shopkeepers, artisans, workers, and peasants were nearly three times as numerous at the July Monarchy Centrale. One suspects that the latter group, for whom a son at either school could mean the sacrifice of a sizable part of a laboriously earned and fairly recent fortune, had not yet acquired the taste for the State servant's status-honor that appealed so much to the comfortable *propriétaire*, lawyer, or colonel; they savored instead the more tangible rewards of private industry.[50]

By the end of the century two important changes had occurred in the pattern of recruitment to the two schools. In the first place, some of the differences in sectoral representation between the two schools had all but disappeared. With classes now about the same size (200 per year), they shared almost equally in the sons of *propriétaires* (a declining category in the population), members of the liberal professions, and large merchants and industrialists. On the other hand, functionaries, employees, and officers of all ranks now favored Centrale, whereas artisans and shopkeepers had switched to a two-to-one preference for Polytechnique. The second change is even more striking: Polytechnique, according to Shinn's figures, now had opened itself to the lower ranks of the popular classes. No less than 11.2 percent of its students were the sons of workers, peasants, and domestic servants.[51] The share of such categories in the Centrale clientele, however, remained as small as it had been under Louis Philippe.[52] With regard to the representation of the least privileged social categories, the pat-

Table 3.8
Social origins of Polytechnique and Centrale students

Category	1830–1847 (percent)		1881–1917 (percent)	
	ECAM (N = 707)	Poly. (N ≈2,300)[a]	ECAM (N = 478)	Poly.
A. Rentiers, *propriétaires*	20.4	36.8	10.7	13.3
B. Liberal professions, upper category[b]	9.9	15.5	9.4	11.2
C. High officials and military officers[c]	7.4	14.5	16.1	9.1
D. Engineers	1.3		10.8	
E. Large merchants, industrialists[d]	34.6	12.7	23.2	25.2
F. Large merchants, bankers	13.4		7.3	
G. Smaller merchants	4.2		2.3	
H. Middle functionaries, lower military officers	10.7	12.0	17.6	9.8
I. Petty functionaries, enlisted men	4.2	4.3	14.4	9.8
J. Primary school teachers	0.3		5.0	
K. Artisans and shopkeepers	9.5	4.0	5.8	10.4
L. Shopkeepers	1.4		0.6	
M. Artisans[e]	8.1		5.2	
N. Laborers, workers	0.3 ⎫		0.0 ⎫	
O. Peasants	1.7 ⎬ 0.2		2.0 ⎬ 11.2	
P. Domestic service	0.3 ⎭		0.2 ⎭	
Q. Unclassifiable but known	1.0		0.5	
	100.0	100.0	99.9	100.0

Sources: Shinn, *L'Ecole Polytechnique*, 185; Student Dossiers, A.ECAM.
a. Shinn gives no total, but an earlier study by Adéline Daumard gives a "known" total of 2,215 for the period 1831–1847.
b. Shinn does not distinguish between upper and lower categories, as suggested in Daumard's Code, but it is clear from his text that almost all his cases fall in the higher category. Lower categories for Centrale were grouped in H.
c. Major and above.
d. Includes manufacturers, entrepreneurs and high-level managers.
e. Includes *ouvriers spécialisés*.

tern that appeared at the Academy of Sciences, in the *secondaire* science teachers' corps, and at the Polytechnique was not repeated at the Ecole Centrale.

The relatively slower changes in its social recruitment meant that Centrale would continue throughout the century to ensure that the French engineering profession would be forged, in large part, from *haut bourgeois* metal. International comparisons suggest that this particular relation between schooling and the social composition of the profession may not be an exception to some purported general pattern in which the growth of educational institutions has brought "democratization," but rather a clue that the general pattern is not so simple. In the case of nineteenth-century Britain, where engineering drew its members not from secondary schools or universities but from apprenticeships, the evidence (fragmentary, but uniform in direction) indicates that the profession continued to recruit large numbers of self-made men of humble background.[53] In the United States, on the other hand, where John Rae has made a comprehensive survey of the backgrounds and education of members of the engineering profession born before 1860, the proportion of the sons of the "popular classes" (farmers, laborers, and craftsmen) decreases, and the proportion from the "middle and upper classes" (estate owners, "business and finance," and professionals) increases, with the rise in the percentage of those who attended colleges or technical institutes (see table 3.9).

In any case, the argument from technological necessity presents itself: France needed high-powered technical schools to overcome its backwardness with respect to Britain. The kind of school that its engineers received was dictated by the kind of profession that engineering is: Paris alone offered the chance to draw from the best available science, which could only be properly transmitted to future engineers in intensive, extended, and hence expensive education. Since the governments of the July Monarchy were not inclined to underwrite the education of private (that is, non-State) engineers in any important way, the upper bourgeoisie were destined to be the mother lode of the new profession. Only they could afford to bestow upon their sons the long years of *enseignement secondaire*, preparatory classes, and Ecole Centrale. This interpretation is best tested, perhaps, by

Table 3.9
Family and educational backgrounds of American engineers (percent)

Category	Before 1790	1790–1830	1831–1860
	(*N* = 51)	(*N* = 201)	(*N* = 583)
Farmers, laborers, craftsmen	51.4	29.4	20.9
Planters, estate owners, business, and professional	31.3	57.4	64.8
Military and engineers	17.6	13.3	14.5
	(*N* = 92)	(*N* = 335)	(*N* = 1,378)
Nontechnical colleges, engineering and scientific institutes	15.3	26.5	60.2
Self-taught	17.4	12.0	3.0
Apprenticeship	22.8	18.0	14.7
Secondary school	1.1	14.6	8.7
Military academy	11.9	15.6	6.8
Privately educated	14.1	4.8	2.7
Unknown	17.4	8.4	3.8

Source: Survey of the Biographical Archive of American Engineers in the National Museum of History and Technology, published by John B. Rae, "Engineers Are People," *Technology and Culture* 16:3 (July 1975), 414,416. Rae's study does not indicate which engineers received which type of education, however.

considering the experience of those private engineers who did not attend the Ecole Centrale.

Two State-supported trade schools, the *écoles d'arts et métiers* at Châlons-sur-Marne and at Angers, already played important roles in technological education in the years before the founding of the Ecole Centrale. The circumstances surrounding the establishment of the first of these at its original site in Compiègne illustrate the vagueness of both Napoleonic manpower policy and the use then made in official circles of the term "engineer." Compiègne in 1801 served as *une école pré-professionelle* that prepared students for apprenticeships in manufacturing or in army and navy workshops. The pupils studied reading, writing, elementary arithmetic, and drawing until the age of fourteen, when they were placed with local artisans by means of apprenticeship contracts.[54]

In 1802 Napoleon decided to visit the school. In his departing observations he emphasized that it should serve the needs of the economy of the surrounding region:

In the Nord I found excellent foremen . . . but none who could make a layout. . . . This is a real gap in industry. I want to fill it here. No Latin—that will be learned in the *lycées* just organized—but the work of the various trades with the theory necessary for their progress. Here we shall form excellent foremen for manufactures.[55]

Napoleon thus implied that the school's educational program should run parallel to that of the *lycées*.

Explaining Napoleon's views in a report delivered to the government on 4 February 1803, Chaptal stated that one of the most important goals of such an institution was to "multiply that species of precious men, unhappily too rare, who combine that general knowledge on which the (industrial) arts depend with skill in details of execution, and who are alone capable of making manufacturing flourish and giving great development to national industry."[56] The students were to be organized into brigades, sections, and detachments similar to those used in military schools. The best students would be sent on to the Conservatoire des Arts et Métiers to prepare, in two years, the *brevet d'ingénieur des arts et métiers*.[57] After reading Chaptal's report, the government drew up a draft decree in which the instruction at the "Collège de Compiègne" would be given "in such a manner as to form good artisans (*de bon artistes*) or engineers of arts and manufactures (*ingénieurs des arts et manufactures*).[58]

In this draft, then, the graduates of the school were to become engineers with precisely the same title, *ingénieur des arts et manufactures*, that graduates of the Ecole Centrale later received. But when the decree was actually published on 25 February 1803, the school was instead assigned the objective of forming "good workers and heads of workshops (*chefs d'atelier*)." Eight hours of each day were to be devoted to practical exercises, and two hours to "theoretical studies" such as mathematics, descriptive geometry, and drawing.[59] No reference was made to training engineers.

The education at the *école d'arts et métiers* improved steadily throughout the first half of the nineteenth century. Since in 1805 the only condition for admission was being the son of a soldier, most of the students were illiterate.[60] By the Restoration, all entrants were required to know how to read, write, and cipher.[61] To the traditional crafts taught in the workshops, such as furniture making, carpentry, delicate carving, and model making, were added the construction of textile and hydraulic machinery, foundry work, and the fashioning of precision mathematical instruments.[62] In 1832 a year of previous apprenticeship was required of all those who were to receive State financial aid.[63] In that same year, the schools at Châlons and Angers received capable new directors. They strengthened the curricula—which now included descriptive geometry, *mécanique industrielle* (the use of mechanics to explicate machine design), French composition, and "the principal notions of physics and chemistry applied to industry and to construction works"—and installed more modern equipment, including a steam engine, in the workshops.[64]

The schools' origins did not mark them as especially attractive to the bourgeoisie. On the one hand, Napoleon's view of the schools was made clear in a note to the minister of the interior in 1808 in which he objected to the annual expenditure of 540 francs per student; "The pupils are the sons of soldiers and poor artisans; it is against my intention to give them habits and tastes that will be harmful to them."[65] The emperor specially insisted that the sons of officers and of enlisted men should not be placed in the same schools. The enlisted men's sons should be sent to Châlons.[66] Despite the emperor's sense of hierarchy, however, the schools quickly gained a (not unjustified) reputation as hotbeds of Bonapartism, which did not endear them to later governments or to the more cautious of the bourgeoisie.[67]

An even stronger influence on bourgeois attitudes was the association of the schools with philanthropy, with the charity workshops designed to save paupers and foundlings from degradation. Technical education had served this function since the end of the sixteenth century.[68] The connection was symbolized in the person of the Duke of LaRochefoucauld-Liancourt, probably the best-known philanthropist in France, who served as inspector-general of the *écoles d'arts et métiers* at the end of the Napoleonic regime and during the first eight years of the Restoration.[69] LaRochefoucauld's own conception of the schools' role seems to have changed during his inspectorate, and he began to seek pupils not only among the sons of soldiers but also "among the sons of craftsmen (*artistes*), and even well-to-do citizens (*citoyens aisés*)."[70]

Table 3.10, which incorporates the recent research of C. Rod Day, shows the degree to which the duke's hopes were realized in the following half-century. In the period after 1830, the decline in the military contingent among entering students' parents (which came to include, *pace* Napoleon, both officers and enlisted men) was more than made up by the increase in artisans and "businessmen and manufacturers." The latter group remained a distinct minority, however, in a clientele strikingly more "popular" than that of the Ecole Centrale. The State charged 500 francs for tuition, room, and board at the *écoles d'arts et métiers*, and the trousseau cost another 200, but three fourths of the students received scholarships, nearly half of which were for the entire amount.[71] The rapidly expanding *primaire* school system was sending the cream of its students to the annual entry competitions, as is reflected in a report from the prefect of the Gard in 1840:

My previous reports have described the ardor with which the elite of our primary schools hastens to enter the *concours* for the *école des arts et métiers*. This eagerness now resembles a kind of infatuation. Distances, financial sacrifices, nothing discourages the applicants. And yet the majority of them belong to poor families who, perhaps, even borrowed the money to get to the examination site at Nismes.[72]

The competition became fiercer each year. In 1832 300 candidates entered the *concours* for 200 places; in 1840, 800 did so.[73]

Despite their momentary flirtation with the title *ingénieur*, Napoleon and his advisors clearly intended the *écoles d'arts et métiers* to produce foremen, mechanics, and skilled craftsmen. Official descriptions of the schools' purposes continued to use such terms throughout the period under study. The reality, however, was rather different, and that is what gives the *écoles d'arts*

Table 3.10
Social origins of students at the Ecole d'Arts et Métiers compared with Centrale

	Arts et Métiers 1806–1830 (N = 262)	Arts et Métiers 1830–1860 (N = 249)	Arts et Métiers 1806–1860 (N = 511)	Ecole Centrale 1829–1847 (N = 707)
I. Day categories (percents)				
1. Small manufacturers	5.0	5.6	5.3	0.0
2. Small business, shopkeepers	7.3	4.8	6.1	1.4
3. Independent artisans	17.2	26.1	21.5	8.1
4. Company clerks, draftsmen[a]	1.9	2.0	2.0	3.0
5. Merchants, bankers	1.5	2.0	1.8	17.7
6. Businessmen, manufacturers	2.3	12.0	7.0	15.4
7. Managers, engineers	2.7	3.6	3.1	3.4
8. *Propriétaires, rentiers*, nobles	0.0	0.0	0.0	19.2
9. Public employee (prof. level)[b]	2.3	2.8	2.5	21.8
10. Schoolmasters	1.5	2.0	1.8	0.3
11. Military men	22.9	8.4	15.9	5.1
12. Public employees (low level)[c]	23.7	11.2	17.6	1.8
13. Farmers (peasants)	6.9	6.8	6.8	1.7
14. Factory workers	1.9	7.6	4.7	0.0
15. Miscellaneous workers	3.0	4.8	3.9	0.6
16. Unclassifiable but known	0.0	0.0	0.0	0.6

Table 3.10 (continued)

	Arts et Métiers 1806–1830 (N = 262)	Arts et Métiers 1830–1860 (N = 249)	Arts et Métiers 1806–1860 (N = 511)	Ecole Centrale 1829–1847 (N = 707)
II. Weiss-Daumard categories				
Upper bourgeoisie (5, 6, 7, 8, 9)	8.8	20.4	14.4	77.5
Middle bourgeoisie (1, 2)	12.3	10.4	11.4	1.4
Employees, low cadres (4, 10, 12)	27.1	15.2	21.4	5.1
Popular classes (3, 13, 14, 15)	29.0	45.3	36.9	10.4
Military Men (11)[d]	22.9	8.4	15.9	5.1
Unclassifiable but known				0.6

Sources: C. Rod Day, "The Making of Mechanical Engineers in France: The Ecoles d'Arts et Métiers, 1803–1914," *French Historical Studies* X:3 (Spring 1978), 459; A.ECAM. Student Dossiers.
a. Includes for Centrale, other employees of low to middle rank.
b. Includes liberal professions, both levels.
c. Includes employees in liberal professions.
d. Unable to categorize without rank.

et métiers their significance for the present study. The *Gadzarts*, as the schools' graduates were known,[74] became collectively one of the most remarkable cases of upward mobility since the end of the Old Regime. Taken together, the three schools established before 1850 were one of the largest sources of privately employed engineers and other technical managers, a major contributor to government technical services, and an important producer of independent industrial entrepreneurs.[75] At the ends of their careers, only 11 percent of the group remained at the level of "skilled workers, foremen, draftsmen, etc." whereas 53 percent were the owners of enterprises, directors, or engineers, and the remaining 36 percent were in middle-level "white-collar" supervisory positions.[76] X. F. Jourdain (Châlons, 1819) became an important manufacturer in Alsace; Nicolas Cadiat (Châlons, 1824), the designer of the Pont d'Arcole, became director of the Forges d'Aubin; Emile Delahaye (Angers, 1859) was a pioneer in the automobile industry; Hippolyte Fontaine (Châlons, 1851) manufactured industrial dynamos and electrical turbines in the Rhône-Alpes area; Denis Poulot, who had called for a "baccalaureate of work," rose from foreman at the Gouin locomotive factory to become a major producer of machine tools and mayor of XIe *arrondissement* of Paris.[77] The *Gadzarts*, found most often in highly efficient, middle-sized firms, fed French industrialization the lion's share of its machine and motor parts, pumps, railroad parts, and agricultural machinery and played a major role in hydraulic and civil engineering.[*]

The commanding heights of French industry (as well as many enterprises in Spain, Latin America, and elsewhere) nevertheless remained largely in the hands of *Centraux*.[78] *Polytechniciens*, too, sprinkled rather lightly throughout private industry before 1850, began at the turn of the century to appear in greater numbers in the offices of large corporations, a pattern that would continue to the present.[79] In part the division was a matter of particular specialties and particular firms: the *Gadzarts* remained powerful in mechanical engineering, and with metallurgical firms such as the Schneider company at Le Creusot.[80] In most of the larger firms, however, and in fields such as railroads, civil construction, chemicals, and, later on, electrical goods, automobiles, and aviation, alumni of the Ecole Centrale, to a considerably greater extent than those of the *écoles d'arts et métiers*, populated the board rooms and managerial offices.[81]

---

[*] *Ibid.*, 456. Most of Day's *Gadzarts* rose to success in the latter half of the century. The important recent research of James Edmonson has established that in the first half of the century most of the important machine builders of France, who were organized into the Union des Constructeurs, had even less education than a *Gadzart*: "Out of a total of eighty-eight members, only six held diplomas from engineering or trade schools. At least two other members (and perhaps more) attended classes on an irregular basis at the Conservatoire des Arts et Métiers. The remainder, totaling eighty individuals, appear to have received no formal technical education whatsoever. They came to positions of responsibility as engineers and industrialists following some form of apprenticeship and travel." James Edmonson, *From Mécanicien to Engineer: Technical Education and the Machine-Building Industry in Nineteenth-Century France* (PhD Dissertation, University of Delaware, 1981), II-10 (temporary pagination).

It was in its control of the engineering profession as an organized body, however, that Centrale exerted its strongest influence upon the identity of the *ingénieur civil* as a social type. The founding meeting of the Société des Ingénieurs Civils was held on 4 March 1848 in an amphitheater of the Ecole Centrale, with only teachers or alumni of that school in attendance.[82] Neither *Gadzarts* nor other privately employed engineers were excluded; as early as 1862 the former comprised one fifth of the Société's membership.[83] The leadership of the Société, which was France's dominant professional engineering society throughout the nineteenth century,[84] nevertheless remained firmly in the hands of *Centraux*.[85] They brought their upper-bourgeois backgrounds and perspectives to the Société's attempts to enhance its members' professional status and to define the scientific identity and social role of the *ingénieur civil*.* This pattern contributed to the French civil engineer's tendency to adopt the outlooks and orientations not of a parvenu who had risen to the top in the bustling, competitive struggle of early industrialism, but of an *héritier*, an inheritor of wealth who sought to preserve and nurture that patrimony in the socially stable industrial growth brought by the orderly application of Industrial Science.

In sum, the evidence about the clientele of the Ecole Centrale and other technical schools can be viewed from two perspectives. If one examines the individuals who attended, one can find many cases of upward mobility (what the French call social promotion) apparently made easier, or made possible at all, by technical education. If certain twentieth-century models were taken as guides, one might even argue that the upper-bourgeois preponderance at the Ecole Centrale merely reflects upward mobility from past generations: education uplifts, and the longer an educational sys-

---

*The struggle to establish and enhance the independent status and social esteem of the *ingénieur civil* is skillfully summed up in Eugène Flachat's speech at the inauguration of a new headquarters for the Société on 7 June 1872, reprinted in *Memoires de la Société des Ingénieurs Civils* (hereafter, *M.S.I.C.*) (1872), 288–309. Flachat was a wealthy, versatile engineer who employed many *Centraux* on his projects. According to Edmonson, *From Mécanicien to Engineer*, II:3, 1, 6–7, he had tried to turn the Union des Constructeurs into a professional society but had found that body too preoccupied with ensuring the economic survival of its members. He found a more responsive forum in the Société, of which he served six times as president.

tem has to develop, the more individuals get uplifted. Some individuals just happen to arrive before others. On the other hand, if one shifts the focus from individuals to the relation between the configuration of educational institutions and the pattern of social stratification, the role of a school like Centrale appears in a different light. In the first half of the nineteenth century, or even later, Frenchmen of humbler orgins really did perform engineering, entrepreneurial, and managerial feats without any diplomas, as James Edmonson's new research has shown (see footnote, p. 85), or with only the certificate of an "inferior" trade school, as Rod Day has established. But theirs remained the narrower, less certain upward path to professional recognition and high social status. They had to walk up the slope; those Frenchmen born closer to the summit found in Centrale a chair lift to ensure that they reached the top more easily and in greater numbers. By its very existence as an efficient mechanism to inform, discipline, and find jobs for the sons of the *haute bourgeoisie*, a mechanism that was located only in the capital city, that was expensive, and that extracted a special measure of prestige from the Polytechnique model and the glitter of the Parisian scientific world, the Ecole Centrale helped to ensure that the pattern of stratification in the society of the July Monarchy would be reproduced later on in the hierarchies of industrial enterprise and in the structure of the engineering profession.

4

La Science
Industrielle

The Engineer as Scientist:
The Problem

The above analysis of the recruitment patterns of Centrale students reveals some of the most important ways such a school contributed to the stratification of nineteenth-century French society. For the members of the *haute bourgeoisie*, who formed the vast majority of the school's clients, Centrale equipped their sons with the technological knowledge needed to transform wealth acquired through commerce, agriculture, the delivery of professional services, or manufacturing into streams of income stemming directly from the activities of industrialization. For a significant number of sons of the "popular classes," attendance at Centrale opened the possibility that a man who began with little property could attain an important and lucrative position in the sector of private enterprise. For a few men such as the Vicomte de Bellefonds, an education at the Ecole Centrale offered the means to improve a family's finances so that it could adopt a style of life more in keeping with its previous social status.

Even if these three social functions of an education at the Ecole Centrale were clearly perceived at the time, the process retained a certain ambiguity. The difficulty was that the school claimed to be creating not just entrepreneurs or industrialists but *ingénieurs civils*. These latter were new animals in the French social, political and cultural zoo: by the content of its curriculum, by the images evoked by its name, and by its position in relation to other institutions of technological education, the Ecole Centrale helped to determine the social status of its graduates even more directly than did *enseignement secondaire*.

Decisions made at the Ecole Centrale about what to teach and how to teach it shaped the social position of those who attended the school in at least two ways. In the first place, of course, the quality of the students' education helped to determine their competence as engineers and industrialists. To measure precisely the contribution of an Ecole Centrale education would require the identification and isolation of a number of other influences upon individuals' careers—aptitude, personality, family connections, the openness of opportunities for advancement—as well as the construction of an acceptable, unitary measure of "success." In any case, the negative hypothesis seems unlikely: the private records of the Centrale and the public debates about its condition contain very few claims that its graduates were technically incompetent. On the contrary, in 1867 Sir Lyon Playfair wrote that the excellent French products displayed at the industrial exhibition held that year in Paris were the creations of *Centraux* "in the great majority of cases."[1] In 1872, at a time when the defeat by the Prussians brought a strong criticism of many French institutions, the *Revue des deux mondes* printed an article on Centrale that claimed that "from the very first this school placed itself at the summit of professional education and has maintained itself there."[2] The *patrons* of leading industrial firms continued to send their *héritier* sons to the Ecole Centrale throughout the nineteenth century, as did many prominent engineers.[3]

Whatever may have been its success in instilling technical competence, however, the teaching program of the Ecole Centrale helped to determine in what ways and to what extent the *ingénieur civil*'s identity would be associated with the statuses of artisan and of scientist. Despite the "arts and manufactures" in the name of the school, the founders clearly did not intend to confine themselves to transmitting the customary methods and private formulas of the industrial trades as they were then constituted, nor even to the systematic selection of current "best-practice" technology. In their view, the Ecole Centrale would break completely with such artisanal traditions and with the apprenticeship system of education—which, incidentally, was still producing almost all British engineers. Nor did they seek to produce *savants*, scientists devoted to wide-ranging fundamental research and theoretical formulations in the manner of the members of the Académie des Sciences. The realm of the engineer lay somewhere "between" these two types of activities.

The nature of these "intermediate" activities—or, more precisely, the way they were described—was the crucial determinant of the scientific component of the engineer's identity. The relation between science and engineering may take at least two forms. On the one hand, one may apply particular discoveries and theoretical principles—lawlike statements concerning the operation of the natural world—to practical problems. On the other, one may employ scientific methods in solving engineering problems by such actions as carefully controlled experimentation, a search for lawlike statements about technological processes, or a concentration upon precise quantification. In the former case, the activity is dependent upon some prior set of scientific discoveries or formulations. In the latter, the engineering activity is an autonomous enterprise. In its independent advance, the "technological" science fostered by the engineer's research and experimentation nevertheless acquires the characteristics of a "basic" science: a systematically organized, empirically verifiable set of facts and lawlike statements.[4]

The process of educating engineers can incorporate both these approaches simultaneously. The history of engineering during the eighteenth and nineteenth centuries suggests, however, that the act of combining them encountered certain difficulties. The gap between a "basic" science and the art to which it seemed to refer was often too great to bridge. Two books published by James Renwick, professor of natural experimental philosophy at Columbia, illustrate this problem. The first, *The Elements of Mechanics*, published in 1832, was a conventional exposition of the science of mechanics that "followed a well-trodden path in treating systems in equilibrium by the principle of virtual velocities."[5] His second book, *Applications of the Science of Mechanics to Practical Purposes*, published in 1842, was essentially a survey of existing practice in the fields of civil, mechanical, marine, and mine engineering.[6] During the course of his discussion of clocks, ships, and various types of machinery, Renwick occasionally referred to the principles discussed in his first book, but understanding them was not essential to the achievement of the "practical purposes" described; an engineer wishing to construct a waterwheel could have learned much from Renwick, but ignoring the references to the Science of Mechanics would have done him little harm. America experienced a period of extraordinary mechanical inventiveness during the generation preceding the

Civil War, but little of it was due to the application of Newtonian mechanics.[7]

The history of British and French engineering helps to clarify the second form of relationship between science and engineering, the development of "autonomous" technological sciences. John Smeaton, one of the first men to call himself a civil engineer, used careful experimental methods to establish a number of law-like statements—which he called "maxims"—concerning the construction of waterwheels. Layton cites the following examples:

In a given undershot wheel, if the quantity of water expended be given, the useful effect is as the square of the velocity.

In a given undershot wheel, if the aperture whence the water flows be given, the effect is as the cube of the velocity.[8]

Such "maxims" were neither laws of nature nor logical deductions from the science of mechanics. As Layton suggests, they "constituted the germ of a new technological science."[9]

As early as 1729 the French inspector-general of mines, Bernard Forest de Belidor, had written about technological science in his *La Science des ingénieurs*, in which he complained about the lack of interest in practical problems among *"les scavans."*[10] What theories he did derive from their work often turned out to be erroneous when tested by systematic experiments. In first attempting to determine the correct design of a simple beam to carry a given load, he arrived at the correct result, despite incorrect mechanical assumptions, by a proportional comparison of tests using a similarly loaded beam.[11] Louis-Marie-Henri Navier made a similar complaint that theory and practice were too far apart. In prefacing the first edition of his *Résumé des leçons sur l'application de la mécanique à l'établissement des constructions et des machines*, written in 1826, he claimed that research on the strength of materials by men such as Jacques Bernoulli, Buffon, and Coulomb had until that time been "more useful to the progress of mathematics than to the perfecting of the art of constructions." Most *constructeurs* still determined the dimensions of buildings and machines by following "established usages or existing works."[12] Navier's own attempts to bring mathematical precision to the calculation of elasticity were not without their problems. Like so many of his contemporaries, his work started from the Newtonian conception that the elastic properties of bodies can be explained in terms of attractive and repulsive forces between their ultimate particles and that these forces diminish rapidly with the increase of the distance between molecules. This led him to unnecessarily complicated calculations to determine the equations of motion of elastic bodies. Cauchy's introduction of the notions of strain and stress greatly simplified such calculations,[13] but as late as the 1830s, the theory of elasticity "had no rigorous solutions for problems of practical importance."[14] By this time, however, Charles Dupin's careful experiments in the bending of wooden beams had established a set of empirical results that agreed closely with the deflections to be predicted later by Navier's theories.[15]

The list of examples could be extended, but the main point should be clear: what actually took place in the history of nineteenth-century technology was something far more complex than the application of scientific theories or discoveries to custom-bound practical arts. As Navier suggested—and later scholars were to emphasize—the reverse process occurred as well: technological developments were applied to science to produce theoretical advances.[16] In Alfred North Whitehead's view, the crucial fact was that in the nineteenth century "a *developed* science wedded itself to a *developed* technology."[17] In an obituary notice on Navier written in 1837, Baron de Prony, director of the Ecole des Ponts et Chaussées, preferred to interpret the changes as a progressive extension of the domain of technological science.

The system of knowledge that currently constitutes the science of the engineer (including architecture) embraces, from the theoretical, experimental, and practical points of view, almost the entirety of diverse branches of the physicomathematical sciences. This system of knowledge has not always been so extensive—far from it; but in comparing the monuments created during various epochs, not according to their size or the amount of work required but according to the degree of instruction necessary to draw up the plans and direct their execution, one can see an evident correspondence between the scientific march of the human spirit and that of the art of constructions.[18]

In Prony's view, the sixteenth century, in which appeared the first systematic Italian treatises on hydraulics, was the period during which "the science of the engineer began to experience the great development that prepared its current state of transcendence."[19]

Not all of his contemporaries shared Prony's concept of an all-inclusive, "transcendent" engineering science. For many others the sharp division between theory and practice seemed a more accurate reflection of reality, and one not without implications for moral and social hierarchies. In *Democracy in America*, published two years before Prony wrote, Alexis De Tocqueville associated the theoretical with the aristocratic. In aristocratic ages, he argued, "vast ideas are commonly entertained of the dignity, the power and the greatness of man." Such opinions led those who "cultivated the sciences" to "the highest regions of thought," causing them "to conceive a sublime, almost a divine love of truth." After asserting that "in aristocratic ages science is more particularly called upon to furnish gratification to the mind; in democracies, to the body," he stated that because they possessed education and freedom, men living in democratic ages could not fail to improve "the individual part of science." "All the efforts of the constituted authorities" ought therefore to be directed to support the "highest branches of learning and to foster the noble passion for science itself."[20] It is likely, however, that the response to such phrases among Tocqueville's French readers would not have been the same as within his American audience. If Americans, as good Republicans, would have been less than eager to be associated with aristocratic pursuits, to many French bourgeois such association remained the summit of their social ambitions.[21] Under the Restoration, after all, the award of a marquisate to Laplace and a barony to Georges Cuvier showed that the pursuit of "pure" science could be literally ennobling.

Perhaps the most influential pronouncement on the question of the engineer's scientific identity came from Auguste Comte, who later sent to the Ecole Centrale many of the students he met as admissions examiner at the Ecole Polytechnique from 1836 to 1844. In December of 1829, one month after the Ecole Centrale opened its doors, Comte began his course on "positive philosophy" at the Athénée, where Dumas was still lecturing on chemistry.[22] In his second "lesson," Comte took up the question of the relation between science, "the totality of theoretical knowledge," and art, "the totality of practical knowledge."[23]

Given the degree of development obtained thus far by the human intellect, the sciences do not immediately apply themselves to the arts, at least in the most perfect cases. There exists between these two orders of ideas a middle order, the philosophic character of which is as yet poorly determined, which becomes more perceptible when one considers the social class which is specially concerned with it. Between scientists properly so called (*les savants proprement dits*) and the effective directors of productive enterprises an intermediate class has begun to form itself in our day, that of the *engineers*, whose special mission is to organize the relations between theory and practice. Not at all concerned with the progress of scientific knowledge, they consider it only in its current state of development in order to derive from it the industrial applications that it is capable of providing. Such is, at least, the natural tendency of things, although there is still much confusion on the matter.[24]

No matter how poorly understood, the order of ideas with which engineers concerned themselves was thus clearly distinct from "the sciences," and implicitly subordinate to them, just as engineers themselves seemed cut off from the possibility of contributing to the "progress of scientific knowledge." Nor does Comte suggest, as Belidor, Dupin, and Prony might have argued, that a common adherence to scientific *methods* might serve to link the two orders of ideas, to unite the activities of the class of *savants* with those of the class of engineers. In the *Cours de philosophie positive*, after all, his principal goal was to establish a clearly articulated hierarchy of what he considered to be the various fundamental sciences, one which would progress in orderly fashion from the sciences of greatest generality and independence to those of greatest complexity—from mathematics and astronomy through physics and chemistry to physiology and, finally, sociology. This scheme was not just a taxonomic device; it was a crucially important program of study. Comte was claiming, for example, that one could not understand physiology without first mastering astronomy.[25] This orderly progression could not be maintained in the study of the arts, however, since each incorporated several sciences:

Each art depends not just on one corresponding science, but on several at once, so that the most important arts borrow directly from almost all the principal sciences. Thus the true theory of agriculture—to take only the most essential case—requires an intimate combination of physiological, chemical, physical, and even astronomical and mathematical knowledge: it is the same with the fine arts. One can thus easily see why such theories have not yet been possible, since they suppose the previous development of all the different fundamental sciences. Such a consideration offers another reason to exclude this order of ideas from a course on positive philosophy since, far from being able to contribute to the systematic formation of that philosophy, the general theories proper to the different arts must on the contrary emerge later as one of the most useful consequences of its construction.[26]

Comte's presentation thus holds an ambiguous message for the engineer. Only through the full development of the fundamental sciences can the arts be understood, a development that cannot benefit from the investigation of industrial or agricultural practice at this time. To the extent that such an understanding determines the progress of the useful arts, then, the engineers' efforts would seem to be misplaced. Their goal as a class is to perfect some "intermediate" order of ideas in order to arrive at the "rational organization of the relations between theory and practice."[27] But why should an engineer bother with such a technological science if the real victories can only come through carrying out the fundamental research organized by positive philosophy? Nor was it certain that the esteem the engineers received for their efforts would be anything more than the secondary recognition given a worthy subordinate. In alluding to the moral connotations of scientific pursuits, Comte's language was no more neutral than that of Tocqueville: whatever might be the value of the services rendered to industry by engineering activities, scientific theories had "a more elevated destination," the satisfaction of that "fundamental need felt by our intellect to know the laws of phenomena."[28]

Perhaps the essence of the situation lay in Comte's admission that there was "still much confusion" about the engineer's relation to science and about the general nature of the science-technology interaction, a confusion that his own pronouncements amply demonstrated. The greater the confusion, the greater the opportunity for the founders of the Ecole Centrale to define a new professional type, a new Technological Man, and to suggest that his profession constituted an elite. The classificatory schemes, the associations between science and social status, had become no more clearly defined in the public mind than had the actual pattern of interaction between fundamental scientific advances

and technological innovation. If Tocqueville and Comte prescribed a society in which theoretical ponderings alone captured
the heights of esteem, Prony and Navier proclaimed an equally
exalted but more comprehensive "science of the engineer." As
has been mentioned in chapter 2, moreover, the hierarchy of Tocqueville and Comte found itself challenged by the proponents of
even broader concepts. The *Journal des connaissances usuelles*
and other proponents of the diffusion of useful knowledge devoted themselves to the sanctification of a far less sytematically
organized and more heterogeneous body of "positive knowledge."

To the founders of the Ecole Centrale, the phrase "industrial science" (*la science industrielle*) best expressed the unity and coherence of their educational program. In every prospectus of the
school and in most other documents in which the founders gave
an account of their activities, the same maxim appeared, usually
in italics: "industrial science is one, and every *industriel* must
know it in its entirety or suffer the penalty of remaining inferior
to his task."[29] The study of the subject was to be no mere cataloging of processes and principles but an organically integrated, unified experience. The unpublished history of the school asserted
that "the very principle of our attempted creation" was that "all
the courses in the new school formed in reality only a single
course."[30] One would "search in vain" to find "arbitrary limits with which to circumscribe the instruction of the students
according to the directions they were to take upon leaving
the school."[31] The courses were all connected to a single *plan
d'ensemble*, a "bold new conception" that had "never before
been put in practice in the teaching of the sciences."[32] In a speech
at the fiftieth anniversary of the founding of the school, Dumas'
rhetoric made the most ambitious formulation: *la science industrielle* embraced "the study of matter and of force, of everything
that has weight, that vibrates, that changes."[33]

The rejection of the primacy of specialization in engineering education thus became one of the intellectual cornerstones of the
Ecole Centrale, and, consequently, of the entire French civil engineering profession. On this point the teachings of Lacroix,
Comte, and the founders of the Ecole Centrale converged: cap-
itulation to the demands of specialisms could only "deform

the minds" of a technological elite.[34] Dumas celebrated the double nature of *la science industrielle*: the "special character" of the Ecole Centrale lay in its delivery of "an education at once general and technical."[35] There was no suggestion of a paradox. His audience probably had no more difficulty believing in the possibility of a *culture générale technique* than did those nineteenth-century bourgeois who believed that a humanistic general culture was imparted by the study of a small number of narrowly focused classical texts.

The Presentation of the Physical Sciences

Let us investigate, then, how the Ecole Centrale served as the tabernacle of *la science industrielle*, how its teachings attempted to give a unified structure to the diversity of scientific and technical fact, how it taught Comte's "new social class" to organize the relation between theory and practice. The next three chapters will take three different approaches to answering these questions: (1) an investigation of the way certain key subjects were presented in syllabi and, especially, textbooks; (2) an examination of the structure of the entire teaching program and the methods used to carry it out; and (3) an analysis of statements made about the course of instruction by those who sought to justify it and to explain its larger significance.

The three physical sciences begun in the first year—Mechanics, Physics, and Chemistry—constituted, along with Descriptive Geometry, the core of the Centrale curriculum. Only a scattering of course notebooks showing what was taught each day are available, but certain other materials can help us to sketch the general outlines of how these sciences were presented.

In the case of Mechanics, the unpublished history of the school provides the most detailed description. In the course proposed by Philippe Benoît when the initial plans for the school were being made, a brief survey of simple machines (the familiar classifications of Reuleaux: lever, pulley, inclined plane, and so forth) was to be followed by the study of friction and the strength of materials, "but their study would only have limited itself to the results of experiments."[36] Benoît intended to deal next with the "transformation of movement" as accomplished by machines for moving heavy objects, pressing or acting on bodies by shock, raising water, or pumping various gases. A short section on weighing machines and chronometers would be followed by a discussion of "machines powered by water, steam, and air," with mention of the applications of these machines to "drainage and different methods of transport." Finally, he proposed to survey "special apparatuses" used in agriculture, milling, winemaking, sugar processing, brickmaking, ropemaking, cotton spinning, wool spinning, printing, papermaking, wood drawing, wood planing, and the construction of machines. In the opinion of the *Rapport à présenter*, Benoît's proposals would have produced "not a course on applied and theoretical mechanics but a vast essay on mechanical *technologie*[37] taught from the point of view of the transformation of the raw materials used in industry."[38]

When Benoît withdrew from active participation in the life of the school even before the doors were opened, Lavallée and his colleagues requested that Gustave-Gaspard Coriolis teach the basic Mechanics course. Coriolis, who also served as *répétiteur* in the Mechanics course given at the Ecole Polytechnique (the course in which Comte served as *répétiteur adjoint*) and who later became director of studies at that school, offered a course much more to the liking of the Centrale founders. He dealt first with the composition of forces, the theory of their moments, and their conditions of equilibrium. He then studied the composition of velocities, the proportionality of forces in accelerations, centrifugal force and tangential force, and the "work of forces." He then gave equations of kinetic energy (*force vive*), for the application of braking to measure the work of machines (Prony had invented the crucial device for this in 1826),[39] for the calculation of momentum (*moments d'inertie*), and for the theory of flywheels (*volants*). He then took up the theory of water flow and hydraulic wheels. After next showing how to calculate the work accomplished by the expansion of gases, he "passed in review the work (in the physical sense) of men and of animals, of air resistance, and of windmills." He then showed how to assess the efficiency of machines, how to economize in their use through minimizing friction and the "losses produced by shocks to nonelastic bodies."

He concluded with "notions concerning the pulling of vehicles and the theory of the hydraulic ram."[40]

Such a course, organized around such general procedures as the measurement of moments and the calculation of the physical work accomplished by various devices, left the founders completely satisfied. The authors of the unpublished history had "no need to point out the superiority of this program over the one adopted previously."[41] Not all the theoretical procedures learned in this course could be applied to the cases of particular machines and mechanical problems encountered in subsequent courses at the Ecole—or, for that matter, in engineering practice—but certainly Coriolis covered many that *were* useful. Perhaps the clearest example would be Chézy's famous *formule hydraulique,* an empirically derived statement of the relation between the velocity of flow in an open channel, the dimensions of the channel, and the character of its surface. Perfecting this "law" of technological science became a major preoccupation of French engineers during the first third of the nineteenth century.[42] One of the most important contributions to this refinement was made by Joseph Belanger, who would later serve in Coriolis's post as professor of Mechanics at the Ecole Centrale.[43]

In establishing his Mechanics course at the Ecole Centrale, Coriolis drew upon his own pioneering study of the theory of work. He was, in fact, the first to introduce the term "work" (*travail*) into mechanical theory.[44] In the opening paragraphs of their report to the Académie des Sciences about Coriolis's treatise, Prony, Navier, and Girard emphasized its importance to both scientists and engineers. The "theoretical considerations" that governed the calculation of the effects of machines were "most worthy of the interest of scientists (*savans*) both for their own sake (*par elles-mêmes*) and for their influence on the progress of the arts." Instead of treating Coriolis's study as merely a contribution to the long-established field of *mécanique rationelle* (which dated back at least to Newton's time), the authors suggested that it was part of a newer, technological discipline:

The science of machines, considered in all its ramifications, is extremely vast and embraces almost all the arts. Even if we limit ourselves to that part of the science that belongs to Mathematics, we must recognize that it borrows essential notions from Geometry, Statics, and Dynamics.[45]

At the same time, they pointed out, *Du calcul de l'effet des machines* filled a gap between two fields. Theoretical "notions" concerning the use of motors (*moteurs*) had been absent from the teaching of *mécanique rationelle*, while at the same time they received no full development in "specialized treatises on machines." As the three *rapporteurs* recognized, however briefly, Coriolis's study of the measurement of work was thus an important step in the creation of a set of intellectual links that would come to constitute *la science industrielle* as it was taught at Centrale.

In Coriolis's hands, the subject just may have been teachable. An elementary grasp of calculus is required to understand the mathematical sections of Coriolis's study. Although it is true that calculus was not required of Centrale applicants, many of them had studied it in their advanced preparatory classes. At the same time, Coriolis himself probably introduced his students to calculus as he taught his courses on Mechanics. His *Du calcul de l'effet des machines*, intended for a far more advanced audience, carefully and clearly establishes the derivation of all his equations from the most elementary operations.[46]

On the other hand, Joseph Liouville, who took over the course on Applied and Rational Mechanics after Coriolis left in 1832, encountered considerable difficulties. The report of a special commission appointed during the 1837–1838 academic year found that "few students" were learning much in the course: (1) Many of the students had a weak mathematics background. (2) Individual interrogations were infrequent. (3) Certain students felt "considerable indifference" toward mathematical studies, "which they judge useless for the special career for which they are destined." Not swayed by this last observation, some members of the commission had suggested prefacing the course with lessons on "notions of analytic geometry, and even differential and integral calculus." Liouville himself opposed this, claiming that he "intercalated" analytic geometry where appropriate and that his own (unspecified) methods obviated any extra teaching of calculus. The majority of the commission concluded, in any case, that the school had to make its teaching "appropriate to the capacity of the weak [in mathematics] but intelligent students that it has admitted and will continue to admit."[47]

The commission's recommendation to hire a *répétiteur* to conduct more frequent interrogations was implemented immediately. At the end of the year, moreover, Joseph Belanger replaced Liouville, who returned to the Ecole Polytechnique. Belanger's solution was to offer first-year students a simplified and shortened *mécanique rationelle* "with immediate applications," especially apparatuses for the transmission of motion and the effects of friction on their various parts, and to second- and third-year students courses solely in applied mechanics, especially machine construction and hydraulics. Neither calculus nor any concern with the theory of work appears in the surviving student notebooks from these courses.[48]

The honor of inaugurating instruction at the Ecole Centrale was given to Eugene Péclet, whose *cours* on *physique générale* officially opened the school on 8 November 1829. The *Rapport à présenter* gives far fewer details concerning Péclet's courses than it does for Mechanics, but the expected incantation is not omitted: "Theory and practice were therein skillfully united."[49] The *Rapport à présenter* does state that in his general physics course and in his two-year course on "industrial physics," the first such course ever given anywhere, Péclet gave special attention to lighting and heat "considered from an industrial point of view," subjects on which he also published treatises.[50] In the absence of notes from Péclet's lectures, then, his books seem the best guide to his contribution to the promulgation of *la science industrielle*.

Péclet's first work on lighting, the *Traité de l'éclairage*, began with a tribute to technological change. Methods of lighting "had remained stationary," confined to wax, candles, and lamps, until the "discovery" of the double-jet burner by Argand in 1786.[51] Since that time, progress had come to the "art" of lighting "with astonishing rapidity": hydrostatic lamps, lamps with clockwork movements allowing the regulation of the wick position and the reservoir, lighting by hydrogen gas "discovered by a French engineer and applied on a wide scale in England," and Fresnel's improvements in lighthouse beams by the introduction of a system of concentric wicks and specialized lenses. Péclet therefore felt that the time had come for a general work that would present the details of this "long series of discoveries, made in such a short period."[52] The purpose of his book was to "give an exposition of the theory of lighting and the current state of its different branches."[53] In order to accomplish this, the work was divided into eight chapters: (1) a "résumé of the laws of optics that have applications in the art of lighting" (27 pages); (2) an examination of the different sources of light, principally the varieties of combustion (29 pages); (3) lighting produced by solids (26 pages); (4) lighting produced by liquids (73 pages); (5) lighting produced by gas (67 pages); (6) comparisons of different methods of lighting (73 pages); (7) a discussion of apparatuses that modify light by reflecting it or focusing it (20 pages); (8) a discussion of devices for instantaneous illumination, such as flint wheels and tinderboxes (14 pages).

In his first chapter, Péclet promised to give an account only of those "phenomena" that could be applied to the arts, ignoring those that had primarily "philosophical" value. His readers thus received the barest of essentials. Péclet subscribed to the wave theory of light, which by this time had been fairly well established by Young and Fresnel, although the corpuscular theory still had a small but prestigious band of adherents.[54] The rather elementary level of Péclet's discussion is illustrated by the following discussion of radiation and the speed of light:

2. *Rectilinear radiation of light.* Whatever the nature of the body from which light emanates, it radiates in a straight line. Everyone knows that when an opaque body is placed on the line that joins the eye and the luminous body, the light is intercepted. One can also observe directly the path of luminous rays when a beam is let into a dark room through a small opening in the window. The dust suspended in the air traces the path of the rays.

3. *Speed.* The speed of light may seem infinite because when one lifts an opaque screen from in front of a light source, one can perceive no gap between the instant when the path is cleared and the manifestations of light on the body that had previously been screened, no matter how far that body is from the source of the light. However, the transmission of light is not instantaneous; it is only so rapid that the distance that separates terrestrial objects is not completely perceivable. Only by the observation of celestial phenomena has the speed of light been successfully mea-

sured. It has been found that the light reflected upon the planet Jupiter covers 70,000 leagues [173,880 miles] per second.[55]

Bits of information such as the speed of light were of little practical use to Péclet's readers in Restoration France, but his general introduction also included a rough method for the determination of the intensity of a light when measured against a standard source, as well as such basic rules of optics as the equivalence of the angle of incidence and the angle of refraction and the way concave mirrors could focus a beam of light.[56]

Five of the book's eight chapters, however, were devoted neither to the discussion of general scientific principles nor to the application of these principles to the arts. Instead, Péclet gave a detailed description of each of the various types of lighting devices much as might have been done in some eighteenth-century compendium of the arts. The bulk of the *Traité de l'éclairage* was, like Benoît's proposed course on mechanics, a "vast essay of lighting *technologie*," a cookbook more than a scientific treatise. It offered students only the most technical kind of education, a fruitcake of particularities with only the thinnest icing of general notions about the state of optics as a scientific discipline.

One might nevertheless argue that even in the *Treaté de l'éclairage*, of all his works the one least concerned with general scientific principles, a kind of embryonic *science industrielle* could be found. He did mention the most dramatic relevant case of the application of scientific knowledge to the useful arts, Humphrey Davy's invention of the safety lamp. Until this invention miners had been faced with a choice between the risk of often fatal explosions and working in the feeble light of a flint and steel spark wheel.[57] But the *Traité*'s main contribution to the development of an "engineering science" lay in the chapters in which Péclet compared the different methods of lighting. In chapter VI he gave a detailed account of his testing procedures, including the problems he had encountered in measuring the intensity of the light produced from a given source.[58] In measuring the light output of various candles and lamps, Péclet not only compared the duration and intensity of the illumination per unit of weight but also gave the current price for the various candles and lamp fuels and commented on the quality of the light (such as its color) and the defects of the various devices. Chapter VI, in fact, reads like a set of articles from some modern product-testing periodical.[59] This testing and certifying function became, in fact, one of the most important links between the scientific community and the industrial sector during the early nineteenth century. The Académie des Sciences, the Société d'Encouragement, and the associates of the *Journal des connaissances usuelles* spent a good deal of their time sponsoring reports on various new products and processes. Despite its cookbookery, then, Péclet's treatise on lighting did contribute something to *this* aspect of *la science industrielle.*

For all its scientific approach and its opening tribute to the recent march of technological progress, however, the *Traité de l'éclairage* betrayed the dimmest of visions of the future of that progress. In reading it, one gets little sense of where to look for the next innovations in lighting or how to discriminate between methods promising further development and those likely to become antiquated. Péclet was apparently not prepared to make such judgments. Perhaps it is not surprising that he made no mention of the Drummond light (which used incandescent lime) since it had been invented by an officer of the British Royal Engineers only the year before Péclet published; the Frenchman Nollet had not yet made his proposal to modify it for use in lighthouses.[60] But the brilliant light generated by an arc between carbon electrodes had been well known since Davy first demonstrated it in 1802; only the development of a cheaper source of electricity prevented him putting the discovery to practical use.[61] The *Traité* gives no hint that this, too, might be made part of *la science industrielle.*

In Péclet's *Traité élémentaire de physique*, on the other hand, science and technology became more clearly open-ended enterprises. Although Péclet did not teach the course on General Physics during the latter part of the period under study, his textbook, constantly updated, continued to be the assigned reference work at the Ecole Centrale.[62] The introduction gave Péclet's fundamental views concerning what physics was and how it advanced. The object of physics was the study of the "general properties of inorganic bodies that we can see, touch, and weigh; as well as of several fluids that seem to be without weight, of which the existence is as yet only a probable hypothesis: the fluids that produce heat, electricity, and light." All the phenomena associated with these bodies depended upon "a very limited number of general facts, of which they are the necessary consequences."[63] If these facts were known, one could deduce, by rigorous reasoning, all the complicated phenomena to which they gave birth, "just as in geometry all the properties of various figures could be deduced (se déduisent) from a small number of axioms." The difficulty was that these "first causes" were unknown; for centuries philosophers had sought them in vain. But science made rapid progress only "since the time when the only method of investigation that we can employ" was known and practiced:

One observes a great number of phenomena of the same nature, into the series of which one attempts to vary only one of the circumstances on which they depend; one measures with precision its influence, and one searches by experiment (par des essais) the law that accords with the numerical results obtained.[64]

Whatever its limitations as a description of the way scientific advance occurs—many questions are left unanswered, such as the way one selects and orders the phenomena to be observed, or the role actually played by reigning paradigms, hypotheses, and pure intuition—this was a fairly conventional statement of what ordinary mid-nineteenth-century scientists thought they were doing.[65]

Péclet was quick to point out the limitations of this approach; when the laws governing the phenomena in question were "too complicated to be perceived," one had to resort to attempts to "tie together facts" by algebraic formulas that contained a certain number of constants determined by experiment. Yet these experiments themselves "presented great difficulties" because of the difficulty of measuring with precision "a length, an angle, or a temperature." Hence the "laws deduced by observation" were only valid within the limits of error contained in the experiments themselves and within the range of conditions that the experiments covered.[66] When the "observed laws" failed to permit the deduction of the "cause" of a phenomenon, it became necessary to create hypotheses; only when a single hypothesis could explain the observed phenomena would one reach "complete certainty."[67]

Such was the nature of progress in research in physics. The

march was "slow but sure"; well-observed facts and "laws established by exact experiments" became "definitive acquisitions" that formed new points of departure. At the end of his introduction Péclet stressed that much remained to be done: "physics, like all other branches of the sciences," was far from perfect; the list of things not known would be far more extensive than the "catalog of established knowledge."[68]

However small might be the realm of "established knowledge" when compared to the size of the unknown, the *Traité élémentaire de physique* was nevertheless an extensive catalogue. The fourth edition of 1847 was a two-volume work containing 1401 pages. In the introduction Péclet pointed out that textbooks on the "sciences of observation," unlike those designed to teach mathematics, could not proceed in a "perfectly logical order" in which each new item was based upon those which preceded it; whatever method of presentation one adopted, one was always obliged to make "petitions of principle."[69] Péclet thus divided his *Traité* into two parts: the first dealt with the general properties of all bodies, then with the particular properties of solids, liquids, and gases, and with acoustics; the second dealt with the "imponderable fluids": heat, magnetism, electricity, and light.

Although in this work Péclet was writing a general introduction to physics, not a survey of "industrial physics," his approach differed little from that used in the earlier treatise on lighting. The principal change was the addition of a hagiographical element. Since Péclet intended to be "especially careful to cite the scientists who have contributed to the advancement of science, because it is primarily by their inclusion in books designed for teaching that their names became popular and they receive the most noble recompense," the book became a tissue of names and reports of experiments.[70] Mathematical statements of the scientific principles examined were regularly set apart in smaller type so that they could be systematically ignored by his less advanced readers.

Despite such devices as smaller type for the more complex mathematics, Péclet's treatise imposed a considerable editorial burden upon a teacher attempting to use it with first-year engineering students. Even in the course of sixty ninety-minute *leçons*, only a fraction of the material could be covered. Péclet was at times excessively repetitious[71] and often too concerned to give the laurels to as many *savants* as possible, with the result that his book reported closely similar experiments, many of them clearly of tangential interest. As in the *Traité de l'éclairage* and in his book on heating, the *Traité de la chaleur*,[72] he constructed the work as a handbook, an encyclopedia, rather than as a systematic introduction to the general principles governing the field. The great weight he gave to reporting the details of experiments that established the various properties of matter he was discussing led him to include mention of a number of misguided or inappropriate experiments and to spend considerable space describing the functioning of apparatuses of measurement. Under the properties of gases, for example, he included a seventeen-page discussion on barometers that detailed the operation of nine different types. His explanation of this choice of presentation illustrates the approach he used throughout the *Traité*:

Barometers with a wide bulb—those of Gay-Lussac and of Fortin—and those with a dial face (*à cadran*) are the only ones currently in use. But, since the discovery of Toricelli [which Péclet had described in detail], barometers have been modified in an infinity of ways. . . . But all these attempts have been fruitless, and almost forgotten, because they involved inconvenient consequences. However, several of them contain ingenious arrangements that might receive other applications; and besides, it is useful to know them in order that those who try to perfect the barometer will not fall back into arrangements previously found to be unproductive. We shall thus discuss briefly the most notable of these previous barometers.[73]

Even though Péclet called his work an elementary treatise, then, he seems to have envisioned a significant number of his readers as future scientists who would continue the search of new techniques and new knowledge. Strangely enough, Péclet did not mention the meteorological uses of barometers, but his *Traité* did not confine itself to techniques of use only in "basic" research. He mentioned a range of technological devices, from steam engines (under "Heat") to daguerrotypes (under "Light") to the telegraph. That the telegraph should receive only two pages of commentary whereas the electric eel received four (not to mention seventy-seven for barometers) may indicate a preference for "pure" research, but a better explanation may be that Péclet's

approach was too strongly historical, too thinly woven with reflections about the possible future importance of various discoveries. He might have argued that it was pedagogically more sound
to describe instruments and procedures that had reached a relatively high level of development, such as the Gay-Lussac barometer, than to spend much time on rapidly advancing technologies
for which the specific devices were clearly in an "experimental"
stage, such as photography; but by the same token, the elaborate
attention he paid to superseded barometers seems even less justified. Perhaps Péclet was too much influenced by the structure
of the scientific literature: the number of reported experiments
concerning a particular problem in physics seemed to exert an
influence upon his editorial choices far greater than could be justified in a general survey of the field. In short, the *Traité élémentaire de physique* tried to be too many things: a general introduction, an intermediate survey for those with some
mathematical skills, an encyclopedic account of current pure and
applied research, and a handbook for experimentalists.

The fourth edition of Péclet's treatise, published by Hachette in
1847, also gives certain indications of his position in the French
scientific community. On the theory of heat, the study of the
creation and propagation of which became his primary specialty,
Péclet abandoned only with reluctance the rear guard of believers
in caloric. The second half of the *Traité élémentaire de physique*,
which dealt with "imponderable fluids," opened by noting that
two "systems" had been prepared to explain the nature of heat.
The oldest, "which had been almost universally accepted until
very recently," held that all space was occupied by an imponderable fluid whose molecules possessed great "expansive force" and
more or less "energetic affinity" from different bodies.[74] This theory
also held that all bodies constantly emitted caloric, which tra-
versed both empty space and gases with "a prodigious speed" and
produced varying degrees of reflection and absorption upon striking intervening objects. The second theory, "to which numerous
discoveries, only several years old, seem likely to assure the preponderance, if it is not already acquired," Péclet expressed as
follows:

In this system one admits, as in the first, the existence of an imponderable and expansible fluid, and one regards a phenomenon

relating to heat as the result of molecular vibrations that are propagated in the caloric fluid as are sound waves in the air—the hottest bodies would be those where the vibrations have the most rapidity and amplitude.[75]

Péclet then claimed that neither of these hypotheses was sufficiently demonstrated by known facts. "Happily," however, the "laws of heat" were "independent of these hypotheses." Nevertheless, "the first being more convenient for the explanation of most phenomena, we shall adopt it, even if we shall question its value later on."[76] Such statements published in a textbook in 1847 left their reader rather far from the forefront of scientific knowledge, especially when Péclet did not, in fact, explicitly question the value of the caloric theory in later passages.[77]

If *la science industrielle* is to be seen as a coherent intellectual structure unified by generalizations about natural phenomena that cut across the particular compartments of that science, the *Traité élémentaire de physique* stands open to the charge that it failed to lay the foundation for such a structure. Equally as serious as Péclet's adherence to the outdated caloric theory was his failure to introduce the concept of work into his discussion of the fundamentals of physics. These two lapses left his readers with no conceptual framework for understanding the great unifying theories of thermodynamics then emerging: the equivalence of heat and mechanical work is the linchpin of such theories. These laws might have given unity to *la science industrielle* just as they gave it to pure science. As a teacher at the Ecole Centrale, Péclet could hardly have been closer to the tradition of scientifically oriented French engineering that Thomas Kuhn sees as one of the keys to the development of the law of the conservation of energy (the first law of thermodynamics).[78] Of the two men most responsible for introducing the concept of work, Coriolis and Poncelet, the first was Péclet's colleague at the Ecole Centrale (if only briefly) and the brother of his wife, and the second taught at the Polytechnique and the Sorbonne. Emile Clapeyron, a prominent civil engineer, first presented a mathematization of Sadi Carnot's theories on the motive power of heat in 1834 in the *Journal de l'Ecole Polytechnique*, a journal that any serious physicist would have been expected to consult.[79] In 1839 Marc Séguin, a pioneer railroad engineer who had collaborated with Péclet's colleague

Perdonnet, discussed the equivalence of heat and mechanic work in his treatise on railroads and locomotive construction.[80] The publications of J. P. Joule, meticulous reports on the results of experiments that established the precise quantitative relation between heat and mechanical work, had begun to appear as early as 1843.[81] Without being given any discussion of such matters, Péclet's readers would have had no basis for appreciating the way other "conversion processes" then becoming better understood— electrical to mechanical, chemical to electrical, thermal to electrical—could become one of the great intellectual achievements of the nineteenth century.[82] Yet the transformation of energy was perhaps the single most promising organizing principle for *la science industrielle.*

On the other hand, in Péclet's defense it could be argued that even if information concerning the equivalence of heat and work had been published in various sources, the law of the conservation of energy, framed in its full generality to cover mechanical, electrical, chemical and magnetic processes, had not yet been articulated in such a way that one might expect it to appear in a general textbook. Attaining this generality was a complex process. Only in 1847 did Joule finish tracing the "connections" among the various energy conversion processes revealed by his own experiments and the work of others.[83] Again, only in 1847 did Helmholtz publish his *Ueber die Erhaltung der Kraft,* which established his claim to the discovery and first statement of this key principle.[84] Not until the 1850s did energy conservation become fully and universally established as scientific canon.[85]

Nor is it entirely fair to say that Péclet did not seek unifying principles for his treatises, however much the results seemed merely compendia. At the end of his discussion of electricity in the *Traité élémentaire de physique,* he, too, sought the general "connections" among phenomena:

In the long series of facts presented by the study of dynamic electricity, we have seen that movement, heat, chemical actions, and magnetism were sometimes the source of electrical currents, sometimes the effect produced by these currents. Electricity is thus intimately linked to the principles of heat and that of magnetism, and to the elementary constitution of substances. It is similarly linked to the principle of light since electricity produces light and exercises a direct influence upon light. . . . Hence electricity is involved with all phenomena, and we shall not discover its nature and its method of action without discovering at the same time the nature of heat and of the composition of substances.[85]

If the passage falls short of articulating the concept of energy, it at least demonstrates that Péclet hoped to point the way to future discoveries of greater generality. If *la science industrielle* is to be sought in the use of such discoveries as guides to technological endeavors, then, perhaps final judgment upon Péclet's contribution should be sought in his last work, a three-volume study entitled *Traité de la chaleur*. The third edition was completed in 1859, just before Péclet's death. Whatever the great practical usefulness of this handbook on the technology of heating devices (it became something of a classic in its field), it suggests strongly that the search for generality had flickered out. Neither the principles of thermodynamics nor anything resembling a concept of energy can be found in its pages.

If during the first twenty years of the Ecole Centrale's history physics groped with difficulty toward establishing certain of its most important unifying principles, chemistry's search for unifying structures and general laws produced considerably more noise and smoke without yet bringing forth a steady light that could command agreement as a guiding beacon. In the period before the construction of the periodic table or the beginning of the structural understanding of carbon compounds—both of which came well after 1848—the most promising unifying concept in chemistry was probably the atomic theory. After its first presentation in its modern form in John Dalton's *New System of Chemical Philosophy* in 1808, however, a long period of experiment and continuing scientific controversy ensued in which the atomic theory encountered both support and resistance, and in which experimental results several times seemed to call the theory into question. Leonard Nash characterizes the thirty years following 1827, when Dumas' work on vapor density first began to provide results that did not seem to support atomic theory as then understood, as a "period of intense scientific activity and a rather confused response to the atomic theory."[87] In his history of chemistry, Léon Velluz describes this period as *l'âge ingrat*, "the beginning of adolescence when forms are still imprecise and un-harmonious."[88] The discovery of important new elements such as chlorine and iodine added more grounds for speculation, and chemists responded with a "host of new theories," many of which were in fact directed to the reestablishment of some "universal principle" analogous to that held by phlogiston in the eighteenth century.[89]

The best evidence about the way the formulators of *la science industrielle* attempted to give a unified structure to a body of knowledge so much in flux lies in the pages of J.-B. Dumas' *Traité de chimie appliquée aux arts*. Stimulated by the state-of-the-art investigations and original research Dumas carried out in preparation for the *cours* he began to give at the Athénée in 1826, the *Traité* began to appear in 1828. New volumes were completed every two or three years until 1846. The *Traité's* publication thus coincides closely with the years during which Dumas taught Industrial Chemistry at the Ecole Centrale.[90]

Dumas presented his readers—and his auditors at the Athénée and the Ecole Centrale—with a grand tour of basic chemistry and industrial chemistry that gave little hint of the *Sturm und Drang* prevailing in chemical science during the second quarter of the nineteenth century. The *Traité* achieved this sense of clarity and unity principally by adopting three methodological strategies:

1. Dumas specifically excluded theoretical speculation about the physical phenomena that affected chemical reactions. He acknowledged that almost no movement of "molecules," and hence no chemical reaction, could occur without the application of light, heat, or electricity, but he refused to burden his readers with an investigation of the nature of these three agents since physicists were "of divided opinion" on the matter:

Some consider them to be distinct, imponderable fluids that by their absence, presence, or accumulation produce the phenomena we observe. Others admit the existence of only a single imponderable fluid permeating the natural world the various movements of which produce all the results that we attribute to light, caloric, and electricity. Whatever one may make of these two opinions, in chemistry we need only envisage these three sources of action as forces, as powers of which the actual nature matters little to us as long as we can observe the laws that they obey or at least appreciate what is general in their influence.[91]

In his discussion of electrolysis and of the relation between electrical and chemical "affinity," he adopted the views of Ampère, speaking of positive electrical fluid and negative electrical fluid, each surrounding a molecule of opposite charge.[92] He nevertheless warned his readers against accepting electricity as the single basis for the explanation of chemical reactions. He pointed out that there were a "host of phenomena" that remained inexplicable in electrical terms.[93] Factors such as the number of molecules, their relative position, or "perhaps even other circumstances" could introduce modifications that one could not yet predict or explain.[94]

2. Having warned his readers against electrical monism, Dumas then proceeded to subscribe to the most influential chemical monism of his day—the atomic theory. This theory, "accepted today by the majority of chemists," rested on "such simple principles" that it could be presented in a few words. For Dumas, the term "atom" referred to "the very small particle of body that gives birth to a combination by simple juxtaposition with the particles of another body.", Each atom thus underwent "no real alteration" in forming compounds; the properties of these compounds arose from "the grouping of atoms of various sorts." When a compound was destroyed in such a way that the elements (*corps simples*) composing it were isolated, the atoms of these elements reappeared with their original properties and "probably" with their original forms and dimensions, without any alteration. The atom of an element was thus "the smallest particle of that body that undergoes no change in chemical reactions."[95] At this point, however, Dumas revealed one of the ways in which his interpretation of the atomic theory differed from the modern conception: "The atom of a compound (*un corps composé*) is in turn the small group formed by the union of the simple atoms that constitute it."[96] Dumas' *Traité* thus never distinguished clearly between an atom and a molecule.

Dumas' version of the atomic theory soon ran into grave complications, however, especially in his work on vapor densities; his notions of chemical affinity prevented him from envisioning the union of two or more atoms to form a single molecule.[97] Yet he continued to use the language of atomic theory. As late as 1835, in the fifth volume of the *Traité*, he discussed the "determination of the number of atoms contained in an organic material."[98] In the following year Dumas delivered a series of lectures at the Collège de France that culminated in his famous declaration that if he were given the power, he would "erase the word 'atom' from science" because he was persuaded that the existence of such entities had not been established by experiment.[99] Dumas is thus often represented in the history of science as one of the principal opponents of atomic theory.[100] Yet even in his Collège de France lectures he never said that he did not believe in atoms, and he did retain a belief in the existence of some ultimate particle. For the purposes of the present investigation of the construction and propagation of *la science industrielle*, however, the important point is that Dumas did not feel compelled to change his language because of his doubts about the experimental fruitfulness of "Daltonian" atoms. The *Traité* contains no hint of these doubts. The last three volumes of the *Traité* were devoted entirely to compound-by-compound descriptions of properties and analytical procedures free of any general or theoretical statements about the nature of chemical reactions or elemental structures. If asked what gave a unifying conceptual structure to his applied chemistry, then, a young engineering student of Dumas' would have been able to offer without much hesitation the suggestion that one of its foundations was an interest in the precise way each of the compounds represented a particular combination of atoms of elements.

3. Dumas insisted on the unity of theory and practice. This was the guiding theme of the work, which began with a general discussion of the fundamentals of chemical theory and ended with a compound-by-compound discussion of properties, processes of manufacture, and analytical procedures. Each item was to be seen as an integral part of a work addressed to "*les jeunes industriels*" and to "young chemists now spread out in almost all the cities of Europe." The "ideology" of *la science industrielle* was perhaps best summed up in a passage from the preface to the *Traité*:

Many readers will find that I have given too many details of pure chemistry (*chimie pure*), that I have been mistaken to treat questions of [technical] art in a theoretical manner, and finally that I should have avoided the use of atomic language (*l'emploi des atomes*). To all that I shall reply that this work addresses itself to young men and not to manufacturers already established; that my intention has not been to describe the practice of the arts, but to clarify the theory of them, and that those scientific details which frighten away manufacturers "of a certain age" will be the easiest of games for their children when these sons will have learned in *collège* a little more mathematics and a little less Latin, a little more physics or chemistry and a little less Greek.[101]

In this paragraph Dumas neatly summed up the case for envisioning the (chemical) engineer as a scientist while simultaneously stating some of the basic premises of *la science industrielle*. There could be no separation between "pure" chemistry and the work of the engineer, entrepreneur, or agronomist whose activities involved chemical processes. Each of the chemically based practical arts in fact had a "theory" underlying its activities, a theory that the engineer needed to understand. Even if much of current practice seemed based on tradition and trial-and-error methods, future generations would adopt a different approach, based upon scientific knowledge. Classical education would inevitably be *supplanted* by (not augmented by) a more mathematical and scientific education.

Dumas was also suggesting that applied chemistry was a unified field. Nevertheless, the *Traité* does not give a completely clear picture of the degree to which he expected *Centraux* engineers to attain the total knowledge implied by the school's motto that "industrial science was one, and every *industriel* had to know it in its totality or suffer the penalty of remaining inferior to his task." He did not seem to demand such a totality of knowledge from the readers of the *Traité*. He admitted that "whatever one might say" about the improvement of scientific education in the *collège*, in order to draw profit from the "precise notions of chemistry" in industrial applications, it was "indispensable to study them in depth since the smallest details become of major importance when operations take place on a large scale."[102] Since it "would be difficult for each manufacturer to accord the same attention to all branches of pure chemistry," however, Dumas divided his work in order to group those arts that had "certain common bases." The most simple scheme, "after many attempts," contained four groups:

1. Nonmetallic materials, and the products and arts based upon them: water, the principal acids, ammonia, air, coal, heating and lighting.

2. "Alkaline earths" metals (*métaux des terres*), calcium, potassium, and magnesium compounds, and alkalis; "here as applications one finds the fabrication of certain important salts such as the potashes, soda, alum, saltpeter—and hence the preparation of gunpowder." This group, according to Dumas, had the "precious advantage of joining, among others, closely related manufactures such as potteries, glassworks, straw, paste jewelry, and finally, chalks and cements."

3. "Ordinary" metals: iron, copper, lead, zinc, tin, gold, silver, and platinum. Dumas stated that the "essential and necessarily preponderant" part of his presentation would be devoted to the extraction of these metals and the making of their alloys, although "less important products" had not been neglected.

4. All organic "products of nature" and the numerous applications derived from them: dyeing and bleaching, papermaking, sugar, soap, alcohol, cheeses, tanning, hatmaking, "etc., etc."[103]

Although the processes and the substances in Dumas' four groups may have had some unity according to the categories used in chemistry, these common ties had rather less significance in the world of industry and agriculture. The metallurgical processes in group 3 certainly shared certain procedures, but the connections between dyeing and cheese production seem more remote. In any case, the organization of Dumas' presentation into four groups is less important than the approach he intended to take when he examined each process:

In addition, I hope that by placing certain generalities at the head of [the section treating] each important manufacture . . . I will be able to furnish each manufacturer with the means to study the chemical principles of his industry, even though he has not studied general chemistry in any depth.[104]

As long as one understood some of the scientific principles upon which his own activity was based, therefore, the specialization produced by the technical demands of his work was nothing to be feared.

By the time Dumas had finished his *Traité*, the work had expanded to eight volumes divided into thirteen books totaling more than 6,000 pages. Despite the claims of the preface, the *Traité* did not confine itself to the application of chemistry to the arts or to the elaboration of the scientific principles underlying industrial processes. In the first place, the work contained discussions of many substances for which no practical use existed, such as the salts of yttrium, strontium, and selenium. They were of interest to "pure chemists" alone. In the second place, in a large number of cases the *Traité's* treatment of a process was indistinguishable from that of any careful description of the state of the art; no scientific principles were educed, no chemical language was employed. The section on ceramics, for example, contained little but the recipes for making the various kinds of pottery, with the ingredients described not in chemical terms but in the nomenclature traditionally used in the art. The description of the making of ironstone china, for example, was essentially a recipe giving the correct proportions of altered feldspar, Devon argil, flint, and flintglass to be fired at a temperature "lower than that of Chinese porcelain."[105] In other cases, scientific chemical analysis had a role to play, but only a limited one. In his discussion of breadmaking, for instance, Dumas offered a careful but nonanalytical description of the preparation of the ingredients and the procedure for baking. The new science of chemistry entered the picture only in an auxiliary role through Frédéric Kuhlmann's new test of copper sulfate contamination.[106]

Taken in its entirety, however, the *Traité de chimie appliquée aux arts* gave its readers ample reason to believe that they were consulting something that was neither a potpourri of philosophical speculation nor a cookbook of randomly chosen recipes. Neither discussion of the mysteries of the imponderable fluids nor doubts about atomic terminology disturbed its pages, while at the same time every effort was made to integrate "theory" and "practice." Certainly one could question the structural clarity, thoroughness, and coherence of Dumas' professed unity of theory and practice; at times the two seemed merely juxtaposed, not truly interconnected or independent. But Dumas had reason to argue that his successes were more important than his failures— and all the more remarkable in view of the encyclopedic range of the work. If there were certain omissions, such as the chemistry of photography, and peculiar inclusions, such as a long, somewhat pointless (even to a breeder) report on the character of seminal fluid,[107] there were rather more successes. From the summary of general chemical theory in the first volume to the clarification of the agricultural function of nitrogen and the carbon dioxide-oxygen cycle in the eighth, Dumas sought every opportunity to introduce "theory" into his discussion.

In this regard Dumas' task was somewhat easier than that of Péclet. Whereas all the products with which Dumas dealt could be seen as combinations of the same limited set of chemical elements, the physical common ground underlying, say, the behavior of gases, optics, and electromagnetic phenomena was more difficult to perceive. A second advantage for Dumas was that chemistry's practical applications had been publicly recognized since at least the Revolution; the rapid expansion of its accomplishments into new technological fields, producing new products and processes, was a phenomenon widely proclaimed and easily demonstrated. Finally, chemistry was a more accessible science; the mathematics required to understand most of its theory consisted of no more than elementary algebra, a situation rather different from that in physics and mechanics.

In one crucial respect, however, Dumas' contribution to the creation of a *science industrielle* did not differ so much from that of Péclet. The *Traité de chimie appliquée aux arts* presented a scientific *approach*, a *method* of systematic investigation, applied to industrial and agricultural processes. The analytical procedures repeated in Dumas' treatise were in this sense analogous to the careful testing and comparison in the *Traité de l'éclairage* and the *Traité de la chaleur*. The concern for precise measurement, the careful control of the conditions under which analysis and experiment were carried out, the attempt to standardize and purify reagents, the consultation of the full range of scientific literature, and the use of comparative techniques all received the sanction of Dumas' authority and the illumination of his clear, effective writing style.

Dumas' *Traité* also furthered the "deprivatization" of the chemical-using industries. Much of what Dumas reported had previously been part of the domain of individual manufacturers. If modern science had become defined in part by the fact that it was "public knowledge,"[108] it followed that *science industrielle* could not include trade secrets. Although Dumas did not pay tribute to other scientists in the hagiographical way that Péclet did, he usually acknowledged his debt to them, and he customarily began a section with a short bibliography of the most important recent journal articles.

Despite these positive contributions to the advance of chemical technology, Dumas' treatise did not provide a complete answer to Comte's question about the nature of the engineer's "special function" to "organize the relation between theory and practice" and to create the "genuine theories of the different arts." His *Traité* uses chemical theory to explain technological processes, but it never really specifies what constitutes the "theory" of the art in question, or even how such a theory might be organized. Such a theory is not inconceivable: as one way to begin building it, the practice of a variety of arts from metallurgy to agronomy could be described with reference to the lawlike statements analogous to Smeaton's "maxims." Since Dumas chose not to take this approach, one of the most important questions in the establishment of the scientific identity of the engineering profession, the autonomy of *la science industrielle*, was not fully answered.

The *Traité de chimie appliquée aux arts* also had its weaknesses as a guidebook for technological or scientific progress. Despite Dumas' frequent use of the phrase "history of chemistry" where merely "chemistry" might have been more appropriate, the work contains surprisingly little information about the *dynamics* of change, the way "chemistry" came to be "applied to the arts" in the past or the way it might be applied in the future. The reader learned nothing of the way new processes were invented, the way hypotheses were constructed and verified, or the conditions under which innovations were accepted or rejected. Dumas was hardly ignorant of such matters, but he chose not to discuss them. Neither the scientific victories of Lavoisier nor the technological victories of Fourcroy and Monge during the Revolution, neither the chastening story of Leblanc's business failure after his invention of the salt-to-sodium carbonate process nor the dramatic story of Dumas' own intervention on behalf of Daguerre[109]—

all of which might have revealed both the adventure and the inner workings of *la science industrielle*—ever appeared in his *Traité*.

Although it constitutes the principal sources of evidence about the substance of Dumas' teaching during this period, the *Traité* has certain limitations as a clue to the nature of *la science industrielle*. If one suggested to Dumas that in it he seemed to sanction far more specialization than seemed acceptable in his more programmatic statements, that he seemed to grant the impossibility of grasping the *"ensemble"* that the school's unofficial motto posited as the goal of its studies, he might have replied that his work was not addressed to Centrale students alone. The intended readership of *jeunes industriels* was much larger and more varied in its training. Even with the expected improvement of scientific education in the *collèges*, these men needed both a general introduction and a technical reference work. Legitimately immersed in the details of their particular specialties, they should be afforded the opportunity to understand some of the scientific principles underlying it while perhaps gaining inspiration for expansion into fields of endeavor which the *Traité* showed to be related.

Whatever arguments about the needs of his readers one might impute to Dumas, however, one cannot escape feeling that such needs were not the stimulus to which the writing of his treatise responded. The *Traité* appears more as the working out of a private agenda, published according to Dumas' vision of how the theory and practice of chemistry might be united. During the twenty years when the volumes of the *Traité* were being written, Dumas was still a productive scientist, one of the few men in the history of chemistry who made contributions to all its important fields, from the study of chemical substitutions (metalepsis) to physiological chemistry.[110] This was probably the last period in which a single man could claim a knowledge of the important aspects of all fields of chemical science as well as all the industrial and agricultural processes related to it. When such chemical encyclopedias were attempted later in the century, they were invariably collective endeavors. Although his Athénée audience's hunger for novelty may have helped to launch Dumas on his project, the drive to continue the survey seems to have stemmed from the urge to establish that the whole of chemistry could be harnessed to practical purposes. Whether this would be done by engineers, *industriels*, or the entire population seems in the *Traité* an issue of secondary importance.

The *Traité de chimie appliquée aux arts*, after all, was neither a prospectus for the school nor a public statement concerning the role of science in the education of engineers. In other words, it was not a primary location for an ideological statement concerning the status of *la science industrielle* and its practitioners. As has been suggested, when he addressed the public as a representative of the Ecole Centrale, Dumas made more explicit claims to be educating generalists who understood the *ensemble* of current technology. When dealing more immediately with the complexity and enormous variety in the substance of just the chemical part of this *science industrielle,* he more easily accepted specialization.

5

How They Were Taught

The search for *la science industrielle* now shifts to the structure of the teaching program and the methods used to carry it out. At least one of the difficulties raised by the previous chapter's study of engineering education through inferences drawn from general treatises and textbooks thus disappears: whereas it is not entirely clear for whom Coriolis, Péclet, and Dumas wrote, the curriculum at the Ecole Centrale was designed only for its students. The following discussion first considers teaching at the Ecole Centrale during the years 1829–1848 taken as a whole, then deals with some of the changes that occurred during this period.

During the early years of the Ecole Centrale, the educational requirements for admission were not especially stringent. The students had to demonstrate competence in the subjects taught in the *mathématiques élémentaires* class: arithmetic, elementary geometry, and elementary algebra up to and including quadratic equations. In addition, they were required to write a short composition on an assigned subject and to copy the outline of a human head.[1] This was the same set of accomplishments required of candidates for admission to the Ecole Polytechnique in 1794. Since that time the Polytechnique had added more complex requirements: trigonometry, conic sections, elementary statics, the use of logarithms, and the copying of an entire human figure (*une académie*).[2] As a result candidates who were judged *admissible* in the Polytechnique competition were exempted from the requirement for an additional Centrale examination.[3]

Unlike the system used for the Polytechnique, admission to Centrale was not organized as a *concours*. Instead of public examinations by examiners external to the locality and a strict system of ranking, the Ecole Centrale enlisted mathematics *professeurs* in the *collèges* or professors at foreign universities. Usually these teachers actually examined the students, many of whom they had taught themselves, but often they merely certified that the candidates had reached the necessary minimum level of attainment.[4]

Such admissions procedures, coupled with the wide range in age permitted (as young as fifteen, with no upper limit), produced a student body with widely varying capacities for profiting from the more advanced instruction.[5] The first year's curriculum thus had a double purpose. On the one hand, it was a "preparatory" year designed to fill the gaps in the students' knowledge and to "bring them as much as possible to the same level," ready to profit from the advanced courses of the two remaining years. On the other hand, it was supposed to provide the "general knowledge" that would be needed by all the students before they adopted one of the four specialties offered in the later years: machine building; the construction of buildings, roads, and bridges; mines and metallurgy; and chemistry.[6]

In response to this situation the school constructed a pedagogical system closely resembling that of the Revolutionary Ecole Polytechnique, which had been faced with a similarly heterogeneous student body. Four main methods of instruction were used: (1) The *cours*, sets of formal lectures broken into *lécons* of 90–120 minutes. Attendance was required for all students. (2) The practical exercises, both *manipulations* in laboratories and workshops and classroom practice in *dessin*. Part of these, considered of general interest, were required of all students; part were completed only by those enrolled in the relevant specialty. (3) Examinations and individual interrogations. (4) Individual projects completed during the school year or during vacations.

From early November until early August, students arriving at the school first gathered in the main courtyard at 8 A.M. for the reading of various announcements. In later years this gathering became a military formation and roll call with the students arranged by "division" (class).[7] The students then entered one of the main lecture halls to receive their principal lesson of the day.

Exactly what went on in those amphitheaters morning after morning is the subject of considerable importance not only for evaluating the technical competence of the *Centraux'* education but also for understanding the way they came to think about themselves and their place in society. Along with their parents and their professors in *enseignement secondaire*, these lecturers were their most important role models. Although few details about this teaching have yet appeared, the consistency of those that are available argues for the acceptance of a small but significant set of generalizations. In the first place, a certain minimum standard of pedagogical competence was successfully maintained. I have discussed how Benoit's proposed Mechanics course,[8] a rather poorly connected collection of topics, was quickly transformed into a logically progressive and coherent treatment by the skill of Coriolis. In general, poor lecturers did not last long. Parent-Duchatelet, a noted authority on hygiene but a timid, disorganized lecturer, lasted only two years before the school dropped him.[9] Antoine Raucourt, more interested in managing his own business affairs than in improving his lectures on civil construction, lasted no longer. Joseph Liouville, who was later to win election to the Académie des Sciences for his contributions to the theory of determinants, was asked to give up his teaching at Centrale (after six years of lecturing) because he lacked any interest in "applications."[10]

By contrast the men who formed the core of the teaching staff during these years were accomplished, effective instructors. Charles De Comberousse, who attended the lectures of three of them, has left us an account in his history of the school. Péclet, who taught General Physics to the first-year students and Industrial Physics to the advanced classes, "made himself both loved and respected." He had a straightforward, candid style that helped to temper the formality demanded of the *leçon* with a degree of spontaneity. He displayed considerable modesty, often describing his own work without attributing it to himself. According to De Comberousse, it was only after the lecture, "always listened to with pleasure and profit," that the students fully realized the "subtlety of Péclet's insights."[11]

The lectures of Jean-Baptiste Dumas were as different in style as they were similar in pedagogical effectiveness. The immense confidence that Dumas communicated to his listeners emerges in the choice of phrases in De Comberousse' assessment:

M. Dumas combined lucidity, elegance, breadth of vision, and richness of detail. The way he demonstrated the connections between apparently disparate facts removed all the aridity from a chemistry basking in the light of a new day. The most complex phenomena seemed to find order and simplicity. Thanks to that limpid speaking style, as sure of itself at the last minute as at the beginning, when you left the amphitheater you felt you knew everything.[12]

Dumas was in fact one of the most successful and inspiring lecturers in the entire French scientific world during a generation when such lectures, given at institutions such as the Athénée, the Sorbonne, and the Collège de France, played such a prominent role in French intellectual life. On 9 December 1842 Louis Pasteur wrote to his parents

I am attending the Sorbonne course given by M. Dumas, the celebrated chemist. You cannot imagine the size of the crowd in this course. The lecture hall is immense yet always filled. You have to go half an hour early to get a good place, just like at the theater. Similarly, there is much applause. There are always six or seven hundred persons there.[13]

According to his biographer it was at these chemistry lectures that Pasteur "became the disciple of the enthusiasms that Dumas inspired in him."[14]

In most of the ways that men of the July Monarchy measured success, Dumas had succeeded.[15] He had risen from obscure origins in the small southern town of Alais to become a member of the Académie des Sciences at the age of thirty-two, an extremely productive chemist engaged in internationally acclaimed research (at least until the end of the 1840s), and a *notable* whose brilliant marriage to the daughter of Alexandre Brongniart had assured him the highest rank in Parisian bourgeois society as well as the wherewithal to support a large private laboratory.[16] Nor was he the only figure at Centrale who could have inspired the emulation of the intellectually and socially ambitious. The first generation of the school's faculty included other men whose membership in the Académie des Sciences reflects the recognition of their achievements: the physicists Coriolis and Victor Regnault; the mineralogist Alexandre Brongniart; and the chemists Eugène Peligot, Jules Pelouze, and Edmond Fremy.

Théodore Olivier, the third member of the founding group to lecture at the Ecole Centrale, was also a forceful teacher. He spoke in a ringing voice, punctuating his sentences with carefully measured gestures. He solved one of the special problems in teaching descriptive geometry, the need of three-dimensional diagrams, by using cork-covered frames connected by hinges. Wooden rods inserted into the cork represented lines projected into space on various planes. According to another of his students, Francis Pothier, his favorite saying was that descriptive geometry was the handwriting of the engineer. Anyone who knew how to "read in space" could visit a factory and, without taking notes, prepare himself to reproduce a complicated machine after he returned home.[17] It was De Comberousse, however, who best captured the enthusiasm with which Olivier explained the subject perfected by his idol Monge: "Olivier raised descriptive geometry to the level of a religion."[18] As Monge had served Napoleon, so Olivier paid tribute to the emperor by enhancing his already remarkable resemblance to him: he maintained a lock of hair in the middle of his forehead in the same manner that Napoleon affected in the later years of the Empire.[19] Olivier was also known for his habit of engaging in familiar chats with his students after his lectures,

Théodore Olivier

talks in which he gave out advice on matters both technical and moral.[20]

The most detailed account of what went on in the Ecole Centrale's lecture halls appears in the memoirs of Daniel Colladon, who taught at the school from 1829 to 1835. Although he was acknowledged to be a promising young scientist, Colladon could neither teach in the *Université* nor gain election to the Académie des Sciences because he was unwilling to give up his Swiss citizenship.[21] He could teach at a private school, however, and Dumas had invited him to attend some of the final planning sessions before the opening of the school.[22] Originally employed as a *préparateur* for the physics and chemistry courses, he briefly succeeded Coriolis in the Mechanics chair in 1831, then developed a new course on steam engines.[23]

In our description of the most influential "core" professors of the school, those whose values and achievements would be most likely to inspire emulation, it would be a mistake to view Colladon merely as a young scientist in the forefront of pure research. To a greater extent than any of the professors mentioned, Colladon was personally familiar with the world of mechanical engineering then flourishing in the shops around Paris. The others were all strong advocates of the unity of theory and practice, but none knew better than Colladon what engineering practice really was. When he was asked to take over the Mechanics *cours*, Colladon imposed a single condition: that he be allocated the sum of 100 francs to pay for the cost of transporting to the Hotel Juigné items of machinery he would borrow from various shops. In successfully requesting Lavallée's support, Colladon claimed that he already knew "most of the important machine builders in Paris." The owners of the machines had to agree to allow their property to be dismantled and reassembled as often as required by the needs of the course.[24]

Colladon used the machines both as devices to illustrate the general laws of Mechanics and as subjects of study in themselves. In the section of the course dealing with the dynamics of pump action, for example, he collected fifteen different kinds of pumps and allowed the students to observe the action of each as it transferred the water between two tubs installed at different heights.

While each pump was in operation, Colladon pointed out its individual "secondary qualities," such as the ability to pump without interruption water containing sand and rocks. In the interval between the Monday and the Wednesday *leçons*, the students disassembled each of the pumps and made drawings of its interior.[25]

Colladon's lecturing demonstrated a flair for the dramatic. In order to show the "effects of weak but prolonged resistance to the flight of an object," he obtained a powerful "air cannon" with a relatively long barrel. He gave it a good charging, then pushed his thumb firmly against the muzzle as his assistant pulled the trigger. His students could hear the cannonball advance, then drop back down the barrel without having struck his thumb. Without recharging the weapon, Colladon then pierced a two-centimeter board with the same ball. At the end of the demonstration, Colladon warned his students that "a man who presses forcefully at the end of a cannon that has a certain length can do it without danger, but he takes the risk of seeing his thumb blown off if the barrel is too short."[26]

Even in the absence of details about the lecture-hall comportment of other professors, the main point in question seems established: the Ecole Centrale had a core of lecturers who could equal any other institution in their pedagogical skill and their force of character. If the formation of an engineering elite were simply a matter of responding to the charisma of a few major professors, the attempt to understand the contribution of a Centrale education to that formation would not need to extend much further than vignettes about these men. The process of education is obviously more complex than this. Without trying to get some idea of what the students themselves actually did, what tasks they performed other than listening to magisterial lectures, the search for an understanding of the French engineering student unearths but a shadowy figure.

During the *leçons* of the principal *cours*, the students took notes in specially supplied notebooks stamped with the school's crest. Each professor was supposed to check a sample of these notebooks at the end of each lecture and to report the results to the director of studies on a special form. Each *répétiteur* was supposed to examine the student's notebook during his individual interrogation; the result counted for one fourth of the grade for that particular exercise. Grades were assigned for the upkeep of notebooks when all were collected at the time of the final examinations.[27] Although the teachers' reports on student notebooks have not survived, the requirement to conduct such checking does not seem to have been a dead letter: in 1840 Victor Regnault's repeated failure to carry out this duty was mentioned in the normally circumspect minutes of the *Conseil des études*.[28]

Colladon's account of his course on steam engines, first given in 1831, illustrates the way that the school linked lectures with other exercises. This course was designed for third-year students, which meant that by this time they had all taken at least the first part of Péclet's Industrial Physics course and were familiar with "all that concerned the heating of boilers, hearths, radiators, and chimneys" as well as the laws of the vaporization and condensation of liquids. Colladon could thus assume such "theoretical" background and advance directly to the construction of steam engines. His course covered the following:[29]

1. Summary description of the principal machines with cylinder
at high pressure, without condensation,   with and
at high pressure, with condensation,   without
at medium and low pressure, with condensation.   *détente*

2. Work done by a given quantity of steam.

3. Special study of cylinder machines: dimensions, resistance of the cylinders. Metallic pistons or pistons covered in tow. Leaks and friction. Apertures and circulation pipes. Distribution apparatuses with and without *détente*; procedure used to move them. Condensation chamber, injection, air pump.

4. Recoil and transformation of movement. Oscillating cylinders. Parallelograms. Beams. Relative speed of piston and crankshaft.

5. Work (*travail*) absorbed by the interplay of the parts of a steam engine and by leaks. Disposable power. Practical results.

6. Systems adopted by the principal machine builders of England and France: advantages and disadvantages of each. Possible im-

provements. Choice and purchase of a steam engine. Installation, start-up, and maintenance of a steam engine. Annual costs. Accidents and explosions. Examples of steam engines applied to the pumping of water, work in mines, factories, and spinning mills.

7. Steam vehicles. [He refers to locomotives.] Steam navigation.

8. Little-used types of machines. Engines in which the steam acts by impulsive force. Direct use of steam to raise water. Atmospheric machines. Engines of immediate rotation.

This was the first time in France that a full course on steam engines was given in which full-scale machines had been used. When steam engines had been dealt with before—at the Ecole des Mines or the Ecole des Ponts et Chaussées—only small models (Colladon called them "toys") had been used.[30]

Before coming to Paris Colladon had spent considerable time in the spinning mills and machine shops in Alsace, where he had close relatives in the entrepreneurial class. He had continued this sort of inquiry in Paris. In the company of two collaborators, he had conducted "numerous experiments on all the principal steam engines in the *département* of the Seine." His recollections about the reaction of Parisian manufacturers to his request to conduct the experiments suggests that the division between theory and practice carried with it considerations of relative social prestige:

I had connections with the principal machine builders of Paris: Calla, Moulfarine, Saulnier, and the directors of the enterprise at Chaillot. They were flattered to have the theoretical advice of a scientist (*homme scientifique*) who knew workshops and who did not consider those who worked in them to rank below the theoreticians of pure science.[31]

As a result he was able to borrow whatever machines he wanted free of charge.

Because Colladon felt that no classroom demonstration, even with working machines borrowed from shops, could completely replace an investigation of steam engines as they functioned as prime movers in actual industrial situations, he arranged two workshops and factory visits each month for those students who planned to specialize in machine construction. During these visits, he carried with him workers' smocks, which he offered to students who hesitated to dirty themselves by a close inspection of the machines.[32]

Perhaps the most important aspect of the visits, however, was the training they gave in the art of reproducing a machine from memory. After a shop visit, Colladon divided his students into two or three groups. Each drew a sketch of a mechanical device that they had just observed. Colladon then compared their results to his own sketch. This was far more than an academic exercise. At a time when the leading industrial nations attempted to restrict the export of machinery and the emigration of mechanics, the information about particular machines often could be obtained only by brief visits to British shops and mills. Even if they received permission to visit the shops, foreigners were usually forbidden to make sketches during their tours.[33] In Colladon's course, therefore, *la science industrielle* came to include training in industrial espionage.[34]

Visits to industrial sites and scientific field trips soon became a regular part of most Centrale instruction. Auguste Perdonnet took his students to the Chaillot works to study locomotive building and to the Séguins' installation on the Ile St.-Louis to learn how a suspension bridge was constructed; Colladon added a tobacco factory to his itinerary; and Léonce Thomas, Colladon's successor in the steam-engine course, went as far as the works at Saint-Ouen.[35] Constant Prévost took his mineralogy students on trips to the sulfur springs at Enghien and the limestone quarries at Montmorency, and Brongniart took botany students to the hillsides of Sèvres.[36]

Colladon's memoirs reveal that these visits were more than just occasions for the acquisition of the information. He told the proprietors of the various industrial establishments that among his students were "many sons of the principal French and foreign *industriels*." Visits to their factories would then be "in their own interest" since it would give Parisian entrepreneurs and the *fils de famille* a chance to get to know each other.[37]

The *cours*, divided into *leçons* of 90–120 minutes, and their associated field trips are the most easily measured part of the Centrale curriculum. As table 6.2 indicates, for example, during the first semester of the year 1834/35, the first-year students received ten *leçons* per week, two each in General Chemistry, Descriptive Geometry, Mechanics, General Physics, and English, while the second-year students received fourteen *leçons*, two in Analytic Chemistry, Industrial Physics, Theory of Machines, Machine Construction, Public Works and Architecture, and Mineralogy and Geology, and one each in Stonecutting, Carpentry, and Gears and *Histoire Naturelle* Applied to Industry. Almost every year brought some slight change in this program, but its general pattern remained the same.

Along with the *leçons*, in the main *cours* the students were given *conférences*—lectures or sets of lectures, usually shorter than *leçons*, on more specialized topics. In addition, the school began to hire young tutors (*répétiteurs*) who conducted weekly individual examinations and acted as teaching assistants in the drawing and laboratory exercises. In principle, moreover, each *leçon* was concluded with an *interrogation* in which any student might be called on to answer a question or work out a problem on the blackboard.[38]

The distinctive characteristic of the Ecole Centrale's education, however, lay in the variety and quantity of the practical exercises that occupied the remaining hours of the school day—and probably much of the evening as well. For the first-year students, the projections and working drawings associated with the course in Descriptive Geometry were especially important. Two afternoons a week were spent on these and other board exercises that came under the heading of *dessin*. Each class also spent one afternoon each week in laboratory "*manipulations*" associated with the relevant type of chemistry. The variety of the remaining exercises reflects the diversity and complexity underlying *la science industrielle*: problem sets in physics; sketches (*croquis*) and working drawings (*dessins*) in Architecture and the machine courses (as early as 1832 thirty-six of these were required in Architecture alone); laboratory assignments in Stereotomy, General Physics, Industrial Physics, Mineralogy, and Mechanics; topographical exercises carried out in the environs of the school; and, especially for the advanced divisions, individual projects (*projets*) that required the integration of knowledge from a number of disciplines.[39]

Students about to enter their second or third years were expected to continue their training during the September–October recess. Not surprisingly, an excursion to England was the most desirable way to spend this time, after which the students' sketches, notes, and collected documents were evaluated and placed on file at the Ecole. Those who could not reach England were encouraged to seek temporary employment (a *stage*) in some industrial enterprise.[40] If this were also impossible, a student was required to submit drawings of selected buildings and machines.[41]

The climax of a student's career at the Ecole Centrale came not so much with the general examinations of the various *cours* (as was the case at the Polytechnique) but with the project executed for the final year's graduation competition (*concours de sortie*). After the completion of the course examinations, usually in the middle of July, the third-year students were given an order of the day listing projects for each specialty. In 1832, for example, the following assignments were given:

Chemical arts competition: Design a factory for the refining of gold and silver.

Metallurgy competition: Design a coke-using blast furnace, a hydraulic motor, and a blast-engine (*machine soufflante*).

Construction competition: On a given site, construct an Ecole Centrale des Arts et Manufactures for 150 day students. Design an aqueduct-bridge. Draw up a general plan for a medium-sized maritime and commercial city.

Mechanical competition: Design a cast-iron, piston-driven bellows set in motion by a wood-and-cast-iron hydraulic wheel with a force of 30 horsepower. Design a hydraulic wheel that can transmit 60 horsepower to an English-style forge (*forge à l'anglaise*).[42]

The students were then placed in a room for eight hours with only their own notes for reference, at the end of which time each was to produce a memorandum containing the "principal conditions, the elements, and the bases" of his assigned project. These memoranda were then deposited with the jury of the *concours* after the student had made a copy for his own use. Each competitor then had one month to study all the details involved in the execution of the project, to prepare finished diagrams and draw-ings, and to compose a memoir defending his choice of design. During this period the students could consult their professors, their comrades, and various textbooks, but they could not alter their initial statement of "principal conditions, elements and bases." Finally, in the middle of August, the students defended their projects before professors and classmates in an oral exami-nation taking as long as two hours.[43]

The development of this kind of teaching program at the Ecole Centrale helped to give structure and substance to the *science industrielle* that the school's prospectuses proclaimed. The atomic theory as expounded by Dumas (at least until the mid-1830s) might have only tenuous links with the execution of in-dustrial chemical processes; the principal of energy conservation may not have been part of a student's education before 1848; the use of the concept of work may have been confined to the course in Mechanics rather than used as a unifying thread—neverthe-less, a characteristic method, spirit, and orientation can be glimpsed in the daily activities of students at the Centrale. If Péclet's works emphasized the importance of systematic testing, so the classes of Colladon reinforced the point with each dem-onstration. The visits to industrial sites, the trips to England, and the *stages* gave training in precise observation that could be me-thodically communicated to others by the "science" of descrip-tive geometry. The final *projets* embodied a kind of "engineering science" in a number of ways: (1) The assignment's structure re-quired the student to present formally his analytical procedure. In depositing a list of "general principles," the "elements" and "bases" of his proposed project, he had to demonstrate familiarity with the scientific laws that set the limiting conditions of the processes involved. In refining gold and silver, for example, the physical properties of the two metals, as well as their combining proportions with other substances used in the process such as copper, lead, and mercury, would have to be stated.[44] (2) The na-ture of the assignments usually forced the student to demon-strate his abiliy to combine knowledge from several of the sciences in Comte's scheme: thus a blast furnace was simulta-neously a chemist's retort, an example of materials stress under high temperature, and a machine.[45] (3) It is likely that in establishing the "elements" and "bases" of such projects, the students employed what Layton calls technological science, lawlike statements derived not from nature but from the action of human creations. Even the less than completely successful *projets* indicate that students could move in this direction. The 1835 *concours* in the *constructeur* specialty called for the plan for a suspension bridge over the Loire. In his *projet* François Colinet first dealt with the decision whether to use iron chains or spun cables in order to support the deck. In addition to pointing to practical matters such as the easier fabrication of cables and the fact that single strands were more easily tested than links of a chain, he also mentioned the (rather approximate) formula that the tensile strength of a given thickness of iron wire cable was double that of the same thickness of wrought iron link. With more sophisticated "laws," however, he had more difficulty. A misunderstanding of an equation of Navier's relating sag, tensile strength, and deck weight led him to the conclusion that longer bridges were less subject to oscillation than were shorter ones! He then went on to claim that this "remarkable result" was confirmed by experiments: the engineer's science had to be empirically grounded.[46]

What is more difficult to judge is the degree to which students were willing to innovate. In the above-mentioned example Colinet conceived of the problem as largely one of adapting existing designs and formulas. His design for the deck anchoring, for instance, was copied from the one judged best by Vicat in his survey of Rhone suspension bridges. This was the typical pattern of the *projets* although not all students cited their authorities so carefully. Because the pressures for conservatism in *projet* drafting were so high, it is difficult to find in them the source of the small changes and adjustments that scholars such as David Landes view as vital determinants of the shape and speed of a country's industrialization.[47]

A better answer to the question whether the Ecole Centrale's program encouraged innovation and gave guidance on how to conduct it lies instead in the information about experiments carried on at the school itself. For twenty years Eugène Péclet conducted there his experiments on the transmission of heat and the flow of gases that would be incorporated into his *Traité de la chaleur*.[48]

He was assisted by a number of students, one of whom, Marc Daniel, later became his successor in the chair of General Physics.[49] In fact, Péclet brought his teaching into close touch with his laboratory work. As early as 1831, he had begun grouping together during the second semester the second- and third-year students who were taking his *cours* on Industrial Physics. He then adopted the method used when he had been a student at the Ecole Normale Supérieure: teaching in *conférences*, which in this case became laboratory seminars. The students were at times assigned *projets* similar to those in the final *concours*, at other times given limited experiments to conduct. Péclet oversaw the execution of the tests and drawings involved in the *projets*, giving short introductory talks, pointing out the relation of the activities carried out that day to the more general points covered in the first semester's *cours*, and discussing possible ways to improve the apparatus employed.[50]

During the years he taught at the school, Colladon conducted a wide-ranging series of experiments, from studies of electric eels to modifications in the design of high-pressure steam boilers. He often used students as collaborators in his work at the Ecole Centrale itself (where he was one of three resident professors), in his *atelier* at the rue Popincourt, and on his voyages to England and the provinces. In 1832, for example, he finished the construction of the engines for the steamboat *La Seine* at the rue Popincourt. In the company of three Centrale students he then sailed the steamboat to the quai Voltaire, near the Institut. Next he circularized the members of the Académie des Sciences, inviting them to inspect the craft. Among others, Poncelet, Navier, and Dupin accepted his invitation. His report of their inspection suggests that to these eminent *Polytechniciens* steam boilers of this type were still a novelty: "I was astonished to see how little they knew about the construction of machines. I was obliged to demonstrate every piece to them." His own students were "far superior to these distinguished scientists in their knowledge of both the theory and the practice of steam boilers."[51] Two years later, Colladon was assisted by another Centrale student, Luc Championnière, in a series of tests that established that for certain purposes the Savary steam engine, which had all but disappeared in France, was as efficient as more recent models.[52] Louis Véret, another Centrale student, accompanied Colladon on his first visit to Birmingham and Manchester in 1835.[53]

The archives of the school also contain the records of "research laboratories" that operated between 1832 and 1848. The work of these laboratories consisted almost entirely of the chemical analysis of various ores, alloys, and industrial compounds (such as slag). The reports suggest that this work was a routine part of the activities of students in the chemical specialty, performed, perhaps, at the behest of industrialists linked to the school. On the other hand, many of these operations, such as the extraction of uranium from pitchblende performed in 1839, had no immediate practical importance. The reports of these experiments often merely listed the results of the analyses in the manner of an assayer's report, giving just the amount of each constituent substance found; but others described fairly elaborate procedures.[54]

There is some evidence that the Ecole Centrale encouraged, or at least made possible, independent technological experimentation by students. Olivier's personal register for the year 1832–1833 tells how two third-year students, Léonce Thomas and Charles Laurens, began an attempt to use superheated steam to augment the power of steam engines. From previous experiments in Colladon's course, they calculated that the optimal effect would be reached at a temperature of 300 degrees. When they showed their results to Coriolis, who had been kept informed of their work, he advised them to publish their results immediately because an English physicist was known to be working on the same problem. The two continued to collaborate after graduating from the Ecole Centrale, where Thomas later succeeded Colladon.[55]

Such, then, were the activities comprising a student's education at the Ecole Centrale: lectures, laboratory exercises, problem sets, projects, factory visits, *stages*, voyages of observation, examinations, and experiments. The visible sign that a student had demonstrated his mastery of this composite "industrial science" was the school's diploma (*diplôme*) certifying his competence in all fields and his special achievement in one of the four specialties. Until 1838, when grades in previous work received equal weight, the graduation competition (*concours de sortie*) was the sole criterion for the decision to grant the *diplôme* or the lower-ranking certificate (*certificat*). Students were judged either "capable" or "incapable" of "directing industrial projects (*travaux industriels*)." Those judged incapable left the school "without diploma or certificate," and the professors were forbidden to give them private recommendations.[56]

The above paragraphs have described the general characteristics of the Ecole Centrale during the first two decades of its existence. In emphasizing those elements that were common to the period as a whole, they have slighted the changes that took place. The system established in 1829 had at least as much flexibility as it did rigidity, however, and revisions and experiments took place constantly. Some of the most important of these changes should thus be mentioned.

Once the school had been established as a going concern, the admissions requirements became more extensive. As the authors of the unpublished history admit, a rigorously exclusive policy was not deemed appropriate in the early years: "In order to assure the success of a nascent enterprise and to make it more easily acceptable to public opinion . . . it was deemed wise to admit all those candidates who seemed capable of understanding the *cours*."[57] In 1837 both the procedures for admission and the knowledge required became more elaborate. The admissions examiners in the provinces were required to submit not only the *procès-verbaux* of their oral examinations but also the written compositions, which covered the "solution of a practical proof in geometry with the use of logarithms, the solution of two equations with two unknowns," and the literary and drawing exercises referred to above.[58] At the same time, the prospectus of the school added to the set of subjects that were indispensable for the candidate a list of recommended subjects "for those who had the chance to extend their studies beyond that which was strictly necessary": the elements of descriptive geometry, "rectilinear trigonometry, the analytic geometry of straight lines and planes, and some notions of physics and chemistry."[59]

The year before those changes in the admissions program were introduced, the Ecole Centrale strengthened considerably its instruction in drawing and design (*dessin*). Six ninety-minute sessions were held each week. The 120 first-year students were divided into groups of 12, with three such groups assigned to the drawing boards of any one session; hence each student underwent at least six *dessin* sessions each month. In these sessions they were assigned a detailed list of line drawings and wash drawings (*lavis*) to be executed under the supervision of M. Thumeloup, a graduate of the Ecole des Beaux-Arts specially hired for the purpose.[60]

The variety and refinement of its *dessin* program quickly became one of the distinctive features of a Centrale education. Students were trained to work with a wide range of instruments, in both free-hand and mechanically assisted drawing, from models, written specifications, and—as Colladon's instruction shows—from memory. The rigors of the *dessin* program find prominent mention in the two most extensive accounts of student experiences at the school, those of Gustave Eiffel and Maurice Donnay. Eiffel's preparation for the Polytechnique left him poorly equipped to cope with the demands of Centrale for expertise in drafting. Centrale was to form "not mandarins of science, but practical engineers." He wrote to his mother that *dessin* was "almost the most important occupation of the first year." He told her that his other marks were so good that he would have been first in his class if it were not for this "*diable de dessin* where I am always very weak." It became Eiffel's "only nightmare"; he was convinced that the professor gave him bad grades "out of habit."[61] Maurice Donnay, a playwright who became one of the two nineteenth-century *Centraux* elected to the Académie Française, was himself the son of a graduate of Centrale who had been a *répétiteur* in the Railroads course from 1859 to 1874. Despite this background, Donnay was no more than mediocre in the exercises in descriptive geometry and *dessin*. According to his own account, this fault failed to block his engineering career only because of the special treatment given him by friends of his father.[62]

Changes in the school's *cursus studiorum* did not all move in the direction of stiffer requirements and more elaborate syllabi, however. There continued to be room for experiments that failed. In response to their teachers' frequent exhortations to familiarize themselves with British technology, the students requested in 1832 that a course in English be established. The school promptly hired a M. Spiers, who also taught English at the Ecole des Ponts et Chaussées. The course remained outside the required core of offerings, however, and in 1836 it suffered the fate of so many such optional courses when it was discontinued for lack of interest.[63]

The most frequent changes came in the last two years of the school's program, when the students began to pursue their specialties. In 1844 Michel Alcan, an 1834 graduate of Centrale who had rapidly built a successful career in the construction of textile machinery, initiated a third-year course in textile manufacturing.[64] In the same year, Alphonse Salvetat, an 1841 graduate who had become managing director at the Sèvres porcelain factory, established a course in ceramics.[65]

The most important new course established in the period under study dealt with the most dramatic technological change in the nineteenth century, that which became the central symbol of industrialization: the railroad. In 1832, only one year after he began teaching the course on Mining and Metallurgy to second-year students at the Ecole Centrale, Auguste Perdonnet began to add material on railroads to his lectures. In 1837 he divided the course in two and thus established the first full course on railroads given anywhere in the world.[66] Perdonnet's lectures and laboratories, along with the visits he led to the site of the Paris-Versailles line (of which he was engineer-in-chief) soon became the most important single spawning ground for a whole generation of railroad engineers. If any one subject could claim the title, this was Centrale's bread-and-butter course.[67]

Although courses continued to be added and dropped throughout the period under study, the year 1839 can be taken as a kind of curricular watershed. Until that time the school's organization had been considered "provisional." No detailed set of statutes describing the governance of the Ecole and prescribing the content of its education had ever been promulgated. The changes were made through informal consultations between Lavallée, the *professeurs-fondateurs*, and other important faculty members. They were made known to the public when they appeared in the annual prospectus. In early 1838 the need was felt for a "definitive organization," and a committee was appointed composed of

Péclet, the chairman; Belanger, Professor of Mechanics; and Walter de Saint-Ange, who taught the third-year course on machine construction.[68] In the fall of 1839 Lavallée presented the committee's report to the *Conseil des études*, which then issued a detailed set of regulations incorporating most of the committee's findings.[69]

The new *règlement*'s stipulations concerning the organization of the school will be discussed below. For the purposes of this section's investigation of the role of the curriculum in the elaboration of *la science industrielle*, the value of the *règlement* lies in the way it crystallized certain aspects of the school's educational philosophy. Most clearly established were the priorities for the first-year students. The "four principal *cours* and their dependent practical exercises (*travaux*)" constituted the core of the curriculum: Descriptive Geometry, General Chemistry, General Physics, and Mechanics were each allotted sixty *leçons*. Each student was to receive from eleven to fifteen private examinations (*examens particuliers*) by the *répétiteurs* of the *cours* as well as a single general final examination by the professor. The *dessin* exercises were divided into twenty sessions of line drawing and twenty-four sessions of wash drawing.[70]

Three observations may be made about the *règlement*'s first-year program: (1) The regulations clearly stipulated that the first-year students' time must be *equally* divided among the four principal *cours* and their associated practical exercises. The minutes of the *Conseil des études* for 1838 and 1839 suggest that whatever claims might be made that this division reflected the quadrilateral symmetry of the theoretical core of *la science industrielle*, relations among key staff members prompted the emphasis upon curricular parity. The account of the meeting of 9 February 1838 suggests that Olivier, the only *fondateur* giving a first-year course at that moment, was thought to have usurped a bit too much of the students' time. "A member" of the *Conseil* stated that one of the purposes of the newly appointed committee's work would be to "restore the harmony" of the first-year program threatened by the "increased importance" of the drawing exercises, which added more hours to the "already considerable" time spent on *travaux* relating to Descriptive Geometry.[71] (2) This problem of balance was partly solved by the creation of standing committees empowered to suggest changes in the programs of the "four principal courses." Each standing committee included the professor of the first-year *cours* as well as two others from advanced and first-year courses. The committee on Descriptive Geometry thus included Olivier, the professor of Architecture and Public Works, and the professor of Machine Construction; the committee on General Chemistry included the teachers of second and third-year chemistry (hence a place for Dumas). The committee on General Physics included the professors of Mechanics and of Industrial Physics (Péclet); and the committee on Mechanics included the professors of Machine Construction, General Physics, and Industrial Physics.[72] Not only did this arrangement assure that curricular planning would always have to consider the relation between the general-theoretical and the "applied" subjects; it also guaranteed that a *fondateur* would be part of all decisions regarding first-year courses and that none of these courses would find itself bureaucratically isolated with only its instructor as advocate. (3) The subjects that did not find themselves designated as "principal courses" remained in a problematic relation to *la science industrielle*. The most striking example is *Histoire naturelle*, taught in 1839 by the distinguished zoologist Henri Milne-Edwards, who had just been elected to the Académie des Sciences: it never found a stable home in the curriculum, but moved back and forth between the second and the third year. Parent-Duchatelet's course on Public and Industrial Hygiene had also been planned for the first year, but it was not reinstituted after his departure.

The regulations of 1839 also show how the Centrale staff confronted one of the most perplexing problems in *la science industrielle*, the question of specialization. All the students in the second and third years were required to attend all the lecture courses, which varied in length from eighteen *leçons* (for *Histoire naturelle industrielle*) to seventy *leçons* (for Dumas' course on Industrial Chemistry). In addition, all the second-year students were required to participate in a range of practical exercises and field trips that included material pertaining to all four specialties. They also took all the final examinations. In the third year, however, students were allowed to concentrate on their specialties. During the first semester, from 15 November to 15 March, each carried out a "specialty project" that was criticized by the professor in a private conference. In the second semester each student received another such project while taking part in practical exercises confined to those in his specialty: the metallurgists and chemists were given special laboratory *manipulations,* the *constructeurs* and *mécaniciens* made building layouts and designed machines, and the metallurgists took a geological field trip. At the same time, the third-year students were exempted from examinations in courses not relating to their specialty: all three nonchemical specialties did not take the Industrial Chemistry examinations, the metallurgists did not take the examinations in Railroads, and the chemists did not take the examinations in Machine Building and in Railroads.[73]

This system did not provide a complete answer to the question so important to Comte: whether there could really be a separate "doctrine" half-way between pure theory and direct practice. The only second- or third-year course that claimed to lay out the general principles for two of the four specialties, Walter de Saint-Ange's Theory of Machines, disappeared in the new regulations.[74] The vague precepts of *la science industrielle* gave no basis for deciding that metallurgists should be exempted from the examination in chemistry, or, for that matter, that there should be any particular number of specialties in the final years.

Nor did an understanding of the sources of technological innovation emerge from the published curricula any more clearly than did the character of the autonomous principles of engineering science. As many recent studies have shown, the sources of such changes were varied and complex—in their relation to scientific advances, in their institutional origins, and in their economic motivations.[75] The "application of an existing body of scientific knowledge to the industrial arts" hardly describes the only—or even the predominant—manner in which these innovations originated. Those who guided the Ecole Centrale may have been dimly aware of this: they spoke of the *unity* of theory and practice more often than they did of the "application" of theory to practice.

In any case, given the state of French industrial development in the first half of the nineteenth century, the Ecole Centrale was well advised to maintain a program less specialized than Ponts et Chaussées or Mines. Unless he were guaranteed a position in the enterprise of a parent or relative, the Centrale graduate could not be certain what part of his program would prove most useful. Gustave Eiffel took the chemical specialty, intending to become the director, then the inheritor, of his uncle's chemical factory at Pouilly. When his father suddenly quarreled with his uncle during Eiffel's last year at Centrale, the possibility closed forever.[76] Postgraduation shifts in intended specialty were common throughout the nineteenth century.[77] The Eiffel Tower thus stands as only one of many tributes to the versatility permitted by the educational reality that lay beneath the insistence that *la science industrielle est une.*

6

The Curriculum
Justified

During the years from 1829 to 1848, life at the Ecole Centrale was marked by remarkably few of those ceremonial occasions so popular in other French schools. If any speeches were given when the successful students received their diplomas, nothing about their content has survived in the archives or in published references. Since there were no prize competitions, the historian of Centrale must do without the regularly published Prize Day speeches that have provided insights into the changing values promoted at the secondary schools over the course of the nineteenth century.[1] Extended reflections upon the meaning of a Centrale education or the "philosophy" of *la science industrielle*, which might have appeared in such speeches, are consequently less than abundant. Instead of presenting a statistical distillation of the common themes of many documents, then, this chapter examines at length the two most detailed attempts to explain the structure and signficance of Centrale's educational program: a speech by Charles De Comberousse at the fiftieth anniversary of the school's founding and remarks by Theodore Olivier in the preface to his major work on descriptive geometry.

De Comberousse: The Case for
Generality

De Comberousse's speech at the *cinquantenaire* banquet borders on being a secondary source: he attended the school during the years 1858–1859, a decade after the end of the period under consideration here. His explanation of *la science industrielle* could have been influenced as easily by the concerns of the 1870s as by those of the 1840s. And one has to expect a certain amount of idealization in this kind of speech. On the other hand, there is reason to believe that De Comberousse could not have made a speech that was heavily idiosyncratic. Alumni and professors who had known the school in the earlier period were in the audience, most prominent among them being Dumas himself, with whom De Comberousse had worked closely. Nowhere in his speech does he suggest that his understanding of the purpose of the school's curriculum, its relation to *la science industrielle,* or its contribution to the creation of the scientific engineer is at variance with the general view. It may thus be argued that De Comberousse's statement articulates the philosophy of *la science industrielle* as most of those associated with the school understood it.

Although the anniversary banquet was an internal affair, De Comberousse found that he had to take care to make himself especially clear to one part of his audience: the women. He apologized if at times he seemed to put "quite arid pages" before their eyes, but he welcomed their interest in the school and their desire not to "abandon their husbands, brothers, and children on the doorstep of the school that gave them their professional knowledge and their distinctive stamp (*marque distinctive*)."[2] It was to this wish for enlightenment that De Comberousse directed his explanation of the rationale for the special *formation* that shaped the *Centraux* as a social and technological elite.

He began with a statement about the overall structure of the Ecole Centrale's program. According to his interpretation of the curriculum, two subjects stood as the "poles" of that education: Mechanics and Chemistry. The "intermediate sciences" all borrowed "more or less" from these two:

Take, for example, the construction of machines. It is founded on Mechanics for the design of moving parts, for the calculation of the resistance of materials, and for questions of form and dimension. It interrogates Chemistry, in its turn, to learn the intrinsic qualities of these materials, to predict their transformation and deterioration under external influences, or to create them from a variety of substances if this becomes necessary.[3]

Much the same could be said of "civil constructions," public works, the exploitation of mines, and metallurgy:

The blast furnace of factory and foundry is only the crucible of the chemist's Lilliputian laboratory transported to Brobdignagian country, but always heated by Gulliver, the human spirit, who has no need of a larger brain to spawn his miracles.[4]

Such connections between the two "master branches," Mechanics and Chemistry, meant that the two subjects had preoccupied the school's founders from the very beginning. Thus De Comberousse claimed that he must treat in some detail the history of the two sciences that "perhaps one day, by means of the mechanical theory of heat, will become joined in a majestic unity."[5]

According to De Comberousse, Mechanics had originally been "less favored" than Chemistry, but the "struggle for recognition" was soon won, thanks to the efforts of teachers such as Coriolis and Belanger. The latter, whose "works were now in the hands of everyone," had given the Ecole Centrale a complete program in Mechanics, from kinematics by hydraulics. Belanger was one of the most important creators of the science of the engineer:

It is thanks to him that the theory of the resistance of materials, an essential tool, has been disseminated so widely among us. . . . He has familiarized us with difficult notions the knowledge of which is indispensable to all those industries which involve Mechanics. He thus enabled engineers of the Ecole Centrale to create for themselves a kind of specialty in these important subjects.[6]

In the perfection of this speciality lay the key to the liberation of men from physical labor.

In contrast to the early struggles of Mechanics, Chemistry immediately found itself launched on a successful career by the talents of J.-B. Dumas. After fifty years of discoveries and theoretical advances Dumas' original three-year plan was still recognizable beneath all the additions: their arrival had been foreseen. In the first year of General Chemistry were taught the "fundamental principles," the "philosophical part." In the second year the students learned Analytical Chemistry, in which they "descended" more to details, especially those covering "research procedures" and "useful methods of measurement." In the third year, Industrial Chemistry appeared. De Comberousse claimed that in this course, "none of the major applications of chemistry has been left aside: they are envisioned from all aspects and described with minute care." To justify this approach he cited the passage in Dumas' *Traité*, already quoted above, about how the smallest details became of great importance in large-scale chemical operations.[7]

At this point De Comberousse arrived at the most problematic part of his exposition: the "filling of the interval" between the two "primordial" sciences. He took first the course in the Construction and Establishment of Machines, which had been widely imitated abroad. Since the number of machines "necessary to know" grew every year, Centrale instructors had begun to distribute to students lithographs of all the "relevant sketches." Lectures could then "confine themselves to describing the principal types," leaving it up to the students, with the aid of the lithographs, to study "all the derivations and transformations." This method had soon been adopted by all the other courses "invaded by descriptions of new inventions," preventing both professors and students from "succumbing under the burden" of new detail. To De Comberousse, the essence of the course on machine construction lay in a small number of general principles of design:

Proportion economically the dimensions of the parts so that there is always enough material to resist, never enough to weigh down or fatigue unnecessarily; make the choice of material according to the goal to be obtained; seek simplicity without sacrificing elegance; make machines both light and powerful, well balanced, and as silent as possible, by a skillful fitting (*ajustage*) and mounting (*montage*).[8]

The courses in Architecture, Civil Constructions, and Public Works applied the same deductive procedure to the construction of private buildings and other *travaux d'art*. The graduate of the Ecole Centrale was thus ready to make the blueprint for "a guardhouse, a railroad station, a hospital, or a school, just as well as he could lay out the plan of a road, a canal, a port, a viaduct, or a tunnel."[9]

De Comberousse next discussed the sequence of courses dealing with metals:

Mineralogy and Geology in the first year, then the Exploitation of Mines in the second, form another group in which knowledge of the nature of the globe serves as the basis for the examination of its internal riches and the procedures for extracting them.

The final course in the sequence, Metallurgy, included the study of combustibles, metallurgical apparatuses and procedures as well as the different methods of fabricating pig iron and steel.[10]

The only discordant note in De Comberousse's presentation sounded during his discussion of the physical sequence. The advanced part of this set, Péclet's course on Industrial Physics, though one of the most innovative and useful in the curriculum, had been little imitated elsewhere. Those outside the Ecole Centrale "doubtless saw in it, quite mistakenly, a superfetation of isolated lessons." Yet the "industrial study of heat, its modes of employment in all their forms, from the apparatuses that imprison it to the chimneys of our apartments and the heating devices in our homes, [was] of the highest importance." The study of ventilation naturally accompanied it. In short, Péclet's lessons constituted "the baptism of a new science."[11]

At the close of this survey of the school's major courses, De Comberousse attempted to explain how they fit together as a whole:

These courses dealing with a single science, chemistry, mechanics, or geology, extending themselves over the whole three years of study and transforming themselves, from year to year, according to the point of view one wishes to predominate, are one of the special characteristics of an Ecole Centrale education. In this manner they form a rationally organized encyclopedia (*encyclopédie raisonnée*) of increasing and continually renewed value in which the same philosophical principles illuminate the different questions treated and lead the student to rise, by a progressive assimilation of which he is scarcely conscious, to the heights necessary to understand the connection of all the parts he has been taught.[12]

For De Comberousse, then, full appreciation of the unity of *la science industrielle* arrived slowly, somewhat mysteriously, over a long period in which specific facts and general principles (*within* each science) accumulated, became absorbed, and gradually blended within the student's consciousness to produce the mind of a true generalist, encyclopedic yet coherent.

At this point in his talk, perhaps calculating that the attention of the women in the audience would begin to flag, De Comberousse made his treatments of individual courses considerably more brief, and his descriptions of the way courses fit together even more vague. He mentioned the courses on steam engines and railroads merely to pay tribute to their best-known instructors: Colladon, Léonce Thomas, and Perdonnet. Then, "in order not to prolong excessively this *tableau*," he limited himself to the statement that "mathematical Analysis, that indispensable tool; *technologie*; the various bodies of knowledge (*connaissance*) relating to agriculture, newly introduced into our programs; and industrial legislation are not neglected at the school but instead form a truly imposing ensemble."[13]

Only after this seemingly climactic generalization about the "imposing ensemble" presented by the curriculum did De Comberousse decide (as an afterthought?) to give "special attention" to the courses on Descriptive Geometry and Biology (in the school's terminology "The Natural History of Living Creatures"). Descriptive Geometry, "which extends our students' knowledge of geometric forms and their inexhaustible properties, as well as their numerous applications in the arts and in constructions," was the domain of Olivier, to whom De Comberousse duly paid tribute. The Biology course was intended to inform the students about "all organisms, animal and vegetable . . . used in the industrial arts." The "precise notions about the organization, function, and classification of animals, and the internal structure of plants" gave the course a "scientific thrust." This course was in fact the linchpin of the curriculum:

It was an eminently philosophical thought which guided the founders of the school when they made a place for the natural sciences in their program. The introduction of the subject broke down, so to speak, the aridity of the mathematical, mechanical, and technical sciences; it realized that conception of unity that had so much inspired our *Maîtres* from the very first, and that characterized their creation, and which must be developed and not diminished.[14]

De Comberousse thus suggested three forms of unity for the school's curriculum, the first organized around the "poles" of chemistry and mechanics, the second tying together sequences of related courses along progressions from general to specialized and theory to application, and the third somehow encompassing the natural and physical sciences and the arts (not to mention agriculture and industrial legislation, which were not taught during the earlier period). Rhetorical flourishes at ceremonial banquets have their limits as historical documents, however. De Comberousse's scheme had many loose ends. The relation of descriptive geometry to these three forms of unity was left unspecified. Was it a link to more theoretical mathematics, a science capable of continued advance, or merely a set of perfected techniques and formulas? Nor was it clear what engineering practice had to do with the concerns of biologists, even if the animate and the inanimate were all part of the natural world upon which the engineer worked his technological transformations. Should a student in Alcan's textiles course wrestling with the problem of weaving wool concern himself with the question of phylogenetic recapitulation in sheep embryos? During the period under consideration in this study, moreover, the *Histoire naturelle* course was an uncertain quantity, moved from the second to the third to the first year as if no one could decide where it belonged.[15]

The plan of study De Comberousse described had not gone without criticism, and he next addressed himself to these charges. The main complaint was that the curriculum was overloaded. He admitted that even the most enlightened and sympathetic observers thought that Centrale students were becoming overworked and overstuffed with badly digested information. When students were taught less, they would know better. The goal should be not to teach them everything, but to teach them to learn (*apprendre à apprendre*). De Comberousse had to acknowledge the changes in the curriculum since 1829: "Oh, *Mon Dieu*, I am well aware that when the school opened, the students only had nine *cours* to attend and that today, in their three years of study they must take thirty." He claimed that he was thus *theoretically* in agreement with the school's critics but that in practical terms he felt obliged to continue to apply the same programs and the same method, for this was the only way that Centrale could conserve its "superiority" and the "public utility" that would recommend it to the government of the Republic.

The principal justification for the current encyclopedic program was the fact that graduates had no place guaranteed them in State employment. Most did not know where they would start their careers, nor could they expect to stay in the same type of activity in which they began. Of 100 students, only 10 "stayed in the path of their first steps"; hence the students needed not a single key but a "trousseau of keys."[16]

De Comberousse's second justification for the curriculum revealed more clearly the psychological assumptions implicit in its structure:

Undoubtedly, for a profound mind the fundamental principles of a science, presented skillfully and methodically and followed by a few characteristic applications, suffice for the later solution of all problems. But average minds (*les esprits moyens*) would not be capable of following the logic that unites principle to details and hence would remain indecisive when faced with unexpected questions. We thus arm such minds as best we can by multiplying examples . . . by reproducing the guiding idea in its transformations and principal modifications.

Yet the emphasis upon these guiding ideas underlay the superiority of a Centrale education:

The school . . . is thus at the antipode to those technical institutes, so numerous abroad, where students can themselves choose the *cours* they wish to take, following no overall conception (*vue d'ensemble*), and becoming architects without the slightest knowledge of heating or ventilation, mechanicians innocent of metallurgy, chemists incapable of installing a boiler.

The founders of the Ecole Centrale, in a flash of genius, said that "industrial science is one," and their work is there before our eyes to show how right they were.

What a bad thing specialization is! Of course, it facilitates work, which becomes a habit, a second nature, as Aristotle would say. But such work is less attractive: one's intelligence, less exercised, less solicited for information, becomes stationary. It searches for other stimulations, sometimes damaging, outside the assigned task. What fatigues, what wears down, is not work itself, but the same work endlessly repeated. Man is not made for such a merry-go-round—one must correct the specialization of the profession with the generality of the education. Well, I would say that at the Ecole Centrale there are no specialist engineers (*ingénieurs spéciaux*) but only generalist engineers (*ingénieurs généraux*).[17]

De Comberousse also argued that the Ecole Centrale could produce no specialists because the larger fields within which specializations occurred were constantly changing; never before had "applications and modifications poured forth with such intensity." Learning to keep *au courant* had become a necessary part of an engineer's education: "Because of this, one finds more breadth and less depth, because the intellectual volume must remain the same." Yet in this need to keep up with rapid changes in a number of interrelated fields lay the antidote to boredom: "From this, also, [comes] the possibility to renew oneself at will, to avoid becoming locked into a single specialty."

De Comberousse's anniversary speech thus assembled a collection of explications and justifications of *la science industrielle* that left a number of questions unresolved: (1) He made little attempt to establish the autonomy of engineering science. *All* of the knowledge directing an engineer's actions apparently arose from a single source: the application of "science" to the arts. There was nothing of independent value in the procedures and particular facts encountered in engineering practice; all apparently had their roots in "science," a body of discoveries and laws that unambiguously prescribed courses of action to that practice. (2) His remarks justifying the encyclopedic approach of the Ecole Centrale presented contradictory rationales for the curriculum structure. On the one hand, all the subjects taught lay between the two "poles" of Mechanics and Chemistry, apparently incorporating elements of each. On the other hand, the curriculum structure was determined not by the state of development of those "polar" sciences but by the distribution of employing enterprises and the expected shape of careers. At one end, *la science industrielle* sprang from the laboratory of the *savant*, but at the other, it was a child of the French industrial structure, shaped by its needs, both current and anticipated. De Comberousse implied that since most *Centraux* did not have the "profound minds" required to make the bridge between a compact set of general theories and a host of detailed industrial practices, engineering education had to attempt to reproduce within each individual a knowledge of the current state of the art in *every* field. Whatever may be the ambiguities in De Comberousse's speech, however, his conclusion indicates that the essential purpose of his exposition was not a systematic investigation of *la science industrielle*, but rather the embellishment of a crucial determinant of the Centrale graduate's social status: the portrait of the civil engineer as generalist.

Because the Centrale engineer derived a good measure of his cultural cachet, and therefore his social position, from having communed with "philosophically" pure, unified, "primordial" sciences, De Comberousse had little choice but to make the case for generality. Because the actual possibilities for applying the school's education to economically productive tasks remained unclear and ever changing (a fact equally important for the school and for the individual student), he had to make the case for universality.

The claim to generality appealed to many of the same values that underlay the claim to generality made from the traditional *en-seignement secondaire*, shrine of *honnêteté*, the Good, the True, and the Beautiful, and *culture générale*. Even though De Comberousse's engineers had sat in the *lycées* with the rest of the upper bourgeoisie, he suggested that their liberal education somehow had not bestowed a powerful, penetrating understanding of, and commitment to, generality that would then permit them to specialize professionally without deforming either character or intellect. This is not surprising. As has been shown above, De Comberousse, Dumas, and their colleagues at Centrale were unimpressed by the prevailing classicist *secondaire* ideology; they saw clearly its superficiality, fragmentation, and stalemated cultural visions. It was thus all the more important for them to claim unity and generality for their own *science industrielle*. Yet the *secondaire* curriculum had certain important advantages over *science industrielle*. Its material was fixed, static, controllable, purged of impurities and inconsistencies, the distillation of centuries of pedagogy. The *morceaux choisis* had been carefully chosen for their cultural usefulness and their congruence with the prevailing morality. Even when works of French literature were proposed for inclusion in the canon, they, too, could easily be sanitized and excerpted as part of the process of becoming immortal "classics." Even today educators cannot always distinguish between the immortal and the merely embalmed. Science, on the other hand, was more difficult to manipulate. One could sometimes mask its uncertainties, its mysteries, and its flux—we have seen in chapter 4 how this was done by Péclet and Dumas—but in the nineteenth century, the changes were too rapid, too many, and too fundamental to give teachers confidence that the essence of its unity and generality could be securely mastered. It was as if a protean and neurotic Cicero, sitting at the Sorbonne, were still composing orations to be entered into the textbooks.

The intellectual foundations of *la science industrielle*'s claim to generality were nevertheless far less precarious than those supporting its claim to universality. The groping progress of nineteenth-century science was an orderly march compared to the rapid, chaotic economic, social, and technological changes brought by market capitalism, industrialization, and international competition: in the period of 1872–1879, an annual average of 5,085 patents and 5,579 registered bankruptcies; in the following decade, 6,853 patents per year and 7,412 bankruptcies.[18] As De Comberousse spoke, whole new fields of industrial production (such as artificial dyestuffs) were being created, and others (like steel-making) were being totally transformed.[19] A curriculum embodying *la science industrielle* might still appear *raisonnée*, but it was becoming less and less credible that it could remain encyclopedic. The logic of De Comberousse's simultaneous exaltation of the status and employability of the Centrale engineer nevertheless left him little choice but to make such an argument.

Olivier: The Crusade for Utility

Théodore Olivier wrote his reflections about the meaning of a Centrale education not in the self-congratulatory atmosphere of a jubilee ceremony but in the daze of upheaval during the early months of the Second Republic. The document in question, his *Mémoires de géométrie descriptive, théorique, et appliquée*, appeared in 1851, but the crucial prefatory remarks were completed at the end of 1848.[20]

The Revolution of 1848 received no welcome from Olivier. In his preface, he states that during this "ill-fated year for France", he, like all true lovers of science, withdrew into his research to find a "useful distraction" from the "moral sufferings" inflicted by the "demogogic saturnalias sprawling in the streets and clubs." In his escape from "palaverers" (*parlementeurs*) and their "ambitious parliamentary foolishness" Olivier had found solace in his family and intellectual companionship in the Société Philomathique, to which he had delivered the memoirs collected in this volume.[21]

The preface reflects more than Olivier's immediate disgust with the events of 1848, however. He had been worried about the fate of education in France for a much longer time; hence he included in his preface remarks on the Ecole Polytechnique written in July 1847 since he thought that they were "not without interest for those who concern themselves, but too late, with the current state of public education, and who are worried about the future

of the country in the light of the deplorable situation produced
by a system of education maintained too long against everyone
(*envers et contre tous*) despite many warnings from the wise."

Like Comte and Tocqueville, Olivier provides evidence for the
prevalence of attitudes giving high prestige to theoretical
investigations:

Man, . . . forgetting that he is condemned to live on this earth,
and dreaming only of the place where he will go after his terrestrial exile, places above everything his purely intellectual
achievements. Thence comes the distinction between theoreticians and practitioners that has established itself among scientists. Those theoreticians who take the name of *pure scientists*
consider themselves to form an aristocratic corps having the
right to command and dominate the practitioners, whom they
treat as vassals. The pure scientists thus forget, like nobles of all
species and countries, that *work* is the condition imposed on
man, and that work useful for humanity must be honored and
rewarded more than amusements that men devise as mental
games or circus tricks.[22]

Unlike Comte and Tocqueville, then, Olivier refused to make
any gestures toward the superiority of theoretical endeavors.
Man's God-given task was to work, in imitation of the act of creation. Man had been granted intelligence for a single purpose: to
produce "sublime and continually renewed modifications in the
elements that form the terrestrial globe he inhabits." For an idea
to become useful to men, it must be "united to a body, materialized." But not only was technology the sole sanction of science;
technological needs underlay all scientific ideas:

There is not a single human science that does not owe its birth
to the fact that labor is a necessity of the human condition. It is
always the need to satisfy an earthly need that has led human
intelligence to create in succession all the sciences taught in
schools.[23]

Only an elementary development of the sciences was needed to
satisfy these primary earthly needs, however. In each century
appeared men who launched themselves farther in search of the
truth, producing new ideas which remained for some time "only
ideas," without application and without real utility, until finally
they "materialized themselves" to satisfy new needs that had
come to "accumulate in the head of Man." But this process of
"materialization" should not be slighted:

If scientists who reserve all their admiration for purely philosophical and speculative discoveries were willing for a moment to descend to the workshops and to examine carefully the series of operations through which one must pass in order to materialize an idea, I am certain that they would show less disdain for the applied sciences. They would realize that a practitioner must always possess greater energy and intellectual will, and that he must often master a greater variety of scientific specialties to materialize the idea of a theoretician than the latter needed to put forward the idea.[24]

Even while he challenged the claim of theoretical science to greater esteem, then, Olivier granted an important point to its proponents: technological activities were viewed as the result—at whatever stages of remove—of scientific discoveries made by the theoretically inclined *savant*. That a nonscientific artisanal tradition or tradition of technological science might contribute as much to the intellectual underpinnings of this process of fulfilling human needs did not appear clearly as a possibility.

Olivier's argument contained a second difficulty. He based his argument for the moral superiority of technological activities on the divine command to work. As Saint-Simon had been among the first to point out, however, the moral force of technological activity can be seen as resting instead upon its power to liberate men from the most arduous work while simultaneously elevating the moral relations of society by replacing the government of men by the government of things.[25]

When he moved from general comments on the relation between theory and practice to an explanation of the structure of the curriculum, Olivier offered a test for the accuracy of De Comberousse's interpretation of the school's pedagogical rationale:

Education at the Ecole Centrale is based on the study of geometry, mechanics, physics, and chemistry; every engineer must possess a rather extensive knowledge of these four sciences. Geometry teaches him the properties of representative space (*l'espace figuré*); mechanics teaches him the effects and measurement of forces; physics teaches him the laws that govern bodies according to their composition; and chemistry teaches them the proportions of the elements that constitute those bodies; physics and chemistry have links that unite them, and the same may be said of geometry and mechanics, and of physics and mechanics.

The theoretical teaching of these four sciences, but with theory always taught with regard to industrial applications, forms the course of study of the first year. The links uniting these four sciences become apparent if they are taught with regard to applications, and each student will recognize immediately the need to study each of them with equal care if he wants to be an engineer, regardless of his subsequent industrial speciality. . . .

The professors never forget that the purpose of the Ecole Centrale is to provide well-trained, capable engineers, not scientists (savants). Hence during the first year they examine in detail only those theories that lead to industrial applications. All the other theories are examined summarily so that the students will know what has been discovered. If, later, new industrial applications are based on these heretofore sterile theories, they thus will not be unaware of them and will know where they can study them in their full development.

The professors shall not be pure scientists; they must be chosen as much as possible from among active or retired engineers.

All the courses of the second and third years are industrial courses, in which are shown the applications of the theories expounded during the first year to the work of engineers; the number and duration of the courses and the substance of the lectures vary with the progress and needs of industry. . . . Whatever their specialties, the students attend all the lecture courses, but they only execute those projects and laboratory exercises related to their specialties. Industrial science being one, each student must know it in its entirety. This goal is obtained by attending all the lecture courses. The practical projects develop the spirit of invention and teach the students to materialize theoretical ideas; a limited number of projects can achieve this purpose.[26]

Such was Olivier's rationale for the program of technological education examined in the previous chapter. A single year's lectures gave the future engineer a sufficient general knowledge of the four principal sciences because only theories of proved value in the applied sciences received detailed consideration. A "summary examination" of other theories would enable students to gain an understanding of their fundamentals sufficient to enable them to appreciate the significance of new discoveries that promised to yield beneficial practical results. To "become aware" of machines using electromagnetic induction, then, one did not have to be an Ampère or a Faraday; the previous "summary examination" of electrical theory in the first year of physics at the Ecole Centrale would enable one to appreciate the engineering possibilities of a dynamo or an electric motor. (In fact, a practical electrical-power technology failed to appear for more than a generation after the scientific fundamentals were established, a delay that is one of the more intriguing puzzles in the history of engineering.[27]) Scientific knowledge was still unified to the extent that a meaningful acquaintance with all potentially important theories could be imparted in this single year, at the same time that one acquired a more profound knowledge of those theories that had already convincingly established their utility. Olivier assumed that these two kinds of theory could be clearly separated in order to receive different pedagogical treatment: *sommaire* or *approfondi*. Yet when one considers some of the "theories" to which one might apply these classifications, a vision of discrete sets of propositions that could be examined in turn throughout the year does not emerge very clearly. How does one discuss, say, atomic "theory," caloric "theory," or the formulas for measuring physical work? Should these have the same status as propositions about the pressures exerted upon the inside of a boiler, the way to interpret the stress on a beam, or the proper shaping of the blades of a turbine? Lacking any further explanation of the nature of a theory, Olivier's claims about the structure of the first year's instruction seem more like articles of faith than careful considerations arising from reflections about what actually went on in the classroom. How could one be certain, moreover, that the orientation of the first year's courses toward "theories" of already proven utility would not considerably distort the logical sequence and general comprehensibility of the entire program?

According to Olivier's account, the descriptive geometry taught in the first year seems to be all the mathematics that a student at Centrale had to learn beyond what he had already learned in secondary school. Algebraic methods and the analytical procedures of advanced calculus received no independent attention, whatever development they may have received in the Mechanics course. Nor were the biological sciences of the same importance as the physical sciences and geometry; the engineer's task was to act upon the inanimate world. Olivier does not mention the curriculum's early experiment with a subject that recognized the existence of human beings in engineering activities, Parent-Duchatelet's course on Hygiene.

Certain conditions laid down in Olivier's statement leave some doubt as to the effectiveness with which the unity of *la science industrielle* functioned as both legitimizing myth and guide for practical policy. Not only were men who understood theory the only ones who could actually practice engineering effectively, but only men steeped in practice could teach theory: Olivier wanted no "pure scientists," even for his first-year teachers. Yet he failed to distinguish between those men *interested* in the practical applications of science and those with experience as working engineers. Two of Olivier's most important colleagues, Dumas and Péclet, belonged to the first category but not to the second, and the same could be said for many others who taught at the school during this period.

In any case, Olivier's brief discussion does not provide a fully satisfactory answer to the questions Comte raised about the nature, autonomy, and coherence of the engineer's intellectual domain. On the one hand, the preface to the *Mémoires* suggests that he envisioned the generation and development of scientific theories as an activity uniquely the province of "pure science." Regardless of whether the subject were caloric, the behavior of atoms, the flow of water in turbines, or electromagnetic phenomena, the impetus for change seemed to come from the theorizing described in that first year of general science. *Applying* those theories would thus seem to be distinct from such theoretical calculation, a process separated in time and space from the activities of the *savants purs*. In thus implying that engineers' actions were necessarily derivative, Olivier seemed to vitiate his attack upon the social prominence and aristocratic pretensions of the "theoretical scientists" and their claim to the commanding heights of scientific and, hence, technological creativity.

Olivier can be defended here, however. He might have replied that the substance of the first year's instruction had no independent significance, that *la science industrielle* could only be judged as a whole, as the creation of three years of work, of lectures on the general theory of the sciences and on the construction of railroads, of laboratory exercises and personal projects, short terms of employment and periods of travel. The engineer's claim to high social status would thus reside in his command of all aspects of the process whereby ideas were transformed into useful products and services. Theoretical advances in "pure research" had no social and moral content until the engineer transmuted them into the coin of Utility. Consider the way Olivier referred to the practical projects in the last paragraph of the passage quoted above: by carrying out even a small number of projects, a student could acquire the "spirit of invention" and learn to "materialize theoretical ideas." The single general—and generalizable—method of materialization and a single general "spirit of invention" were thus independent ingredients of the engineer's *science industrielle*, autonomously produced at the Ecole Centrale.

But here, as so often in the study of engineering ideology, one seeks in vain for elaboration. Olivier did not intend to engage in a fully developed investigation of the nature of engineering science. If this process of "materialization" could be learned in a limited number of projects, presumably there were certain generalizable principles inherent in the process of which the nature could be further explored in order to render the teaching of the "process" more efficient and reliably comprehensive. But Olivier chose not to conduct this exploration; nor did anyone else at the Ecole Centrale. In the case of the concepts that supposedly unified the lectures in the first year's general sciences, moreover, Olivier was equally reluctant to specify the binding threads of *science industrielle*: the interrelation among the various subjects somehow "became evident" if they were taught "with regard to applications." Only in practice did one perceive the unity of theory.

If, then, Olivier's descriptions of the curriculum leave many questions unanswered, his comments about the history of French education between 1789 and 1848 reveal some important ingredients of the spirit that infused an education based on that curriculum. These passages may not tell us any more about what *science industrielle* was, but they do help to explain the forcefulness with which its merits were proclaimed, the sense of mission that sustained both professors and students through the intermittent adversities and constant hard work of the school's first years.

According to Olivier the Revolution, in establishing the *écoles centrales* and the Ecole Polytechnique, had created a most promising set of institutions. Both of these "remarkable works" were pedagogically sound because they were based upon "the alliance of theory and practice." The *écoles centrales*, moreover, helped to hold together the strongest threads in the social fabric by returning students to their families every afternoon—Olivier echoes Lacroix here—or, at worst, to a small private *pension* where they received the protection and guidance from a *maître* acting *in loco parentis*.[28] Such an arrangement did not suit Napoleon's purposes, however. He wanted to make France a "military camp" in which the students in each *collège* would undergo an apprenticeship in the martial arts, marching from class to class to the sound of a drum. The life of the barracks replaced the life of the hearth; loyalty to emperor and nation replaced family ties. Olivier gave a bleak picture of the consequences of this policy:

From that day on the familial spirit was destroyed in France. When the Restoration arrived . . . the nation found itself without strength and without energy; family ties no longer existed; each returning French soldier found himself isolated and unaided; no childhood friends, no relatives, no family. France was covered with men separated from each other by political opinions.[29]

The economic effects of the years of barracks life were equally severe. Agriculture and industry languished during the decade when all of the nation's efforts were devoted to the maintenance of French imperial dominance. Life in camps and *lycées* had been a poor preparation for the returning soldiers who undertook to revive the economy: "Who can count the sums of money that were thus swallowed up in badly conceived and badly directed enterprise?"[30]

In Olivier's view the Restoration brought no improvement. "Louis XVIII loved Horace: only Greek and Latin were taught." Royer-Collard, powerful as both deputy and member of the Royal Council on Public Instruction, took up philosophy, which then became a second obsession of the schools. In its passion for philosophy and classical studies, the Bourbon government did not perceive that *collèges* so organized would only disgorge upon France a host of "rhetors and sophists, feuilletonists and young tribunes." No one thought to give the country schools capable of producing *industriels* and scientific engineers (*ingénieurs savants*), men really useful to the country.[31]

The distortion of secondary education was paralleled by pernicious developments at the very highest level, at the Ecole Polytechnique itself. Napoleon had reduced the length of studies to two years, required the students to live at the school, and generally increased its military aspects, but he had not challenged the unity and balance of the set of theoretical and practical courses that had enabled the school to serve as a training ground for all engineering specialties.[32] The visions of Lacroix and, especially, Monge continued to guide the school during its Napoleonic phase.[33] This brilliantly conceived organization was then "destroyed, and with a light heart," by the men of the Restoration:

Thus, and I say it with a heavy heart, it was under the baneful influence of Laplace, Poisson, and Cauchy that the Ecole Polytechnique was reorganized in 1816.

These men, who knew no language but algebra, who thought that one was ready for anything when he knew algebra, who esteemed a man only to the extent that he knew algebra, who were incapable of rendering services to this country other than in algebra, destroyed from top to bottom the original organization of studies at the Ecole Polytechnique. Instead of a school of students destined for the public services, they created a school of mathematicians, an Ecole Normale Supérieure entirely devoted to algebra. They suppressed the applied courses, conserving only a few lessons (*leçons*) of architecture and a few lessons of descriptive geometry, and one no longer did anything in the school but algebra; experimental physics disappeared, and physics became a course in algebra; the mechanics of machines disappeared, and mechanics became a course in algebra. They [the three men mentioned above] cannot deny that their thinking was as I have described it, because they and their disciples have often repeated that the Ecole Polytechnique is designed above all to give the country two or three *savants* of Analysis each year.[34]

Given their obsession with reducing all education at the school to questions of pure mathematics, the "reformers" of 1816 should have changed the name of the school from *polytechnique* to *monotechnique*.

According to Olivier, the change in the Polytechnique's program had several negative side effects as well: (1) *Professeurs* in the secondary schools began to spend the last four months of the academic year preparing half a dozen selected students for the entrance examinations while neglecting all the others. (2) The nature of the mathematics taught at the secondary level changed. During the previous period the mathematical manuals were written with "elegance and simplicity": the arithmetic, geometry, and algebra of Lacroix, the analytic geometry of Biot, and the statics of Gaspard Monge. At that time, moreover, the admissions examiners concerned themselves mainly with whether the students grasped "the spirit" of the principal theories. They did not bother with "all the objections, often trivial, and with all those problems and theorems that were more curious than useful." Professors did not overload their students with material useless for anyone not attending the Polytechnique. Since that time, studying such matters had "warped more than one judgment and obliterated many intellects that would have remained healthy but for the baneful direction imposed upon the professors by the admissions examiners at the Ecole Polytechnique." (3) The dominance of the Polytechnique curriculum had grown as the number of students to gain admission to the school gradually became the principal standard by which a mathematics teacher's success was measured. The authority of the *Université*'s own inspectors-general had thus been undermined. At the same time, the *Université* began to lose students to the cramming schools set up especially to prepare for the *écoles spéciales*, especially the Polytechnique. Such a system sapped the provincial *collèges* of their best students while bankrupting the *Université*'s own educational program—as Dumas had shown in his 1847 report.[35] (4) As the algebraic approach came to dominate the Polytechnique to the detriment of methods and subjects more directly tied to practical applications, education at that school became increasingly similar to that given at the Ecole Normal Supérieure. Yet the students at Polytechnique were to become *travailleurs*, whereas those at Normale were destined to be *philosophes*. For Olivier it was "impossible to understand how the same program of study, carried out in the same spirit, could be appropriate to both at the beginnings of two such different careers." What happened, in fact, was a withdrawal from engineering. At the first opportunity young *Polytechniciens* left their projects in the provinces to return to Paris, hoping to gain teaching posts that would support them while they spent the remainder of their time in pursuit of scientific truth. Because teaching posts were difficult to obtain, many were driven to the publication of "a swarm of elementary treatises on arithmetic, geometry, algebra, analysis, and statistics," all designed to capture the market composed of Polytechnique candidates and all "most harmful to the true advance of science."[36]

Confirming Olivier's claim that "dropouts" from engineering comprised a significant proportion of the authors of the elementary textbooks that did, in fact, appear in significantly greater numbers from about 1830 lies beyond the scope of the present inquiry. One piece of evidence suggests that there may be a kernel of truth to the argument, however. Of the 6,131 students who left the school between 1794 and 1853 and were eligible for assignment to the *services publics*, only 21 were originally placed in the State educational system, but by 1854 250 more had entered *l'instruction publique* at some later time. This group of 4.4 percent of all graduates was the fourth highest in Marielle's tabulation, following field artillery (39.2 percent), military engineering (21.1 percent), and Ponts et Chaussées (19.3 percent), but greater than Mines (3.6 percent).[37]

The more important question is the accuracy of Olivier's portrayal of instruction at the Ecole Polytechnique during the post-Napoleonic years. On this matter the sources do not permit a quick and uncomplicated conclusion. This period saw one major reform of the curriculum, in 1816, followed by a series of partial and selective adjustments in course offerings. The 1816 reform was devised by a five-man commission headed by Laplace.[38] Pinet states that Laplace "wanted a purely theoretical school in which the study of advanced Analysis ["Algebra," in Olivier's terminology] predominated."[39] Furthermore, in his report, Laplace asserted for the first time the Polytechnique's crucial role in the French social hierarchy:

[The Ecole must be considered as] an establishment designed to complete the education of the young men destined to form the elite of the nation and to occupy high posts in the State. . . . We live in a time when the instruction of the upper classes (*les classes supérieures*) can alone assure the tranquillity of the State by allowing the members of these classes to obtain, by a personal superiority of virtue and enlightenment, the influence that they must exercise over others. . . . With regard to the sciences and to all varieties of positive knowledge (*connaissances positives*), the Ecole Polytechnique will permit such generous ambitions to be fulfilled.[40]

In Pinet's account, the reform of 1816 climaxed a decade-long process of whittling down the courses devoted to the building of public works, despite the resistance of Monge, "with whom Laplace had had several agitated arguments." The Polytechnique thus "lost the character of a school preparing for the public services."[41]

Documents listing the programs of study in effect at the Polytechnique throughout this period do not reflect such a drastic change, however. Table 6.1 indicates that changes in the portions of time devoted to various subjects did not always move in a direction supporting the Olivier-Pinet interpretation. For the first year, the percentage of total class and laboratory time devoted to Analysis actually dropped from 29 percent in 1806 to 23 percent (including 1 percent for Analysis applied to Geometry) in 1818, while Descriptive Geometry dropped an equal number of points, from 26 percent to 20 percent. For the second year, Analysis remained constant from 1806 to 1818 (21 percent, with 3 percent and 4 percent, respectively, devoted to the "Analysis Applied to Geometry" category), whereas Descriptive Geometry actually increased from 0 in 1806 to 2 percent (10 *leçons*) in 1818. Table 6.2, taken from the *annuaire* of the Polytechnique in 1837 and from official Ecole Centrale sources, similarly falls to reveal the marked contrast between the two schools that Olivier's strictures would have led one to expect.

Nor did the post-1816 reforms reflect a mentality of pure-science elitism. In 1819 the Polytechnique's Council on Improvements, headed by the Duc de Doudeauville, recommended the addition of two courses, one dealing with "the theory of machines and the calculation of their effects" and the other with "social arithmetic," which their report justified in the following manner:

When one considers the daily developments in French industry, and when one considers the necessary relation between this industry and the form of government established by the *Charte*, one must feel that the execution of public works will tend in most cases to take place in the system of concession and enter-

Table 6.1
Curriculum changes at the Ecole Polytechnique, 1806–1818

	1806		1818	
	No. of *leçons*	Percentage of total hours	No. of *leçons*	Percentage of total hours
First year				
Analysis	60	29	55	22
Mechanics	35	17	38	15
Descriptive Geometry	110	26	70	20
Analysis Applied to Geometry	0	0	12	1
Physics	25	5	30	10
Theoretical Chemistry	36	9	36	14
General and Applied Chemistry	0	0	34	8
Grammar and Belles-Lettres	36	2	0	0
History and Belles-Lettres	0	0	?	4
Dessin	75	12	70	6
Second year				
Analysis	50	18	45	17
Mechanics	60	22	55	20
Descriptive Geometry	0	0	10	2
Analysis Applied to Geometry	20	3	15	4
Fortifications	30	6	0	0
Civil constructions	30	6	0	0
Mining	10	2	0	0
Machines	0	0	15	5
Geodesics	0	0	16	8
Physics	25	7	18	4
General and Applied Chemistry	36	9	36	12
Architecture	50	13	38	7
Grammar and Belles-Lettres	36	2	0	0
History and Belles-Lettres	0	0	34	9
Drawing	127	12	70	12

Source: Fourcy, *Histoire de l'Ecole Polytechnique*, 376–379.

Table 6.2
Comparison of curricula at the Ecole Centrale and the Polytechnique

Ecole Centrale	Ecole Polytechnique
First year	**First year**
Descriptive Geometry (60L)	Descriptive Geometry (68L)
General Mechanics and Analysis (60L)	Analysis and Geometry (48L, 4R)
General Physics (60L)	
General Chemistry	Analysis Applied to Geometry (14L)
Drawing	
	Statics and Dynamics (30L, 3R)
Second year	Physics (32L)
Applications of Descriptive Geometry (36L)	Chemistry (36L)
Theory of Machines (60L)	French Composition and Belles-Lettres (31L)
Construction of Machines (30L)	Topographical Drawing (33S)
Industrial Physics (36L)	Other drawing (78S)
Analytic Chemistry	
Civil and Industrial Constructions (36L)[a]	**Second year**
Histoire naturelle industrielle (18L)[b]	Analysis and Geometry (40L, 4R)
(Mineralogy and Geology, 1830–1831)[b]	Mechanics (37L, 4R)
Drawing (44L)	Machines (22L)
	Geodesy (29L)
Third year	*Arithmétique sociale* (6L)
Industrial Physics (36L)	Physics (30L)
Construction of Machines (30L)	Chemistry (36L)
Steam Engines (22L)	Architecture (30L)
Industrial Chemistry (70L)	German (32L)
Constructions and Public Works (36L)	Topographical Drawing (45S)
Exploitation of Mines (with Mineralogy and Geology after 1831: 36L); included Railroads from 1832 to 1837, when a separate course was established	Drawing of Human Figures and Landscapes (45S)
Ferrous Metallurgy (70L)	Wash Drawing (20S)
(Hygiene, 1830–1832)[b]	
Histoire naturelle industrielle (18L)	
Drawing (6S per month)	

Sources: *Annuaire de l'Ecole Royale Polytechnique* (1837), 242–243; prospectuses of the Ecole Centrale.
Key: L = *leçons* (formal lectures); R = *répétitions* (small class, drill); S = *séances* (sessions).
a. At times included a course on stonecutting.
b. Taught intermittently.

prise. It will therefore become necessary that our engineers know how to regulate and direct such projects. They must know how to evaluate the utility and the particular and general costs (*inconvénient*) of such and such an enterprise; they must in consequence have correct and precise ideas about the general interests of industry and agriculture about the nature and influence of moneys (*monnaies*), about loans, insurance, collective financing, amortization, in a word, about all those things that can serve to assess the probable costs and benefits of all such enterprises.[42]

Although it should be pointed out that all these subjects were to be covered in only six *leçons* (2 percent of the class and laboratory time of the second year), it is also true that no course covering these eminently useful subjects was instituted at the Ecole Centrale.

Another piece of evidence that tends to modify Olivier's stark contrast between a purely theoretical, algebraic, Laplacean Polytechnique and a balanced, geometrical, Mongean Centrale is the substantial number of men who taught at both places during the period under study. Dumas, Coriolis, Edmond Fremy, Victor Regnault, and Jules Péligot taught simultaneously at the two schools (some only for brief periods), and Joseph Liouville and Joseph Belanger moved from one to the other.

On the other hand, an important clue to the explanation of Olivier's highly critical attitude toward the Polytechnique—if not to the validity of his argument about the curriculum changes—may in fact lie in the quality of the individual teachers at the Polytechnique who did *not* lecture at both institutions. Monge had been both a first-rate scientist and a first-rate teacher. The years 1815–1848, however, witnessed the presence at Polytechnique of a surprising number of professors who demonstrated conspicuous weaknesses in their teaching, whatever their achievements as scientists. The Polytechnique's centenary publication departed from its predictably hagiographical tone to point out these weaknesses. Arago, Gay-Lussac, and Thénard (who taught for twenty-five years "with talent, but without spark") seem to have been effective teachers.[43] Ampère and Cauchy, on the other hand, were pedagogical disasters. They both "left their listeners far behind." In 1823 a special commission requested that they submit their lecture notes to it for prior approval or, if this were impossible, tie their teaching directly to previously published texts. Ampère never did comply with this directive, however, whereas Cauchy, although he did submit notes, remained beyond the reach of most of his listeners.[44] Arago's brother-in-law Mathieu replaced Ampère in the Mechanics chair in 1827, holding it until 1838 with a singular lack of distinction.[45] In Descriptive Geometry Hachette, the great collaborator of Monge, was relieved of his post for political reasons and replaced, "one can only wonder why," by C. F. A. Leroy, who delivered thirty-two years of fairly clear but notably pedestrian instruction.[46] After the death of the inspired teacher Neveu in 1816, the course in *dessin* became "increasingly banal" until, in 1822, the professorship was abolished "for reasons of economy."[47] From 1815 to 1830 Grammaire, Belles-Lettres, and the History of Morals were taught by Aimé-Martin and Binet, "with scarcely any success." When the Revolution of 1830 caused Arago to relinquish the chair of Geodesy and Machines, his replacement, Savary, taught for twelve years "with a real measure of success but without effacing the memory of Arago."[48] His replacement, Michel Chasles, on the other hand, "did not possess the gift of capturing the interests of large audiences." Navier and Liouville, successors to Cauchy and Ampère, were more successful as teachers than the latter, but they "continued the tradition of their predecessors by doggedly remaining in the theoretical domain"—a fact not calculated to win favor with Olivier. In 1840 the Swiss physicist Sturm replaced Duhamel, but his bizarre personal habits made him the butt of student jokes and "like Ampère, he was afflicted with an extreme gaucherie in posture and gesture, a great difficulty in speaking, and an insurmountable timidity, rendered all the more salient by his corpulence and his flushed, ugly face."[49] Victor Regnault, who succeeded the chemist Gay-Lussac in 1840, became interested solely in physics (the subject he taught at Centrale) from the moment of his appointment and for the next thirty years contented himself in the classroom with a "monotonous paraphrasing of his *Traité de chimie*, which was already in the hands of all his students." In the Literature course, Aimé-Martin was replaced in 1830 by Arnault, who "bored two *promotions* with his excessive classicism" before appointing the witty and popular Léon Halévy as his substitute. At Halévy's death two years later, however, the chair was given to Paul-François Dubois, "distinguished writer and founder of the

Globe," who nevertheless "lacked prestige, eloquence, and finesse in his lectures at the school." Dubois, who also served as a deputy from the Loire-Inférieure and as director of the Ecole Normale Supérieure, finally stopped preparing his lectures, "which everyone studiously avoided."[50]

In the closely knit world of French higher scientific education, Olivier must certainly have become aware of most of these cases of ineffective teaching at the Polytechnique. He himself served as a *répétiteur* in geometry at the school from 1831 to 1837. Not only, then, was the Polytechnique teaching too much theory at the expense of practice; whatever it was teaching, it was not teaching it well. If the specific content of Olivier's attacks dealt with curriculum and approach, this pedagogical weakness of the school may help to account for the general emotional pitch of his language:

We must add, however, that Laplace and Poisson sincerely deplored their mistake, that at the end of their lives they recognized that they had dragged the teaching at the Polytechnique into a false and dangerous path, not only for the interest of the country but for the progress of the sciences. Let us hope that M. Cauchy will also regret, in his last hour, the baneful influence under which he placed himself when he lent the support of his great mathematical intelligence to the destruction of the useful and beautiful institution that was created by the great citizens and illustrious scientists of the French Republic.[51]

The allusions to final repentance mixed with the image of sorrow carried to the grave. In a passage all in italics, Olivier vowed that "only death will bring an end to the sentiments inspired in the hearts of the students of the old Polytechnique by the acts of the men who destroyed that great institution that France owed to Monge."[52]

What actually occurred at the Polytechnique matters less than what Olivier believed had occurred. The language he chose to express his feeling that Monge had been betrayed allows a glimpse of the ethos that permeated the Ecole Centrale during its early years. The *Centraux* were restoring a sacred tradition of engineering education that had been undermined by the elitism and excessive devotion to pure theory brought to the Polytechnique by the men who guided that school during the Restoration years.

In the panoply of institutions that trained the French technological elite, the Ecole Centrale stood not as a supplement or an auxiliary to the Polytechnique, not as the school that merely accomplished for private industry what Polytechnique did for the State, but as an autonomous and even morally superior institution, the sole repository of the salutary union of theory and practice. One may suggest, then, that the claims made for *la science industrielle* served not so much to sum up the intellectually rigorous rationale of a program of study as they did to assert this faith in a special mission.

From the perspective of international comparisons of the relation between education and social structure, the rivalry between Polytechnique and Centrale might seem to reflect relatively minor differences within an essentially homogeneous professional elite. At one critical point, however, that rivalry helped to shape events of major importance in French history. The National Workshops set up after the Revolution of 1848 were initially headed by the former Centrale student Emile Thomas, who was assisted by between 75 and 200 *Centraux*.[53] The hostility between the *Centraux* and the Polytechnique-trained Ponts et Chaussées engineers, vividly described in Thomas's memoirs, contributed to the failure to develop projects more economically sound and politically acceptable than the blatant makework that helped to produce the Workshops' downfall.[54]

Olivier's attitude was certainly not the least important factor in shaping the circumstances under which Comte's "intermediate class" failed at the crucial moment to gain unity of thought and action.

The School as a Society

The Ecole Centrale was of course more than just a place to acquire scientific knowledge and technological skills. Examining
the mixture of myth and reality in the school's "unified" *science
industrielle* gives important clues to the sources of graduates'
professional competence and, especially, to the way their status
was defined in relation to the scientist, the artisan, and the industrial entrepreneur. Yet the contribution of the Ecole Centrale
to the history of the emergence of the civil engineer cannot be
limited to such matters. Like other schools, Centrale was also a
social institution that in some ways reinforced structures and
values dominant outside the school while in other ways offering
contrasting patterns. The difficulty is that in the case of the Ecole
Centrale, this "secondary" curriculum—the development of
common assumptions and attitudes, the vertical links of subor-
dination and deference, the horizontal ties of comradeship and
cooperation, the impulses of emulation or rejection—reveals itself far more sparingly than the "primary" curriculum of textbooks, lectures, and practical exercises. Documents about these
aspects of the school's life are even more sparse than those delim-
iting the elusive *science industrielle*. As has happened before in
this study, engineers' laconism concerning nontechnical matters
compels the resort to a good measure of inference and indirection.

The Externat

A good many of the differences between the experiences of students at Centrale and those at Polytechnique probably arose from
the fact that whereas the latter slept within the walls of the
school, took all their meals together, and were only allowed to
leave during a small number of specified hours, the *Centraux*
were day students required only to take the midday meal at the
school's refectory.[1] Perhaps one fourth of the students lived with
their families.[2] In the case of the remainder, the names of Parisian
"correspondents" often appear on the dossiers. These individuals

could be merely bankers or attorneys who took charge of the tuition payments and passed important information between parents and school administrators, but more often they were keepers of *pensions* or relatives and family friends who boarded the students themselves.[3]

In the years from 1829 to 1848, the gates opened at 7:45 A.M. Students were allowed to leave at 4:30 P.M., although laboratory and workshop exercises often retained them in the school well beyond this hour. In addition, the school opened a library in 1837 that was open from 7:00 P.M. to 10:00 P.M. and during the intervals between courses. In any case, the school's formal control over its students' movement lasted only until the end of the afternoon.[4]

The school's founders claimed that their clients' status as day students made an important contribution to their formation as an engineering elite. The unpublished history acknowledged that in "all instructional establishments" the boarding students brought the most financial profit, but it added that the need for discipline obliged the authorities to place the students under a uniform and inflexible rule that did not suit "all natures and all aptitudes."[5] Nor did it suit the purposes of an engineering school. Training at the Ecole Centrale was the "first step" in a "delicate career." The earlier one learned to deal with "responsibility," the better. Full-time boarding (the *internat*) only "retarded the blossoming of their individual character" and prolonged the debilitating influence of the *collèges*. The positions awaiting them upon graduation required not only "specialized knowledge" (*connaissances spéciales*) but also a "certain moral firmness" that should be developed during their time at the school by putting them "face-to-face" with liberty.[6] Olivier's account made the same point:

To know how to command, one must have learned how to obey; to lead others, one must have led oneself; to direct the industrial enterprises of another one must have order and economy, regulating expenditures according to assets. . . . Students at the Ecole Centrale, being day students, learn not to abuse the liberty they enjoy.[7]

Olivier mentioned a more immediate consideration as well: the students at the Ecole Centrale were of greatly disparate ages, came from many countries, and represented "all political opinions and all religions." If such a mixture were placed in the same dormitory, "discipline would be impossible."[8]

The argument that the independence of the *externe* bred responsible, self-disciplined behavior has a certain plausibility although, as will be discussed below, there were a number of constraints on that independence. And it is true that compared to the Polytechnique, the Ecole Centrale had a much more heterogeneous student body, both in national origins and in age.[9] On the other hand, it might also have been argued that older students would provide a steadying influence on the younger and that the variety of backgrounds and beliefs would have permitted the school authorities to use a divide-and-rule policy. Certainly Olivier was aware of the troubles caused by the political solidarity of *Polytechniciens* during the July Monarchy (see below). The spokesmen for the Ecole Centrale seem, in fact, to have been making a virtue of necessity: they clearly did not have the funds to construct and maintain sleeping quarters for 300-odd students. Regardless of the real reasons for the *externat*, however, the special maturity endowed by experience as an *externe* became one of the standard ingredients in the image of the Centrale graduate.

To describe fully the political and cultural milieu in which a postsecondary student in July Monarchy Paris might have participated would require a separate work equally as long as the present study. Since such activity could have provided an important link between the formation of engineers as a group, with common interests and a common philosophy, and the life of the surrounding society, the question of the extent to which they became part of that wider current cannot be completely ignored. Although pertinent details dealing specifically with Centrale students are all but nonexistent, certain parallel evidence sheds additional light on the matter.

Contemporary literature suggests that supposedly independent "external" students had to struggle to overcome restraints on their freedom hardly less severe than the gates and guards at the Polytechnique.[10] Balzac's novel *The Lily of the Valley* describes some of these difficulties. At the age of fifteen the narrator, Felix de Vandenesse, was taken from his provincial *collège* at Pont-le-Voy and placed by his father in a pension in the *Marais* that a Monsieur Lepitre had established in the old Hotel Joyeuse. In the porter's lodge of the *hôtel* lived Monsieur Doisy, "a regular smuggler, whom it was in the interest of the pupils to make much of; he was the secret chaperon of our rambles, the confidant of late returns, our intermediary with the agents of forbidden books."[11] He was also the students' principal source of credit, which he extended mainly to cover the cost of coffee and sugar, the consumption of which in Doisy's quarters had become the mark of the most aristocratic students. Doisy assumed that all the students had sisters or aunts who would help the young scholars out of financial embarrassments.[12]

Since Felix's father had given him no money, assuming that "if I could be fed, clothed, crammed with Latin, and stuffed with Greek," all would be solved, the young de Vandenesse quickly fell into debt. When his parents came to visit him after his second year, Doisy forced the revelation of his debt. Felix's father was inclined to be indulgent, but his mother was pitiless, denouncing him with prophecies of ruin. At her urging, Felix's father continued the system that had left him without pocket money. He finished his secondary education, studied higher mathematics for a year, then attended law school, all the while living at the Lepitre *pension* and all the while without funds: "But how can one attempt anything in Paris without money?"[13]

There were other ways in which Balzac's protagonist found his "liberty skillfully fettered." All the boys in the *pension* had been escorted by ushers to the Lycée Charlemagne, but in Felix's case, this procedure continued on through law school: "M. Lepitre had me escorted to the Ecole de Droit by an usher who handed me over to the professor, and came to fetch me back after the class."[14]

It is nevertheless far from certain that such close chaperoning of a postsecondary student was the general rule at the time Balzac wrote, even for young aristocrats. Balzac's emphasis and phrasing suggest that whereas the ushering of students to Charlemagne was a common experience, mentioned in order to evoke a conventional setting, the ushering of a law student performs a different function in the novel: it contributes directly to the psychological portrait of de Vandenesse with which Balzac is principally concerned.[15] The escorting of a law student is nevertheless presented as clearly within the realm of possibility. It should be kept in mind that *The Lily of the Valley* contains many details that correspond closely to Balzac's personal experience. The name of the *pension* that appears in that novel was the same as the first of two *pensions* in which Balzac actually lived; the second, the Institution Gaujer et Beuzelin, was located at 7, Rue de Thorigny, one block from the location of the Ecole Centrale.[16]

In the absence of direct evidence that *Centraux* were escorted to and from the school, Balzac's evidence can only suggest the possibility. Colladon's memoirs do indicate, however, that those who kept *pensions* for post-*secondaire* students could take an active role in restraining the behavior of resident *externes*. Colladon and Sturm took rooms near the Pantheon with an unnamed gentleman (Colladon refers to him only as Monsieur) during the weeks when they were preparing their study of the speed of sounds in fluids for presentation to the Académie des Sciences. Alarmed by their continual absences during the evening, Monsieur finally warned them that they could not remain if they continued their mysterious activities. His "fears for their morals" disappeared, however, when they told him of the scientific work that prompted their absences.[17]

Balzac's depiction of the constraint of limited finances finds corroboration in the case of Eiffel. After receiving his diploma in 1855, Eiffel confessed to his parents that he had contracted a debt of 800 francs with his landlord. As in Balzac's fictional case, Eiffel's mother reacted strongly to her son's irresponsibility. The two were reconciled only with great difficulty.[18]

Polytechniciens also had their financial difficulties. Pinet's account gives the following composite "quotation" from his interviews with those who attended the school during the July Monarchy:

We received from five to eight francs a week from our parents. With this we had to get our meals on Sundays and Wednesdays [when the school's refectory was closed], have our boots polished and our suits brushed, and go to the cafe and the theater. It is true that for the theater the uniform had its advantages: the night before a day at liberty (*jour de sortie*), free passes sometimes were delivered by generous friends and correspondents. We didn't always eat well, but we always went to the theater.[19]

Centrale, however, had neither the uniform nor the constituency to produce the free tickets allowing greater participation in the cultural life of the capital.

Perhaps the greatest constraint on the freedom of the Centrale student, however, was the discipline imposed by the school's work load. The school's guiding spirits clearly perceived a connection between the absence of an *internat* and the need to inculcate habits of seriousness and hard work. After celebrating the "moral firmness" that independent residence helped to induce, the *Rapport à présenter* declared that "as a natural counterweight [to this liberty] and to reassure the students' families, it was necessary to organize a vigilant system of verification (*contrôle*). It was necessary that each oversight, each act of self-indulgence, result in a failure, a lowering of one's marks, a reduction in rank order. And all the various forms of tests (*épreuves*) imposed upon the students contributed systematically to this result."[20] For young Gustave Eiffel the constant assignments and examinations contrasted sharply with the leisurely rhythm of his days in secondary school:

The worst (and the best) was the work load: it was crushing. Gone were the tranquil days at the Collège Sainte-Barbe, the memory of which became all the more enchanting. The Ecole Centrale was a real miser about holidays: The Saturday of Christmas and the Saturday of New Year's Day were deemed to suffice. And finally, the school imposed tasks without regard for the time needed, and it expected them to be accomplished on time. Seated in his little room near a small fire, he learned how to work straight through to the morning when necessary. And it was often necessary, especially before the examinations, which were both numerous and regularly spaced along the three years of instruction.[21]

The unpublished history of the school suggests, in fact, that Eiffel's situation was the most precarious of the various possible living arrangements. The authors made a calculation of the chances of success at the school, measured by the likelihood that an entering student would eventually receive a diploma, which deals with those students who matriculated at Centrale before 1871. If the success percentage of those students living alone in a rented room is taken as 100, those students who had roomed with one or more comrades scored 113, those who lived in supervised *pensions* scored 128, and those who lived with their families scored 210.[22] The tempering of "moral firmness" by residential independence came at considerable cost to academic achievement.

When a student did overcome the constraints of chaperonage, lack of pocket money, and the press of his assignments at the school, he entered a city replete with moral, cultural, and ideological temptations. Paris did not appear to him as a formless conglomeration of buildings but as a set of familiar centers of attraction and social networks within which moved a student population of perhaps 3,000.[23] Some young aristocrats, or others with both money and the urge to spend it adventurously, would succumb to the lure of the gambling dens of the Palais-Royal described in Balzac's novels.[24] For others, the Palais-Royal held *galéries* and gardens in which to walk, to talk politics, to buy the latest political pamphlets—clandestine copies of those recently seized by the authorities were special favorites—and to inform themselves in general of *l'actualité*. According to Pinet's informants, *Polytechniciens* regularly made up part of the crowd here.[25] The Ecole Centrale and the adjacent area that housed most of its students was approximately the same distance from the Palais-Royal as was the Polytechnique.[26] In the early years of the Ecole Centrale, a student might have proceeded from this area to the rooms near the *Caisse hypothécaire* where Enfantin gave his exposition of the Saint-Simonian doctrine or to the Athénée, where Comte presented his own elaboration upon Saint-Simon's ideas.[27] After the dispersal of the Saint-Simonian movement, he might have joined one of the semisecret societies that mixed Republicanism with a variety of social doctrines.[28] Or he might have sought admission to the editorial circles of the newspapers and journals that were then such important centers of Parisian intellectual life.[29]

In any case, the young men who lived and studied in Paris during the years between 1829 and 1848 belonged to a generation in

search of new doctrines to replace traditional ideologies that responded to the challenge of industrialization no better than they
legitimized the rule of Bourbons and Orleanists. It is most unlikely that the *Centraux* were somehow immune to the intellectual *Wanderlust* that produced converts for the "prophets of
Paris."[30] On the contrary, they did not all accept the isolation
from political and intellectual ferment that living by themselves
on the Right Bank was supposed to produce. Although many of
Eiffel's free hours were "safely" spent in the homes of relatives
and former correspondents, he also sought the company of his
old classmates from Sainte-Barbe.[31]

As has been indicated above, many of the *Centraux* had spent a
year in special mathematics classes (*taupe*) preparing for the
Polytechnique. The friendships formed in these classes, among
students from similar social backgrounds, were not all effaced
merely by attendance at different engineering schools. Two clear
demonstrations of links with other students occurred in the early
years of the period: the participation of Centrale in the Revolution of 1830 and the funeral procession of Benjamin Constant
on 10 December 1830 when Centrale students marched beside
Polytechniciens and students from Droit, Médecine, Pharmacie,
Alfort (veterinary medicine), and the Ecole Supérieure de
Commerce.[32]

It is indeed possible, then, that the attachments and attitudes of
the *Centraux* were shaped not only by their families, their secondary education, and their experience within Centrale but also
by their participation in the intellectual and political subculture
of the "student estate" in July Monarchy Paris. The evidence
presently available, however, offers few details about that participation. What is clear is that day-student status did not alone determine its extent. Despite the limitations upon their freedom to
leave the school, the boarding students at the Polytechnique participated fully in this subculture. They established recognized
gathering places such as the Café Colbert, the Univers Tavern,
and the Café Hollandais, the latter serving as the site of their
elaborate, raucous initiation rites (*l'absorption*).[33] During their
Wednesday afternoon *sorties* they swelled the audiences at the
Sorbonne, where Saint-Marc Girardin and Lerminier were their

favorites; the Academy of Sciences, where, according to Pinet's informants, funeral orations were especially popular; and the Collège de France, the pulpit of Michelet, Mickiewicz, and Quinet.[34] The *Polytechnicien's* tricorn headgear became a familiar sight in certain parts of the city. When Barbès, acting on the plans of Blanqui, attempted a *coup d'état* in May 1839, his agents sought out Polytechnique students in the Rue de la Harpe. Claiming that Polytechnique had "always been on the side of liberty," they demanded that the students enlist their comrades in the *coup*.[35]

On the other hand, it is doubtful that the *externat* at Centrale prevented participation in political and cultural life outside the school because Centrale students could not form the special solidarity of boarding-school inmates which provided the moving spirit behind activities such as those described above. Anyone who has attended a public high school knows that links of solidarity, codes of conduct, and common values can be forged within the walls of an institution during the day as well as at night. The influence of his classmates could be just as important in forming the Centrale student as was the "moral firmness" induced by his extramural independence.

The Rate of Attrition

By no means all of the students underwent the Ecole Centrale's influence for a full three years. Complete attendance figures for every year are not available, but table 7.1, compiled from archival records, the *Rapport à présenter*, and published histories, clearly reveals a high rate of attrition. In 1839 student leaders calculated that in the first decade of the school's existence, 208 men had "graduated" by receiving the *diplôme* or the *certificat*, while another 54 left during or at the end of the third year.[36] Approximately four times that number had begun studies at the school before 1837. By the time that the Ecole Centrale became a State institution in 1857, 1,291 students had received a *diplôme* or *certificat*, whereas 2,051 had left without either degree, an attrition rate of 61.4 percent.[37] By contrast, during the years 1829–1852, 3,284 students began the Polytechnique's two-year program, of which 2,922 graduated, an attrition rate of only 11.0 percent.[38]

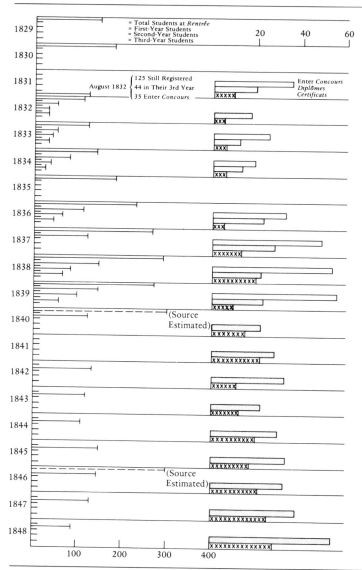

Table 7.1

Enrollments and degrees awarded

Sources: (1) Counts of Dossiers in the Ecole archives (undoubtedly minimum figures, but slightly higher than Guillet's totals for the years 1843–1847). (2) Report from Perdonnet printed in Institution of Civil Engineers, *The Education and Status of Civil Engineers* (London, 1870), 66. (3) *Rapport à présenter*, 6-8. (4) Dumas Papers, Carton 16. (5) Pothier, *Histoire de l'Ecole Centrale*. (6) Guillet, *Cent ans*.

The available records of Centrale have left no systematic accounting for these withdrawals, nor, more surprisingly, does any discussion of the possible reasons for them appear in the archives or in published histories. Death or serious illness probably caused the disappearance of a relatively small number. The sample of 1,093 dossiers for the period 1829–1847 records 10 student deaths, but since the notations were not made in any formal way, some deaths probably went unrecorded. Marielle's tabulation for Polytechnique students entering during the same period lists 50 deaths among 2,882 students, a rate of 1.73 percent, whereas the dossier entries indicate that only 0.9 percent of the Centrale students died during their studies. As will be discussed further below, however, the fear of serious illness underlay many withdrawals, especially during the 1832 cholera epidemic, which nearly wrecked the school.

Expulsion (*exclusion permanente*) for misconduct accounted for an even smaller part of the attrition. The dossier sample produced twelve such cases, which agrees with the figures given by Pothier, who had access to both Olivier's private register and the minutes of the *Conseil d'ordre*.[39]

The high cost of a three-year stint at the Ecole Centrale and the difficulty of its programs undoubtedly account for most of the remaining departures. Even for the relatively prosperous bourgeois who comprised most of the parents, each 2,000-franc annual increment must have caused considerable reflection: the school's archives are full of complaints about the financial burden. Yet it was not merely a matter of high cost. The probable return on the investment in each additional year of study had to be weighed against the income the student might have earned had he taken an industrial post one year earlier. Many fathers probably thought that two years—or even one year—gave an adequate grounding in the "fundamentals"; the rest could be learned on the job, or, more precisely, in father's shop. Thus in 1835 Louis Meiner left after one year, to take over his father's clock and lock manufactory in the Haut Rhin, and Felix Vinchon left in 1837 to join his father, a glassmaker at Bligny.[40] In April 1843 Charles Debray had only four months remaining to complete his third year at the school, but his father instead summoned him to help manage the family asphalt mine at Bastennes.[41] The father of Charles Bouchacourt withdrew his son in 1841 when a friend who directed a foundry at Lyons offered Charles a position there.[42] If more complete information were available about the reasons for withdrawal from the school, a high attrition rate at Centrale might suggest itself as one of the best indicators of the current rate of French industrialization; the higher the demand for technical personnel, the lower the incentive to remain in school. In any case, not all parents shared the Centrale founders' valuation of the organic unity of the curriculum of *la science industrielle*.

After all, the school's two "degrees," its *diplôme* and *certificat*, embodied no special privileges. Unlike the *agrégé* or the *ancien élève* of the Polytechnique or the Ponts et Chaussées, the Centrale graduate had no claim upon the State obliging it to find the holder a post. The members of the Ecole Centrale's ruling council were thus presented with two somewhat contradictory goals. On the one hand, they needed to maintain the total enrollment at a level that would insure the financial health of the institution; on the other hand, they wanted to develop products of recognized value, degrees that would only be granted after rigorous training and that would promote the holder to parity with a *Polytechnicien* or a "Pont." Even when the school was in the middle of its worst crisis, during the end of the academic year 1831–1832, the directors imposed selective standards. In August 1832, 44 of the school's first class, originally totalling 145, presented themselves for the *concours de sortie*. Of these only 35 were allowed to compete, and 26 received either a *diplôme* or a *certificat*.

Grades on written examinations and practical exercises were also used to weed out the weakest students. Entries in the minutes of the *Conseil des études* for the year 1838–1839 permit a more detailed picture:

	First-year students	Second-year students	Third-year students
Present in November 1838	144	83	62
Present at end of year (August 1839)	117	66	

Promoted to higher division	91	50
Allowed to enter graduation competition		53
Diplomas earned		22
Certificates earned		20

Of the twenty-six first-year students not promoted, nine were listed as "action deferred" (*ajourné*), three were authorized to repeat the year because of illness, and fourteen were simply failed (*rayé*). Of the sixteen second-year students not promoted, four were deferred, three were on authorized absences, and nine were failed.[43] At least by 1839, then, the Ecole Centrale was hardly a diploma mill. Only in its failure to copy the Polytechnique in its "external" graduation examiners (*examinateurs de sortie*) did the school miss an opportunity to employ one of the trappings of a rigorous selection process.[44]

Even if the Centrale's high attrition limited the duration of the school's formative influence upon those who initially enrolled, just as the *externat* limited the intensity of that experience, that same attrition may have had a psychological impact on those who remained. They were the survivors, the elect, imbued with the sense of solidarity of men who had overcome obstacles together and the sense of superiority that came from the knowledge that many others had failed to overcome them.

The Years of Crisis

The remainder of this chapter will investigate the immediate social context of that sense of solidarity: how the school functioned as a community; how it developed patterns of order, power, and loyalty; what were its formal rules and informal customs. The story divides roughly into three phases: the three years of infancy climaxed by the cholera crisis of 1832; the reconstruction during the next six years in which Théodore Olivier played a commanding role; and the period of "routinization" from 1839 to the Revolution of 1848, which was marked by the leadership of Bénédict Empaytaz. Running through all these periods, however, were important threads of continuity: in the social background of the students, in the tenure of key members of the staff, and in the program of technical education.

One of the first concerns of the school's founders was the organization of the students. As was the case for the organization of curriculum, they drew their inspiration from the Revolutionary Ecole Polytechnique. Monge and his colleagues had made their choice of *chefs d'études* (or *chefs de brigade*) from among candidates nominated by the students themselves. The Ecole Centrale administration allowed the students even more power: the *chefs d'études* were chosen by the students alone during a single meeting in March of 1830. Each of the four groups who met in the same drafting room chose a *chef d'étude* (also known as *chef de salle*).[45] During the following year the students chose alternates (*suppléants*) and second alternates. The Ecole's founders reserved the right to choose any student to fill a vacancy in the post of *chef d'étude*, but in practice they chose the alternates elected by the students.[46]

In addition to acting as a liaison between the students and the administration, the *chefs d'études* helped to choose other personnel. In 1831 each professor was authorized to hire one *aide-préparateur* (assistant for demonstrations and laboratory work) from among the first-year students and one from among the second-year students. When the final choices were made in public examinations, the *chefs d'études* joined the professors on the juries.[47]

In these early years, at least, the office of *chef d'études* carried real prestige, as is evidenced by the way it could become the first step on a climb to the highest positions in the engineering profession. Camille Laurens, elected a *chef d'étude* in 1831, became a professor at Centrale and a founding member of the Société des Ingénieurs Civils. Gustave Loustau, elected a *chef d'étude* in 1830, also helped to found the S.I.C. When the school entered a period of confusion and conflict in the first years of the Third Republic, Loustau headed the group of graduates who offered to take the school back from the State. He told Pothier that for fifty-six years his old classmates had unfailingly recalled his post at the school whenever they met him.[48] Election to the post of *chef d'étude* augured well for success at the school, too. Of the twenty-six men who received the school's first degrees, ten had been *chefs d'études*.[49]

The *chefs d'études* had no formal disciplinary duties. In 1829 a Monsieur Bideaut was employed "to prevent disorder in the lectures."[50] He was replaced in the following year by a Monsieur Gand, who was charged with conducting "general surveillance" of the entire school while at the same time managing the supply room.[51] As the unpublished history of the school put it, the "methods of repression" were "exclusively moral": a private reprimand by the director, a public reprimand by the Conseil de l'Ecole in more serious cases, and, finally, expulsion.[52]

In any case, these methods did not have to be employed very often. At the end of the first year, the head of the Council on Improvements, Chaptal, appointed a visiting committee composed of the chemists Anselme Payen and Jean Darcet and the Vicomte Héricart de Thury, inspector-general of mines. Their report of 12 July 1830 spoke especially highly of the school's discipline:

Since its opening, perfect order has constantly been maintained. The young men in the school show a degree of assiduity that does honor both to the professors and to the students themselves. The *chefs d'études*, elected by their classmates, and in general chosen from among the best and most devoted, have established both the harmony and the methodical approach so necessary to the productive use of time.[53]

Two weeks later the people of Paris gave a less favorable report on the behavior of the Bourbon regime. During the rioting of 27–29 July that ended the Restoration, the Ecole Centrale, while welcoming the downfall of a government seen as less than friendly to the growth of industry, concerned itself largely with maintaining order. Etienne Véret, a student who was also an officer in the Swiss army, took charge of giving the students some quick training in close-order drill and the use of the musket. The school was given the mission of guarding the La Force prison, three blocks away on the Rue Pavée, to prevent mass escapes by common criminals. Other students were sent to prevent looting at the Picpus convent and to assume sentry duty at the Place Royale *mairie*.[54] Daniel Colladon sought out his friend Sturm in the Latin Quarter, whence they rushed to the apartment of Four-croy's widow to ensure that his collection of experimental appa-ratus was safe from harm. On 29 July two *chefs d'études*, Loustau and Guépin, were delegated by their fellow students to seek the advice of J.-B. Dumas concerning the possibility of manufactur-ing munitions at the school. They made their way through bar-ricaded streets to Dumas' residence at the Jardin des Plantes, only to find him gone. Whatever may have been the risks Loustau and Guépin ran on this mission, however, only one student, Abel Noirot, received official recognition for *"une action d'éclat"* in the struggle to keep order at the Place Royale. Upon receiving a decoration from the new regime, he left the school to make a career in the army.[55]

Such incidents give some clues about the state of mind of the students at Centrale. Whereas few of them were prepared to sup-port the Bourbons, they were concerned not with social revolu-tion but with the consolidation of the new Liberal regime and the maintenance of order. They showed that they believed that Dumas would share their sympathies and that he would be will-ing to cooperate with them in their munitions-making scheme. Thus even Dumas, the most illustrious of the school's teachers, the son-in-law of one of the scientists (Brongniart) closest to the King, seemed accessible in the midst of political crisis.

But for the purposes of the present study the school's later re-sponse to the July Revolution is even more interesting. The Cen-trale prospectus published in the fall of 1830 attempted to capitalize on the changes in career plans forced upon Bourbon sympathizers while simultaneously offering an alternative to place seeking to supporters of the Orleanists:

The mania for government posts (*la manie des places*) should by now have passed out of favor in France: we have just seen an excellent example of the instability of such posts. The young men of the country now realize that there is something worth far more than a post: the capacity to create an independent situa-tion. . . . In England, independent engineers are sought by private companies who wish to undertake public works, establish a fac-tory, or make improvements in an existing establishment. That special type of engineer is missing in France.

The Ecole Centrale had now been established to supply such requests, while also training "directors of factories" and "young men destined to teach industrial sciences."[56]

At the same time that they warned parents that *la manie des places* led to careers both politically insecure and outdated by the greater opportunities offered by industrialization, the authors of the prospectus tried to associate their school with some of the *gloire* then being attributed to the Polytechnique. The founders of the Ecole Centrale "viewed with joy how their students came to rival in courage those of the Ecole Polytechnique," and they observed "not without emotion" that Lafayette had given the title of "*Brave Ecole Centrale*" to the "numerous students whose services he accepted at the moment of danger."[57] To make known their eagerness to share in the spirit of their sister school and to support the new regime, the students voted to adopt a uniform: a dark blue dress-coat with a blue collar bordered in red and adorned with a gold bee, all worn with the *epée civile*.[58]

Even though the July Revolution had supposedly removed all obstacles to industry and liberty, the enrollment at the Ecole Centrale grew far less than the founders had hoped. The political upheaval had disrupted the first year's final examinations, so the first-year students were automatically admitted to the second year of the program, and the summer 1831 examinations covered both years. Despite these measures, and despite the continuing excess of candidates for the Polytechnique, the *total* attendance at the fall *rentrée* of 1830 was only 175. The next year this total was 5 lower.[59]

An even more threatening development occurred in the middle of the school's third year. The outbreak of cholera throughout Paris at the end of March 1832 brought normal routines to a halt.[60] On 4 April Parent-Duchatelet gave a special *leçon* on the characteristics of cholera and the means of its treatment, but it soon became evident that the Ecole Centrale could not continue. Lavallée himself was taken ill, as were his wife and two daughters, the younger of whom died within a month.[61] Students began to leave Paris, while those remaining petitioned the *Conseil des fondateurs* on 20 April to obtain the temporary suspension of all instruction. Lavallée responded with a letter to all parents authorizing their sons to leave the school during the epidemic.[62]

Lavallée's incapacity and the evident threat to the school's future brought by the outbreak produced a flurry of consultations and maneuvers among the founding professors and the other faculty members. The most important proposal came from Raucourt, who wanted both greater power in the administration of the school and a chance to share in the *bénéfice* ("profits") to be expected once the school regained its start-up expenses. He suggested that a *société anonyme* be created in which Lavallée would be issued 384 of the 600 shares, with responsibility for negotiating the sale of the remaining 216. The decision-making power, however, would rest with a *Conseil de direction* made up of all the major professors, not just the *fondateurs* (Olivier, Péclet, Dumas, and Lavallée).[63]

Several students who had remained at the school during the epidemic brought these proposals to the attention of Colladon. They expressed their fear of the uncertainty created by a *société anonyme* (such limited-liability companies were still widely mistrusted) and suggested that the problem of current expenses might be solved if the major professors agreed to forego their salaries for that year. (Teaching at the Ecole Centrale was not the principal source of income for any of them although Péclet and Olivier were considerably more dependent upon it than the others.) Colladon accepted this plan. He next contacted one of Lavallée's relatives in Paris, a M. Lallemand, who agreed to offer his support to Lavallée in an effort to convince the stricken director that he should not "liquidate the school."[64] Colladon convinced Perdonnet to offer his course without fee and to join him in arguing against the *société anonyme* idea. The two then met in turn with each professor, saving the three from whom they expected the most resistance—Raucourt, Olivier, and Péclet—for the last.

The final negotiations produced a compromise settlement. All the professors agreed to accept no fee for courses taught during the current academic year. Lavallée retained the title of director of the school as well as legal ownership, but his duties were confined primarily to financial administration, representation of the school in its dealings with the public, and certain ceremonial functions. All other matters became the concern of the *Conseil des études*, presided over by Dumas and composed of all the principal professors.[65] The four founders had previously made all the important decisions. Now the three *professeurs-fondateurs* became merely the equals of their senior colleagues, but on a council whose independent power was further enhanced by the abolition of the Council on Improvements.[66] The reduction of Lavallée's responsibilities was reflected most clearly in the creation of the new post of Director of Studies (*directeur des études*), which carried the obligation to "ensure the execution of the decisions of the *Conseil*" and to supervise the daily conduct of both academic and disciplinary affairs. Théodore Olivier assumed this post on 3 August 1832.[67]

In the years before Olivier's directorship, the Ecole Centrale conformed rather poorly to the model of a French educational institution described by sociologists such as Jesse Pitts and Michel Crozier, which stresses the aloofness of the teachers and a system of authority relations that creates among the students a "delinquent community" of opposition.[68] Instead of hostility and periodic *chahuts*, one finds collaboration between faculty and students in laboratories, workshops, and research trips. Communication between students and administration functioned smoothly: the *chefs d'études* were intermediaries whose legitimacy was acknowledged by both sides. The frequent student petitions usually received a favorable response. They had been granted their request to adopt uniforms after the July Revolution, to add a third year of studies, to begin instruction in English, and to suspend courses temporarily during the cholera epidemic. The approach to Colladon by those students who stayed at the school during the epidemic was credited with saving the school from the dangers of incorporation.[69] In the summer of 1833 their collective complaints about the teaching of Raucourt helped to bring the resignation of the "villain" of the cholera crisis.[70] According to histories of the school, the careers of these early classes were "more closely followed" by the professors than those of later years. Perhaps the character of the bonds forged between students and teachers during these early years is best expressed by the title given to the twenty-six students who received the first diplomas and certificates in the fall of 1832: *élèves-fondateurs*.[71] One of their number, Jules Pétiet, became director of the school in 1867.

The Years of Rebuilding

It was hardly surprising that Théodore Olivier assumed the new post of director of studies. After leaving the Polytechnique in 1815, he had taught for six years at the artillery school in Metz. In 1821 he received an invitation from the Swedish government to reform their school of engineering and artillery at Marienburg. He spent the next five years recasting it along the lines of the program at Metz.[72] As the only member of the founding group with previous administrative experience—and in view of Péclet's precarious health and Dumas' heavy commitments elsewhere— Olivier was thus the logical person to direct the Ecole Centrale's recovery.

It was also likely that the pattern of social relations described above would undergo some changes during Olivier's tenure. The devoted disciple of Monge lacked his master's affectionate disposition as well as his modesty. Both were dedicated teachers and scientists, but whereas Monge relished the role of statesman and political advisor, Olivier preferred the posture of the military professional. Even the circumspect Pothier, to whom he taught descriptive geometry in 1842, stressed his "great spirit of order, his strictness (*sévérité*), and his nobility of countenance and language, which disposed to obedience those students troubled by the suspension of studies."[73] Colladon, writing half a century after he left the school, mentioned the second weakness:

M. Olivier, a colleague whose merits as a teacher I well appreciated, but who, along with many good qualities, had a fault that sometimes amused the students: a rather childish vanity that made him exaggerate his influence and his discoveries. You can find examples of this in the *Bulletin des sciences mathématiques* of M. Férussac (1826 or 1827), in which he prefaces his theorems of descriptive geometry with the words "problem found in a certain year, on a certain day, and —if I remember correctly—at a certain hour."[74]

Olivier has left only the briefest hints about his vision of the ideal social order, but perhaps there is some subconscious model reflected in the fact that his principal scientific work, on which he labored twenty-five years, dealt with the geometric aspects of cogwheels.[75]

In order to understand the significance of the way Olivier ran the Ecole Centrale, it is important to consider first certain aspects of the extramural environment. There was good reason for parents and teachers to be uneasy about the strength of the "repressive" apparatus at Centrale during those years. The 1830s saw frequent demonstrations, riots, and general unrest among the swelling Parisian population.[76] One could hardly expect a group of students to be unaffected by this effervescence. The clearest evidence of this potential for disorder came from the Polytechnique, where Olivier taught as a *répétiteur*. In the aftermath of the July Revolution, in which they were credited with playing an important part, the *Polytechniciens* won the right to less surveillance and more liberty: they could now leave the school between 2:00 P.M. and 5:00 P.M. on any day.[77] They also forced the withdrawal of their Jesuit Inspector of Studies, J. Ph.-M. Binet. In December 1830 they led 100,000 Parisians (including *Centraux*) in the funeral procession of Benjamin Constant, then the next week joined the National Guard in demonstration marches designed to calm a populace agitated by the trials of Polignac's ministers—an action that won them the thanks of the Chamber of Deputies.[78] Many of the *Polytechniciens* quickly became disillusioned with the new regime, however. A good number sympathized with the Party of Movement, but others turned to active Republicanism. At the school's *carnaval* festivities in February of 1832 the liberty cap was waved on top of billiard cues; 150 students joined the revived *Charbonnerie*.[79] In June 1832 both classes were suspended when they refused to allow the king's grenadiers to enter the school during the pro-Republican rioting, and four *Polytechniciens* were arrested in 1833 for their part in the *conspiration des poudres*.[80] Despite some evidence of a resurgence of sympathy for the Orleanist regime, unrest continued at the school throughout the decade, climaxed by the incidents surrounding Barbès's attempted *coup* in 1839.[81]

Not all the disorder at the Polytechnique had a political motivation. At the *rentrée* of 1834, the students demanded the removal of an especially severe inspector of studies, Colonel Thouvenel. When they continued their demonstrations after the school's commandant refused their request, both classes were dismissed.[82] In 1837 the school's chief physician was driven from his post by a well-organized *charivari* after the students became convinced of his incompetence.[83]

Throughout his tenure as director of studies at the Ecole Centrale, then, Olivier could observe a running duel at the Polytechnique between the forces of order and the students. And it was a struggle in which the opponents were almost equally matched. Not only had the students developed a "culture of solidarity" by their *absorption* rites and collective protection of rule-breakers; the professors, who had nearly complete independence from the commandant in academic matters, did not always support the administration. When the two *promotions* were dismissed in 1834, Thénard, Lamé, Savary, and Mathieu continued their teaching at Dr. Quesneville's laboratories outside the school.[84] The students often found another ally in the opposition press, who took up the defense of the Polytechnique as a way of attacking the government. In an 1837 report to the king, the commandant, General Tholozé, remarked that the *Polytechniciens* had come to consider themselves a fourth power in the State, equal to the monarchy and the two legislative chambers.[85]

With such events occurring on the other side of the Seine, Olivier and his colleagues could not afford to neglect the question of discipline. Just the same, the "apparatus of repression" at their disposal remained limited. In the fall of 1837, for example, the Ecole Centrale staff concerned in whole or in part with the maintenance of order among the 265 students consisted of the Director (in the most serious cases), the director of studies, a single inspector of studies, and a laboratory instructor who was also charged with "general surveillance." At the Polytechnique, this same responsibility was shared by a commandant, a deputy commandant, a director of studies, four inspectors of studies (artillery and engineering captains), four under-inspectors of studies (lieutenants and sublieutenants), two adjutant noncommissioned officers, and a captain in charge of military exercises, arms, and barracks—all for 348 students.[86]

Despite Centrale's relatively small surveillance staff, the school was largely successful in keeping order during the years of reconstruction after the cholera crisis. Much of the credit must go to Olivier. What he lacked in modesty he made up for in energy and resourcefulness. The available records of the school indicate that Olivier gave more attention to individual cases than did any of his successors. He was especially diligent in putting into operation the system of demerits. Unauthorized absences, disturbances, and other violations of school regulations were assigned various demerit values (5 for each day's absence, 5 for a violation of smoking rules); if a student accumulated 100, he was required to withdraw.[87] This was at least in principle the ultimate penalty: I have as yet found no evidence that a student was expelled solely because his demerit total exceeded the limit. One of the reasons that matters never got this far was Olivier's practice of writing to a student's parents after two or three infractions. He made considerable use of this powerful device, regularly informing parents of the progress, problems, or *mauvaise volonté* of any student whom he took to deviate from the normal.[88] After 1834 evidence for a third technique used to prevent disorder appears in the dossiers: the requirement to submit a certificate of good conduct before admission to the school. The request for such an affidavit from a local official seems to have been made selectively, however, principally to foreigners, students who had been privately tutored, students returning after long absences, and those who had prepared for entry in certain smaller, relatively obscure institutions.

As the experience of the Polytechnique showed, the faculty could play a pivotal role in the balance of power between the students and the forces of surveillance and control. Solidarity between faculty and students in welcoming the overthrow of Charles X had become part of the school's folklore, but if the teachers gave even tacit support to student actions opposed to the administration, Olivier would have been left with almost no one else upon whom to rely. He undoubtedly felt that he had no choice but to impose upon his colleagues certain surveillance duties that were not part of the functions of Polytechnique lecturers. On 6 November 1834 the following item appeared in the "order of the day":

At the beginning of each *leçon*, the professor will call the roll. He will make a record of the absences as well as the grades given the students interrogated that day and the subject of the interrogations. This will permit the *Conseil des études* to keep up with the work of the school. These measures have nothing irksome or onerous about them: as a result of their execution studies will improve, order will be more perfect, and the administration will be better informed in its guidance and direction of one hundred fifty students.[89]

On 14 February of the same year Olivier wrote to Dumas, criticizing him for irregular absences: "You did not give advance notice this morning that you would not give your lecture, and the second-year students awaited you in vain. If professors are not reliable, adieu to discipline and order in the classroom."[90]

Olivier constantly tried to improve the quality of the teaching. His concern was partly motivated by a desire to upgrade the school's reputation. The more the teaching methods approached those used at the Sorbonne, the less the Ecole Centrale could claim to offer the same intensity of learning that characterized *écoles spéciales* such as the Polytechnique. The daily order of 6 November 1834 quoted above contained the following passage:

All professors must sign an attendance card at the porter's office upon entering the school. This will enable us to answer those who think or say that the Ecole Centrale follows the same path taken elsewhere, that professors hire substitutes after two or three lectures.

A year earlier, Olivier had asked Jules Pelouze, a chemistry professor, to visit the laboratory when it was used for experiments related to his course.[91] He made the same request to Dumas on 13 November 1933.[92] On the other hand, the cases of Colladon and Péclet indicate that other professors needed no prompting to make the teaching laboratory an important part of their activity at Centrale.

Olivier's insistence that professors regularly make their lectures and supervise their laboratory exercises did not, however, leave him blind to the need of his faculty to keep informed of the latest developments in their fields. When Perdonnet, Walter de Saint-Ange, Colladon, or Ferry requested a delay in the start of his course so that he might prolong an investigative trip to England or Belgium, the request was always granted. Even Dumas, with whom Olivier was not on the friendliest terms, got permission at least three times to attend meetings at the Sorbonne at the time his *cours* was to open.[93] The granting of such leave to allow professors to update their knowledge may not have been as extensive as Lacroix envisioned for his ideal *écoles centrales*, but it was certainly a more liberal policy than that governing the *collèges*.

Taken all together, the Centrale archives, Dumas' papers, and the quotations Pothier chose to make from Olivier's own *registre* permit only the most general conclusions about Olivier's administrative style. They do suggest that despite his military posture, the Ecole Centrale could hardly be depicted as a platoon of sullen churls pushed about by a petty martinet. Olivier did take the trouble to justify his directives:

The regulation forbids students to smoke in the classroom, but the use of tobacco is permitted in the courtyard during lunch hour. This rule is justified because some students may suffer from the effects of smoke in enclosed rooms. It is also true, however, that the students at the school belong to different nationalities, for some of which the use of tobacco is general, for others of which it is exceptional.[94]

When a group of students asked for an extra day of vacation at *carnaval* time, Olivier replied by observing that in the past several students always "lost themselves" during this holiday and failed to return to their studies.[95] In fact, more student petitions were granted than were refused. In December 1834 the third-year students thanked the *Conseil des études* for arranging a special course in differential calculus—which Liouville taught on Sunday mornings. (!)—while recommending that it be moved to the first year of study.[96] In January 1835 three students received permission to complete work on a steam pump project in Colladon's workshop, rather than the school's own, even if he could not supervise them personally.[97] In February 1836 the students were granted several changes in the regulations governing the use of the library.[98]

Olivier sometimes found expressions of student opinion useful in his own attempts to improve the teaching. On 13 November 1833 he wrote to the biologist Milne-Edwards:

Your course is still too theoretical; you must make it more industrial. Students remarked upon this last year, observing that they understood that it took time to adapt the course content to the special needs of the school. Hence try to give an industrial slant (*une couleur industrielle*) to your course.[99]

Of all the teachers J.-B. Dumas caused Olivier the greatest concern. Relations between the two men were never especially warm, but Olivier's opinion that Dumas neglected his teaching duties did nothing to dispel the coolness.[100] Dumas usually visited the laboratories at the Ecole Centrale in the hours immediately following his lectures, at which time he would consult with assistants about the exercises to be assigned to students, the demonstrations to be prepared for his own lectures, and the progress of his own experiments.[101] Despite the entreaties from Olivier, he seldom appeared in the laboratories during the afternoon hours when the students worked there. On at least three occasions Olivier wrote to Dumas enclosing petitions from students that both asked him to visit the laboratories and requested additional lectures on industrial chemistry.[102]

At the same time, then, that Olivier attempted to make his executive influence effective by his methods of "repression" and his mobilization of the faculty, he also tried to keep open the channels of upward communication by responding to students' petitions. In Olivier's eyes, however, the core of his program for avoiding conflict at the Ecole Centrale probably lay not so much in these administrative techniques as in the maintenance of a proper ethos, one based on dedication to work and the exclusion of politics. In September 1833 Olivier received a letter from an Italian named Luchino Valeriani, who expressed the fear that his sovereign, the Duke of Modena, would not look with favor upon the sending of Valeriani's son to politically troubled Paris. In reply Olivier listed in detail the various tasks assigned students during the year, then declared that "as for politics, you know that one must not concern himself with such matters at the school; our statutes in this regard are quite severe. Order and economy: this is our religious dogma; work: this is our political dogma."[103]

The preface to Olivier's *Mémoires de géométrie descriptive* also stressed that the exclusion of religious and political discussion was the key to harmony at the school: It was the principal justification for the *externat*.[104] The first prospectus announced that "works of literature and literary and political newspapers are banned from the school," a phrase repeated each year thereafter.[105] On 20 November 1833 Olivier announced in an "order of the day" that neither professors nor students were to "take any part in political acts and writings."[106]

Olivier probably had less than complete success in his efforts to exclude politics from the school, but it is difficult to judge the extent of his failure. Neither Pothier's excerpts from Olivier's private register nor the available documents in the Centrale archives reveals a single case of disciplinary action resulting from a violation of the prohibition on politics. Whatever the role played by Centrale in the July Revolution, there is no evidence that before 1848 it developed a public political personality analogous to that which enveloped the Ecole Polytechnique. Rioters did not seek out *Centraux* "because they had always been on the side of liberty." In any case, it would have been difficult to identify them: The *Conseil des études* dropped the wearing of the uniform in 1834, citing its expense and the "essentially civil nature of the school's education."[107] After the end of the 1833–1834 academic year, moreover, *étudiants libres* (auditors), a less easily disciplined group, were no longer allowed.[108]

The difficulty of quarantining the school from politics is best illustrated by the case of the Polish refugees. During the years 1834–1836, at least seventeen young men who had participated in the unsuccessful Polish nationalist revolution of 1830–1832 enrolled at the Ecole Centrale after the Marquis de Lafayette wrote to Lavallée on their behalf.[109] At least six had been students at the Warsaw Polytechnic School before the outbreak of the revolution, one other had graduated from the school, one had been a professor of economic science at the University of Warsaw, and three others were listed merely as army officers.[110] All were in serious financial straits, dependent on a stipend of forty-five francs a month provided by the French government. At Lafayette's urging Lavallée agreed to admit them without tuition.[111]

The young heroes were welcomed by their classmates, too. The entire student body combined to subsidize their purchase of graphic supplies and lithographed teaching materials.[112]

Despite their politically dramatic pasts, the Polish refugees caused little trouble at the Ecole Centrale. None is mentioned in any of the available documents dealing with discipline. Since they were dependent upon the French government and the school for their support, they may have been especially reluctant to incur the disfavor of the administration. It is significant, moreover, that Lafayette requested their admission to Centrale, not Polytechnique: at the latter school, the presence of Polish refugees might have provided a dangerous link between the oft expressed pro-Polish sympathies of the Parisian populace and the anti-Russian feelings of the Polytechnique cadets.[113]

As a group, the Poles were reasonably successful in their pursuit of *la science industrielle*. Charles Chobrzynski, in fact, was chosen to serve as a *répétiteur* at the school from 1836 to 1838. At least three others, Felix Szlubowski, Antoine Wolski, and Clutoine Mirecki, also received *diplômes* from the school.

The case of the Polish refugees also demonstrates how patterns of personal patronage and sponsorship played a role in the history of the school. Although Lafayette does not seem to have concerned himself with the fate of the Poles beyond his initial request to Lavallée, the proprietor of the school paid special attention to their progress. He defended them before the *Conseil des études* when "doubts were expressed" concerning the advisability of continued remission of tuition for all who returned in the fall of 1835,[114] he helped to arrange to transfer Eloy Bontemps to the school of military engineering at Metz,[115] he personally recommended Chobrzynski for the *répétiteur*'s position,[116] and, finally, he supported the admission of more Polish refugees in the fall of 1836, two years after Lafayette's death and at a time when the French government was no longer paying a stipend to the Poles.[117]

By means of his position as representative of the school to the public, Lavallée directly influenced the career choices of a considerable number of young Frenchmen. He conducted all the interviews with candidates for admission and their parents. When a candidate presented himself alone, Lavallée occasionally supported him *against* his parents. Jules Pétiet was the best-known example. Pétiet's grandfather had been minister of war under the Directory, and his father, also a military official, wanted his son to enter the Polytechnique. But a copy of the Centrale prospectus somehow reached Pétiet, then barely sixteen, and he was impressed by the emphasis it placed upon "applied science." When an interview with Lavallée reaffirmed his preference for *la science industrielle*, he withdrew his candidacy for the Polytechnique. After receiving one of the first Centrale *diplômes* in 1832, he went on to a distinguished career in the railroads before succeeding Perdonnet as director of the Ecole Centrale.[118]

On the other hand, Lavallée did not encourage all those he interviewed to attend his school even if he thought they had the intellectual capacity to succeed. When Urbain Le Verrier presented himself as a candidate, Lavallée, deciding that his talent for *mathématiques transcendantes* should not be deflected, urged him to apply to the Polytechnique. After graduating from that school in 1833, Le Verrier began the astronomical studies that were to lead him to the discovery of Neptune.[119] Other men who had made successful careers in fields ranging from school administration to literature later paid tribute to Lavallée's influence in their choice of vocation. When Lavallée's son-in-law sought advice from the headmaster of the Lycée Louis-le-Grand concerning the best program of study for his son, he was told that the headmaster was "only giving to the grandson what he received from the grandfather." He had been an unhappy student at the Ecole Centrale when Lavallée advised him to switch to a career in public instruction.[120]

Through the warp of horizontal links among students or staff members ran the vertical weft of half a dozen such sets of personal ties. The most distant, yet probably the most pervasive, was the tie of loyalty to the memory of Monge. Olivier's devotion to Monge has already been touched upon. The special intensity of his loyalty stemmed not only from the fact that Monge's specialty became his own, but also from his unusually long association with his great teacher. Olivier entered the Polytechnique in

1810, but because of illness he could not pursue a normal course of study. He had to stay at the school until 1815, becoming the disciple of both Monge and Jean-Nicholas-Pierre Hachette, Monge's collaborator in the creation of descriptive geometry.[121] The single instance in which Olivier requested permission to cancel a lecture at the Ecole Centrale arose on 18 January 1834, when he was asked by the Polytechnique to give an eulogy at Hachette's funeral.[122]

Nor was Olivier the only disciple of Monge to teach at the Ecole Centrale. Raucourt, who had entered the Polytechnique in 1810 along with Olivier, taught civil constructions and public works at Centrale until 1833; his successor in that position, Charles-Louis Mary, had studied at the Polytechnique from 1808 to 1810. Whereas the Centrale teaching career of Mary's Polytechnique classmate Coriolis was rather brief, that of a third 1808 entrant, Joseph Belanger, stretched from 1831 to 1864. Finally, Achille Ferry, who taught machine construction and ferrous metallurgy at the Ecole Centrale from 1830 to 1864, was the son of Joseph Ferry, who had replaced Monge's brother Louis at the Ecole de Mézières in 1781.[123] After serving as a deputy to the Convention, he taught beside Monge again in the early days of the Polytechnique.[124] At no time, then, were there fewer than four men on the *Conseil des études* who had been strongly influenced by the example of Gaspard Monge.

If all students entering Centrale encountered the influence of Lavallée and Monge, a good number of them had already spent many hours in the care of one of two faculty members at the school, Emile Martelet and William Priestley. Martelet, who had graduated from the Polytechnique in 1826, began teaching applied mechanics at Centrale in 1837. Three years before this, he established a preparatory school on the nearby rue de la Perle. Of the students who entered Centrale between the years 1835 and 1847, no less than 120 (12.3 percent of the sample) attended his preparatory school. Something of a local notable, Martelet later became *maire* of the seventh *arrondissement* and a municipal counselor.[125] Priestley, a great nephew of the famous British discoverer of oxygen, lost his father when a young child. The chemist Francoeur then became the family's patron. He found a position for Priestley's mother as conservator of a herbarium, enrolled him in the municipal drawing school, and, later, helped to support him at the Ecole Centrale.[126] Upon graduating from Centrale in 1836, he was immediately appointed a *répétiteur* in physics. During the next thirty-eight years he served as physics tutor and experimental assistant to Péclet. Exactly when Priestley established his preparatory school is uncertain, but by 1847 the dossiers of twenty-eight students indicated that they had studied with him.

As table 7.2 indicates, the clientele for the Martelet and Priestley schools were somewhat more heavily distributed in the higher-status categories than were Centrale students as a whole. During the period under consideration only these two schools specifically prepared candidates for the Ecole Centrale. Their tendency to recruit significantly more from the upper strata than did Centrale as a whole suggests that their presence reduced the differences in social recruitment between the Polytechnique and the somewhat more "democratic" Centrale that were discussed in chapter 3. In any case, attendance at the two preparatory schools meant that a prospective engineer began at Centrale not as a solitary schoolboy from a provincial *collège* entering a large, impersonal institution but as someone who had both a sponsor within the Ecole—his former instructor—and, perhaps, friends within the student body whom he had met while preparing for entry.

In addition to these more widespread attachments—to the counsel of Lavallée, the example of Monge, and among those from the same preparatory schools—certain other patterns of sponsorship

Table 7.2
Social composition of two preparatory institutions compared to that of the Ecole Centrale as a whole

	Martelet (N = 113)		Priestley (N = 28)		All Centrale (N = 707)
I. Nobles, *haute bourgeoisie*	81.4	(92)	85.7	(24)	68.1
II. *Moyenne bourgeoisie*	12.4	(14)	3.6	(1)	16.3
III. *Employés, cadres inférieurs*	2.6	(3)	3.6	(1)	4.2
IV. *Class populaires*	3.5	(4)	7.1	(2)	10.4
Unclassifiable	0		0		1.0

developed during the course of the Ecole Centrale's first twenty years that were more limited in scope. In the first place, graduates of the school began to be recruited as *répétiteurs*, and, eventually, professors. As table 7.3 shows, the hiring of Bineau d'Aligny in 1832 was the first of sixteen such appointments made before 1848. In 1847 eight of the thirty-three teaching members of the school's staff were *Centraux*, while eight more were *Polytechniciens*. At the latter school, on the other hand, the recruitment of staff from alumni was much more marked: in 1837, for example, no less than fifteen of the twenty-nine members of the teaching staff were *anciens élèves*, whereas only one (the chemist Péligot) had studied at Centrale.[127]

With respect to the quality of the schools' education, the significance of this in-breeding is difficult to judge. The Polytechnique could argue that because it was the preeminent institution of scientific education in Europe, the recruitment of faculty from among its former students was neither surprising nor pernicious. The founders of Centrale argued, on the other hand, that their attempt to unite theory and practice required them to recruit their faculty from among the best of those who had experience in industry, regardless of their performance in a particular school. Whatever the significance of these recruitment patterns for *la science industrielle*, however, they had an important influence upon the nature of group alignments in the school. As has been mentioned above, the rebelliousness of the student body at the Polytechnique was reinforced at crucial moments by the support their protests received from the faculty of the school, especially those who were *anciens élèves*. The smaller number of alumni faculty on the Centrale staff thus reduced the likelihood of such alliances. Moreover, whereas both faculties had acquired considerable independence from their administrations (at the Polytechnique in 1830, at Centrale after 1832), the separation between the two was more clearly drawn at the Polytechnique. Olivier, for example, served as both teacher and chief disciplinary official while enlisting faculty members in order-keeping functions. At the Polytechnique the commandant enforced order almost exclusively through the captains and lieutenants serving as inspectors of studies who, not being *anciens élèves*, were less likely to sympathize with the student opposition.

Table 7.3
Faculty and staff at the Ecole Centrale, 1829–1848

Name	Previous school	Positions, subjects	Interval
Lavallée	Fac Law	Director-Owner,CE	1829–1850+
Dumas, J.B.	Fac Sci	P, CE :Chem, IChem	1829–1850+
Olivier, T.	X	DE, P,CE:Geom	1929–1850+
Péclet, E.	ENS	P, CE :Phys, IPhys	1829–1850+
Colladon	Fac Sci	P:Mec, Steam Eng.	1829–1835
Ferry, A.	Fac + Mines	P,CE:Con.Mach,Metall.	1830–1850+
Walter de St.-Ange	St. Cyr	P,CE:Th. Mach, Con Mach	1830–1850+
Perdonnet, A.	X, Mines	P,CE:Mines, RRoads	1831–1850+
Raucourt	X, P&C	P,CE:Con&T Pub	1831–1834
Mary	X, P&C	P,CE:Con&T Pub	1934–1850+
Bélanger	X, P&C	DE,P,CE:Mec, AMec	1831–1850+
Payen, A.		P,CE:I Chem	1834–1850+
Bardin	X	DE,CE	1839–1840
Empaytaz	X	DE,CE	1840–1850+
Priestley	EC	R,L:Mec, Phys	1836–1850+
Choquet		Pa,:Geom	1821–1831
Ballio-Lamotte	Mines	Pa,R,C:Geom	1830–1834
Laurens	EC	R,P:Geom,Con-TPub	1833–1849
De Paul		R:Geom	1939–1845
Fernique		R:Geom	1845–1850
Martelet, P.-J.-E.	X	P,Ps,R:Mec,AMec,Ana	1836–1850+
Didiez		Pa:Mec	1829–1831
Coriolis	X,P&C	P:Mec	1830–1832
Liouville	X,P&C	P:Mec	1833–1838
Pagès	X	R:Mec	1837–1839
Sonnet	ENS	R:Mec,Ana	1838–1848
Abria	ENS	P:Phys	1836–1840
Regnault	X,Mines	P:Phys	1839–1842
Masson	ENS	P:Phys	1842–1850+
Thomas, L.	EC	R,P,C:Phys,SteamEng	1833–1850+
Fournier	ENS	R,C,L:Phys	1835–1838
Hébert	ENS	R:Phys	1838–1842
Becquerel, A.E.	X	R:Phys	1844–1850+
Laurent		Pa,L:Chem	1830–1833
Pelouze		P:Chem	1833–1836
Péligot	EC	R,P:Chem,AChem	1834–1850+
Fremy		P:Chem	1839–1841
Cahours	X	R,Pa,P:Chem	1835–1850+
Bergouhnioux		R,Pa,C:Chem	1829–1833
Bineau d'Aligny	EC	R,L:Chem	1832–1834

Table 7.3 (continued)

Name	Previous school	Positions, subjects	Interval
Boistel	EC	R,C:Chem	1833–1836
Chobrzynski	EC	R:Chem	1836–1838
Charpentier	EC	R:Chem	1836–1838
Rousseau	EC	R:Chem	1837–1842
Milne-Edwards	(Prof:HenIV)	P:HNat	1834–1845
Doyère	(Prof:HenIV)	P:HNat	1845–1850+
Spiers		P:English	1832–1837
Faure		Ps,P:Mach,Cinemat.	1842–1850+
Devillez	Ec	R:Th.Mach.	1835–1836
Gourlier		P:Con.Civ.	1830–1832
Bussy	X	P:AChem	1830–1832
Boutin	EC	R:Mines,Miner.	1841?–1843?
Knab		R:IChem	1839–1846
Dellisle (sp?)	EC	R:IChem	1846–1849
Poinsot		R:IChem	1849–1850+
Salvetat	EC	P:Technologie (Var.)	1844–1850+
Alcan	EC	P:Textiles	1844–1850+
Prévost, C.	Fac Sci	P:Mineral.	1830–1831
Burat	Mines	Pa,P:Mines	1838–1850+
Lacambre		R:Mineral.	1835–1836
Des Cloiseaux	Mines	R:Mines	1843–1850+
Brongniart		P:Botan.	1830–1831
Parent-Duchatelet	Fac Sci	P:Hygiene	1830–1833
Cornet	EC	R:RRoads	1846–1848?
Jacquelain		L:Chem	1833–1850+
Walter, Ph.	Fac Sci	C:Chem	1836–1846
Scribe	EC	C:Chem	1847–1850+
Fouché[a]		C,L:Phys	1829–1834
Leblanc		C:Dessin,Archt	1829–1831
Chavonhet		C,L:Dessin,Archt	1835–1837
Thumeloup	EBArts	C:Dessin,Archt	1837–1850+
Wurtz		C:Chem	1845–1850+
Sellier		C:Dessin	1829–1830
Leblanc, Ad.		C,R:Dessin	1835–1842
Nouvian		C	1841–1847
Obelliane		L	1840–1845
Sahuqué		L	1837–1840
Daniel		L	1845–1850+
Tronquay		C:Dessin	1846–1850+
Naef		Inspr	1835–1839
Rameau		Inspr	1838–1850
Delsart		Inspr	1838–1842
Bance		Inspr	1838–1840

Table 7.3 (continued)

Name	Previous school	Positions, subjects	Interval
Huet		Inspr	1840–1845
Duroch		Inspr	1845–1850+
Belchamps		Inspr-Sup.	1847–1848
Gand		Inspr	1830–1835
Valton		Inspr	1829–1830
Regnault (?)		Inspr	1840–1848
Cavadino		Inspr (?)	1848–1849
Cance		Inspr (?)	1837–1848
Bideaut		Inspr (?)	1837–1848

Key: X = Ecole Polytechnique; ENS = Ecole Normale Supérieure; P&C = Ponts et Chaussées; EC = Ecole Centrale; Hen IV = Collège Henri IV; EBArts = Ecole des Beaux Arts; DE = Dir. of Studies; CE = Member of *Conseil des études*; P = *Professeur*; Pa = *Professeur adjoint*; R = *Répétiteur*; C = *Chef de travaux*; L = *Préparateur*; I = Industrial; A = Analytic.
a. *Surveillance general*, 1832.

If table 7.3 shows that at the Ecole Centrale, faculty members gradually began to hire their own students as teachers, it also indicates that recruitment patterns in certain subjects were based on common links to other institutions. Péclet made the physics courses the fief of *Normaliens*: two of the three professors added during our period (Abria and Masson) and two of the five tutors (Fournier and Hébert) came from Péclet's own school. The other physics teaching posts were divided between *Centraux* (Priestley and Thomas) and *Polytechniciens* (Regnault and Becquerel). In the case of *Histoire naturelle*, Milne-Edwards was able to insure that Louis Doyère, who succeeded to his chair at the Collège Henri IV, also replaced him at the Ecole Centrale.[128] In the case of the course on mining, three of the four teachers and tutors who taught the subject after Perdonnet split it in two in order to concentrate on railroads had attended the Ecole des Mines, where, like Perdonnet, they had been "auditors" (*étudiants libres*) rather than regular students planning to join the state mining *corps*.

Perdonnet's personal clientele extended far beyond the teachers in the mining course, however. During the Ecole Centrale's first twenty years he served as director of technical services for the Paris-Versailles Railroad, then as administrator of the Compagnie de l'Est. His course at Centrale consequently became one of the richest recruiting grounds for the men who directed the construction, maintenance, and improvement of the French railroad network. Dumas claimed that no other individual had placed so many Centrale students "on the road to work and fortune."[129]

Not all the students who came to Centrale forged personal links with one of its faculty such as those described—preparatory lessons from Martelet, sponsorship from Lavallée, a letter to his parents from Olivier, a trip with Colladon, or a job with Perdonnet. Even the fragmentary biographical evidence available suggests, however, that most did, especially those who played the most important roles in defining the profession of civil engineering. The years when Olivier directed the recovery of the school thus also witnessed the construction of patterns of personal interaction that became key elements in the structure of the French technological elite.

Routinization

By the end of the year 1835–1836, the Ecole Centrale no longer seemed in danger of collapse. The school had been spared reputation-threatening disruptions; the teaching program, constantly strengthened, had proved successful; and no further epidemics had appeared. In May 1836 Lavallée predicted that in the following year the school's revenue would not only cover operating costs but also produce a small *bénéfice*.[130] The next month he learned that State scholarship funds would begin supporting students at the school during the following year.[131] When Olivier announced that his ever precarious health would force him to give up his post as director of studies before the 1836 *rentrée*, the *Conseil des études* rightly thanked him for having placed the school "on the road to prosperity."[132]

The next twelve years on that road have left little more information about changes in daily life at the school than did the first seven. An expanded order-keeping apparatus did appear, but increases in enrollment make this less than surprising. On 18 November 1836 the *Conseil des études* announced the creation of a *Conseil d'ordre* to be composed of the director, the director of studies, and one other professor. It was to meet twice weekly to deal with all infractions of the school's rules; only the most serious cases would be referred to the *Conseil des études*. In 1838 the number of inspectors of studies was increased from one to four.[133]

Joseph Bélanger, who served as Olivier's successor from 1836 to 1838, had the most peaceful term of any director. The most serious infraction with which he was faced came when a group of ten second-year students left the school three hours early on 22 April 1837. Bélanger wrote to each of the families involved, warning them that continued insubordination, especially concerted action, would result in expulsion, regardless of the student's academic standing.[134] The following academic year, that in which the three new inspectors were hired, was one of the most orderly in the school's history: not a single case of indiscipline was brought before the *Conseil des études*.

In the spring of 1838 the Centrale chair in Mechanics became vacant when Liouville announced that he would resign at the end of the academic year. When he succeeded Liouville in that position, Bélanger felt obliged to relinquish his post as director of studies. A selection committee composed of Dumas, Péclet, and Olivier could not agree upon a new director of studies, however, despite five months of discussion. Finally, at the end of January 1839, Olivier reassumed the position on a provisional basis.[135] The issue dividing the three founders was the nature of the position. As he proposed to the *Conseil des études* in a *séance extraordinaire* of 26 January, Dumas wanted to create a new post solely devoted to surveillance and discipline. Like the Polytechnique's military commandant, the new functionary would also have the title of commandant, but in this case he would act as the agent of the *Conseil des études*. In opposing Dumas' plan Olivier argued that "in a private establishment, a disciplinary official would not have the same authority that one's rank gives in a school where the students live in barracks under military discipline; all the prestige of the authority (*le prestige de l'autorité*) would remain with the intellectual supervisor of studies." Oli-

vier clearly felt that his position as professor of Descriptive Geometry was the key to his ability to maintain order. In any case,
the *Conseil* temporized, "accepting in principle" Dumas' proposal but taking no further action.[136]

Olivier's provisional directorship proved to be a stormy one, at
least by Centrale standards. Minor infractions multiplied, perhaps because there were now three new inspectors to report
them. On 18 March Olivier reported to the *Conseil des études*
that whereas the second-year students had been "relatively doc-
ile," those in the third year were becoming "very difficult to manage."[137] As it turned out, the second-year students caused the
most serious incident. At the end of March fifteen of them did
not appear at the school for two days. On 2 April they were ordered to appear before the *Conseil des études*. Dumas declared
that they would all be placed in the special custody of the three-
man *Conseil d'ordre*, which could even expel them if they continued to break the rules. Upon leaving the room, the students
broke into a "disrespectful tumult," objecting that only the full
Conseil des études could pronounce permanent exclusion of an
individual student. Dumas, incensed, proposed the immediate
expulsion of all fifteen students. He apparently carried the majority of the *Conseil* with him, but Olivier refused to carry out
the order.[138] At a *Conseil* meeting the next day, the director of
studies defended the students, calling attention to their good
scholastic records and arguing that a "spontaneous demonstration" was less reprehensible than premeditated actions. He won
a partial victory: the two ringleaders were expelled, but the rest
received only reprimands. Dumas then delivered a warning lecture to an assembly of the entire school.[139]

When the school's "definitive regulations" were published at the
end of 1839, the post of director of studies continued to carry the
responsibility for both the supervision of teaching and the maintenance of order.[140] Yet it was Dumas who finally prevailed. Olivier, shaken by the events of his last term of office, did not object
when in 1840 Dumas and Péclet agreed to hire Bénédict Empaytaz as the new director of studies. In the case of Empaytaz, however, the "prestige of authority" came not from scientific
accomplishments but from more traditional sources. Although a

213

The School as a Society

Polytechnicien like all his predecessors, he was neither a disciple
of Monge nor a scientist, and he taught no courses at Centrale.[141]
He had entered the artillery immediately upon graduation and,
two years later, distinguished himself at the Battle of Waterloo.
By 1830, however, he had become so attached to the Bourbons
that he resigned his commission after the July Revolution.[142] Between the lines of Pothier's obituary notice emerges the portrait
of the perfect bureaucrat:

Most of our comrades knew Empaytaz only through the official
functions, which he performed with justice, moderation, and a
totally military dignity in speech and gesture. He did not like
discussion; when questioned by professors about scheduling matters or by students about his reprimands, he answered merely by
citing the regulations. Under a cold exterior he hid a real benevo-
lence; he had a deep attachment to the school.[143]

Gustave Eiffel also testified to Empaytaz's coldness. In his third
year at Centrale, he returned after Christmas without having finished any "vacation work" (*travaux de vacances*). Awarded a
zero, he sought a meeting with the "pitiless" M. Empaytaz in
order to justify himself, but an *inspecteur* convinced him that
with such a director of studies his case was hopeless.[144]

In this austere old soldier the 1839 *règlement définitif* thus found
a dedicated executor. Few documents challenge the impression
that during the seven years before the Revolution of 1848 the
Ecole Centrale was immersed in an orderly, industrious routine.
The histories of the school, both published and unpublished, find
little to comment upon. Pothier, usually the most detailed, covers the period in three pages.[145] The *Rapport à présenter*'s most
important observation is that the school became an eminently
prosperous enterprise during this period: the mean annual profit
(*produit net*) for the period 1840–1852 was 65,000 francs.[146]

Things ran less smoothly at the Polytechnique. In 1840, when
the school received its first non-*Polytechnicien* commandant
since the Restoration, General Boilleau, the students quickly became barely governable. As Pinet puts it, the students "took back
from him all the liberties that [Boilleau's predecessor] General
Tholozé had been at such pains to suppress." The *divertisse-
ments* (balls, games, and other social events) tolerated within
narrow limits by previous administrations "took on proportions

hardly compatible with the accomplishments of serious studies."[147] (These developments quite probably influenced the decision to hire Empaytaz.) A series of politically inspired disruptions—attempts to demonstrate at the funeral of Jacques Laffitte and in favor of Admiral Dupetit-Thouars' defiance of the English—shook the school in 1843 and 1844. The dismissal of both classes was finally prompted, however, by an issue directly concerning the students' control over their own institutional procedures. The Ministry of War had rejected the candidate for graduate examiner proposed by the Académie des Sciences and approved by the students. Instead it ordered the school's new director of studies, J.-M.-C. Duhamel, to assume the examiner's post as well. In the students' view this created a serious conflict of roles. In response to this affront to their independence and to that of the Académie des Sciences (where they had many patrons), the students refused to take their final examinations. Under pressure from the Right the ministry decided to take strong measures against this latest rebellion and appointed General Rostolan, an infantry officer known for his severity, as the new commandant. He immediately secured the "permanent exclusion" of seventeen leaders of the rebellion, but in Pinet's account, "The iron hand of the new commandant did not succeed in reestablishing discipline until two years later."[148]

The contemporary Ecole Centrale witnessed no such revolt. The available records show only one serious disciplinary case, a second-year student expelled in 1844 after repeated altercations with a tutor: an individual action, not a gesture of student solidarity against the administration.[149] Just the same, the atmosphere at the school could not have been the same as in the early days of the *élèves fondateurs*, or even the years of Olivier's active paternalism. The elected *chefs d'études*, with their prestige and their voice in certain personnel decisions, had become appointed *commissaires*, then *chefs de specialité*, chosen in each course merely to report to professors about those lecture topics on which the students wished further instruction.[150] Student petitions, less frequent even under Olivier, all but disappeared under Empaytaz.

During these years, the students did not refrain from all collective initiatives, however. In the fall of 1845, when Constant Robaut, the son of a printer from Douai, announced that family financial reverses would force him to leave the school, other students, led by Adolphe Cauvet, began to take up a collection for him among themselves. They then decided to create a permanent aid fund (*caisse de secours*). Dumas and Empaytaz argued that the school should control the funds and supply the presiding officers, but the students apparently were able to avoid such a takeover.[151]

They had less success with their attempts to form a civil-engineering society. In 1840 a group of graduates and students tried to gain the approval of the Centrale faculty for an Association des Elèves de l'Ecole Centrale that would "establish a center for meetings of students from all classes" in order to "create and maintain relations of friendship and utility among them."[152] At the same time, the "intellectual and moral influence" of such a "society of civil engineers" would help to "propagate enlightenment" throughout the country. For this purpose the *règlement* laid plans for a library, scientific collections, a technical publication, and the presentation of research papers.[153] Despite the fact that the "founders and professors" of the Ecole Centrale were to be honorary members of the Association, the *Conseil des études* opposed the plan. Because the association would be open not only to *diplômés* but to those who held only the *certificat*, as well as third-year students, standards would be threatened. Even worse, non-*Centraux* could be admitted as corresponding members.[154]

Without the support of the Ecole Centrale itself the efforts of *Centraux* to build their own combination of student-alumni association and professional society met only failure. Between 1840 and 1848, the students tried almost every year to form such a society, with the same results.[155] There is good reason to doubt, moreover, that a concern for standards was the only reason that the Centrale authorities blocked these attempts. In the draft of a note to the Ministry of Public Instruction, Dumas cited a rather different motive for not permitting the formation of such an association:

If we had done so, our students would have been linked in a veritable freemasonry, and who knows what that would have pro-

duced? . . . We have always believed that society would be placed
in grave danger if such a lever was ever placed in the hands of an
impassioned revolutionary. Such a fanatic could then give the
word of command to the entire manufacturing population of the
country at a moment of crisis, the occurrence of which one may
still fear.[156]

The plans for the society had always carefully proscribed any discussion of politics or religion, but this was apparently not enough
to reassure Dumas.[157]

Soon after revolution broke out in 1848, the leaders of previous
efforts to form a society sensed that the new political atmosphere
had given them their chance. A committee composed largely of
those *Centraux* who had led the attempt of 1840 chose a young
civil engineer named Camille Laurens to negotiate with the
school's directors.[158] Calculating, perhaps, that the immediate
protection such an organized group might offer the school in the
current political crisis was worth the risk of encouraging a freemasonry, Lavallée and the professors agreed to support the project. On 4 March 1848 forty men met in the amphitheater of the
Ecole Centrale to form the Société Centrale des Ingénieurs Civils. By the end of the year at least ninety-three more had joined.[159]

Although during the spring and summer months of 1848 Centrale students were in close contact with alumni, they were not
allowed to become members of the newly formed Société. Here,
at least, the wishes of the Ecole's directors seem to have prevailed. Yet one might question the wisdom of the hostility Dumas and his colleagues showed to any form of student organization.
Their view of desirable links between men at the school seems
limited to the personal relations between individual teachers and
students described in this chapter. Yet Empaytaz's preference for
bureaucratic routinization suggests that such relations were
undergoing a certain rigidification, a cooling. Empaytaz may have
wanted his students to feel that they were sons, or protégés, but
he treated them as *administrés*. They were certainly not allowed
to act fully as school citizens. When the Revolution gave the
young *Centraux* the chance to act as citizens of the new Republic, they flocked to the streets to help run the National Workshops.[160] Their eagerness to join such a dangerous political
adventure brought their mentors at Centrale considerable an-

guish.[161] But perhaps this was the price of the years of a calm that excluded organized student activity: the citizen's role is not learned overnight. It was both ironic and predictable, therefore, that the obstreperous, conspiratorial (and thus politically experienced) *Polytechniciens* became much acclaimed guarantors of stability for the new Republic, acting as *aides-de-camp* or bodyguards for the new leaders, serving as messengers for the Provisional Government, and, after the arrest of Emile Thomas, replacing the *Centraux* as administrators of the National Workshops.[162]

8

The Formation of
Technological Man

The Portrait Presented

Despite the general orderliness and profitability prevailing during the first years of the Empaytaz regime, the founders and their colleagues began to be increasingly worried about the future of the school. In the mid-1840s they began to take measures to persuade the French government to assume ownership. On 2 March 1846 Lavallée, Dumas, Péclet, and Olivier wrote to the Ministry of Public Instruction that they were "anxious to see their work perpetuate itself in the face of all eventualities and to endow the country with an institution that they regard as indispensable for the progress of its industry." They pointed out that the governments of Saxony, Baden, Wurtemberg, and Belgium had successfully created schools modeled on the Ecole Centrale after studies of its "constitution and mechanism." Such actions "gave the measure of the political influence that our institution will exercise in the hands of the State." In the hands of Lavallée or his heirs, they argued, the school would be vulnerable to all the vicissitudes affecting any item of personal property.[1] Eleven more years would pass, however, before the transfer was finally made.

For the purposes of the present study, it is fortunate that the takeover did not occur immediately, for during the long campaign to secure the transfer, the men who guided the Ecole Centrale were moved to discuss in writing, if only briefly, the role they thought their school and its products played in the French economy and society. As the dominant political personality in the group—dean of the Faculty of Sciences during the July Monarchy, minister of commerce, industry, and agriculture in the Second Republic, and senator and top educational advisor under Napoleon III—Dumas received drafts of most of these statements before they were sent to the appropriate official: His papers at the archives of the Académie des Sciences, when supplemented by other sources mentioned above, permit the construction of a portrait of the new technological elite as seen by its creators.

The new elite sprang from a new society. In the section called "Industry" of a document outlining possible arguments to use in the campaign for the State takeover, Lavallée suggested that they "explain the two principal causes underlying the growth of French industry in the last half-century: the progress of science and the political movement that destroyed the barriers between the various classes of society." Before the Revolution "the privileged class, which had the power and most of the wealth of the country, a class assured of transmitting to its descendants most of its advantages, could never have given to industry the stimulus it received from the middle class (la classe moyenne), which could maintain its position only by its education (instruction) and its work."[2] Olivier made similar observations in the preface of his Mémoires: the privileges of birth and custom have now vanished; in the new society only two things, "the size of one's fortune and the size of one's knowledge and intelligence, establish the classes into which all citizens place themselves."[3] For educators the principal difficulty arose from the fact that even in this new situation, the possession of wealth did not always correspond to the possession of intelligence and, especially, the will to apply it with diligence. Joseph Bélanger claimed that the two might even be antithetical:

An aptitude for the sciences, a genius for their application, and a love of hard work are the qualities that combine to form a true engineer-industrial (ingénieur-industriel). They are rarely found together, especially among young men whose parents' fortune gives them the advance assurance of easy circumstances. On the other hand, the history of the progress of the human spirit and of civilization is filled with examples of illustrious men whom fortuitous events rescued from the obscure occupations to which they were condemned by birth.[4]

At the same time, the talented and industrious were not alone in wishing to rescue themselves from obscure occupations. The destruction of the barriers of birth had released among the general populace a swell of expectations that threatened to undermine social stability. For Lavallée, the "principal cause" of the current "moral illness" was the "inordinate ambition for fortune and power that today torments society."[5] For Olivier it was "hardly prudent to declass large masses of men by giving them an education all out of proportion and harmony with the needs of each class of society."[6] While seeking to encourage the most capable members of the lower classes, then, educators had to avoid exciting unrealistic hopes among their less-endowed brothers or run the risk of social upheaval. The solution, of course, was a system of selective scholarships, granted by the State as well as by private groups such as the Société d'Encouragement and the Ecole Centrale itself. Unlike previous scholarships, however, these grants would not be designed to "feed the public services," producing only functionaries, but to mold the leaders of the new moral order brought by industrialization.[7] *Boursiers* would thus be the leaven of the new technological elite.

Even while in school, the scholarship students made important contributions to the formation of their future colleagues. In a letter supporting the continuation of the State scholarships to *Centraux*, Dumas claimed that although *boursiers* accounted for half of the "best diplomas" given each year, their real significance lay in their influence on their classmates:

They are poor. Their drive for achievement will never weaken because if they fall below a certain level in the examinations, they will lose their scholarships. In their conduct and their diligence they serve as terms of comparison for all their comrades, because they have something to lose. One can always ask more of them. In matters of discipline, they have always set an example that has made them our best auxiliaries.[8]

At the same time, Olivier saw the scholarship system as a social safety valve that should be closely monitored:

A poor man endowed with a certain intellectual capacity will be restless (*remuant*) if he is deprived of education. To ascend to a higher class, he must acquire instruction that permits him to earn his living by working with his intellect, whereas in a lower class, he can earn it only by working with his hands. . . . We must give scholarships to promising young men who may then pass into a higher class and become useful there. But at the same time, when we become convinced that a mistake has been made, that a young man is not fulfilling the hopes held for him earlier, we must, moved by a severe but equitable justice, bring him to a halt by withdrawing his scholarship.[9]

The student dossiers show that such a policy was indeed carried out at the Ecole Centrale, even for students about to enter their third year.[10] The "restlessness" of a student who narrowly missed capitalizing on his chance for social advancement was apparently a less threatening problem than the frustration of those who felt they had never been given that chance.

If the *boursiers* were the leaven of the new elite, the rest of the dough was composed largely of *haut bourgeois* gold dust. The pattern of social recruitment discussed in chapter 3 reflected the explicit desires of the Ecole's guardians not to "leave without guidance the heirs of the manufacturing domain." Without Centrale's technological education some risked seeing "the establishments founded by their fathers" perish in their own "ignorant hands." Three years at the Ecole Centrale would "assure the perpetuation of manufacturing families."[11]

Nor was the source of the parents' wealth especially important. Despite Bélanger's fears about the lack of industriousness to be expected from those with a financially secure future, the Centrale spokesmen expressed no hostility to the recipients of land rents, functionaries' salaries, or other nonindustrial income. All paying customers were welcome. To the fortunes surviving from the Old Regime had been added the fortunes made after the fall of the Bastille: in land, in government, in industry, or in supplying the armies.[12] (If every private in the Revolutionary armies carried a marshal's baton in his knapsack, filling the rest of the knapsack produced the fortunes that sent sons to Saint-Cyr, Polytechnique, and even Centrale in the days after Waterloo when marshals were made not in battles but in schools.) Whatever the political rivalries among the current possessors of these different types of wealth, the Ecole Centrale promised to unite the ablest of their progeny as the "leaders" of a single "industrial army."[13]

It was inconceivable that the officer corps of this new army could get its training anywhere outside an institution of higher technological education. No longer could men from the shop floor rise to claim the baton; the new officers were entirely different beings:

The manufacturer who formerly entrusted the direction of his enterprise to foremen whose intelligence raised them from the class of ordinary workers knows now that lest he fall behind his competitors, he must seek the advice of a different class of men, who can understand the thousand inventions made each year and can apply them with profit. This class is that of the industrial engineers to which the Ecole Centrale gives both the fundamental and the specialized knowledge necessary for those who really deserve the title.[14]

Only the case of Britain seemed to undermine this argument. Why did France need such an expensive and elaborate school if Britain had become the leading industrial power without any higher technological education worth mentioning? At various times and places Centrale's defenders offered three different answers to this question:

1. Their primary strategy was to claim that Britain's "uneducated" industrialization had been carried out at far too great a cost:

As the founders of the Ecole Centrale immediately perceived, British industry wasted fuel in its furnaces and iron in the construction of its machines. It threw its capital into rash schemes that did not always succeed. Let us spare French industry the dangers of an excessively servile imitation of the prodigalities of its neighbor. Let us learn to make better use of coal and iron, to respect the prudent habits of our commercial enterprises, and to obtain the maximum effect with the minimum force, the maximum product with the minimum expenditure.[15]

The greater efficiency brought to industrial processes by the scientifically trained Centrale engineers would allow France to compensate for her lack of certain resources, such as coking coal, and to close the gap in overall economic development by the systematic choice of only the best British technology.[16] In short, exploiting the principal advantage of backwardness, the chance to borrow only the best of British technology while avoiding British mistakes, implied that one had some way to evaluate the different products and processes one might borrow. The scientific testing procedures of the French engineer permitted selective emulation of the British.

2. They argued that a Frenchman's mind operated differently from that of his counterpart across the Channel:

Everything proves that in France our mental processes are such that we are devoted to starting from principles to arrive at applications, whereas it is on the contrary part of the genius of the English nation to be more willing to begin with the way things are in practice (*la pratique des choses*) in order to build toward pure theory.

English civil engineers, who all began their education with a four-year apprenticeship in the shop and terminated it with a year's stay in the office of an engineer directing a factory, have been formed by a process in harmony with their national character. In France there was nothing to hope for from such a method; it could not grow in our soil, as shown by the failure of attempts to transplant it.

We believed instead that in France, we had to create Civil Engineers of Arts and Manufactures by processes more in keeping with the national genius: a systematically reasoned common education followed by individual apprenticeships in the workshops of industry.[17]

It followed that the government was "compromising the future of French industry" by allowing civil engineers to be "haphazardly educated."

3. Even if the French mentality had been no less capable than the British of turning to profit an inductive, empirical, practice-based learning process, it followed as a corollary of the first argument given above that a school of higher technical education was needed. If purely technological efficiency avoided a waste of natural resources, so the social efficiency of well-organized technical education minimized the waste of human resources. France could not afford the slow, dispersed, uneven system of apprenticeship followed by on-the-job advancement. Selection of the most able scholarship candidates by national competitive examinations would ensure that the Ecole Centrale could quickly locate the most capable of those who could not otherwise afford the school.[18] Even if this general argument were granted, however, an alternative with rather different consequences for social stratification suggested itself: the resources could be put into the *écoles d'arts et métiers*. The Centrale spokesmen seldom failed to claim that the education at these schools was not "on a sufficiently high level" or that they were formed for "an entirely different purpose," the creation of "skilled foremen for certain specialties."[19]

Dumas and his colleagues argued, then, that their new school responded perfectly to the needs of the new society. The technological elite an Ecole Centrale so efficiently produced combined in its ranks the new wealth created by industrialization with traditional fortunes, the new ambitions created by the Revolution with the need to secure positions already won. The school's interpreters' next task was to describe not where their civil engineers came from but what the school had made them.

In the first place, the *Centraux* were the bearers of *la science industrielle*. In the portrait drawn by Centrale spokesmen, the mystique of the Centrale graduate sprang in large measure from the efficacy and coherence of the industrial science learned in his three years at the Hôtel de Juigné. In the generality and comprehensiveness of his curriculum lay the basis of the Centrale engineers' claim to industrial aristocracy, the right to direct large enterprises, the possession of qualities radically different from the "skills in certain specialties" taught in less exclusive schools. If the previous chapters have argued that the pedagogical reality behind these claims supported them rather weakly at certain crucial points, they have also shown that the social myth nevertheless had a kernel of truth to grow on.

Accompanying knowledge of *la science industrielle* came the mystical, exalting ability to unify theory and practice.[20] This capacity meant that the plans and commands of *Centraux* could produce items of great utility, which—as Olivier had argued most forcefully—gave them a second claim to esteem.

The *Centraux* owed their third set of virtues—versatility, resourcefulness, and independence—not only to the content of the curriculum but to the formative influence of the school's environment: the *externat*, strict academic accountability, and, above all, intense effort. De Comberousse drew the most striking portrait of the new man produced by this regimen:

To pass through the Ecole Centrale is to undergo a special kind of tempering. One is now ready for all contingencies. One can leave tomorrow for Asia or America, Japan or Australia, Suez or Panama, sure of being able to face all difficulties, to fulfill all sorts of tasks. One may become a chemist or a mechanician, a constructor or a metallurgist, a farmer or a professor. No matter what point on the immense surface of the applied sciences a graduate of the Ecole Centrale is asked to occupy, he is already acquainted with it. He can, if necessary, deepen his knowledge to the point of becoming a master: none of the parts, either close or distant, is unfamiliar to him. In short, he is a generalist first and foremost; he only becomes a specialist by necessity. This is the most striking characteristic of his intellectual physiognomy.

To build a house, a machine, or a work of art: it is all the same to him. To adapt the available means in the most rational manner to the assigned goal, with neither unintelligent prodigality nor damaging parsimony; to calculate everything, to prepare everything in advance; to forget nothing; to achieve the most useful, convenient, and least expensive solution with the aid of ingenious combinations: such is the Centrale civil engineer, the brain that sets in motion all efforts destined to create the projected ensemble.[21]

Technological Man *par excellence*. By the time De Comberousse spoke, after all, two generations of engineers trained in France had gone abroad to build the harbors, bridges, canals, and railroads that provided the sinews of the modern world economy.[22]

Yet De Comberousse's portrait was as incomplete as it was idealized. In their day-to-day teaching and in their periodic decisions about curricular changes the creators of *la science industrielle* were concerned above all with perfecting the structure of their students' understanding of the physical world. When they presented that structure to the public, they emphasized those aspects that could best be transmuted into the coin of social status: generality and utility. What they wrote about most sparingly was their vision of the special role of the engineer as a member of a particular kind of society with a particular structure and set of values. In short, the Ecole Centrale produced no Saint-Simon. Nor, with the exception of Péclet (at least according to evidence from his youth), did the school's staff contain any real followers of Saint-Simon, or of men like Enfantin and Bazard. One of the more surprising results of the research conducted for the present study was the absence of references to Saint-Simon in the documents left by Centrale's guiding spirits, by their contemporary observers (such as Pothier and De Comberousse), or by the friends and colleagues who wrote their obituaries. That they could employ terms such as *industriel* to refer at once to all strata in the economic (that is, productive) sector of society, just as did Saint-Simon, may indicate not so much the latter's influence as the fact that such a usage reflected a far more widely shared belief that the members of that sector shared common interests.[23] For the most at the Ecole Centrale, the zeal for the propagation of scientific and technological knowledge, for which Péclet claimed Saint-Simon as the inspiration, sprang directly from the precepts of Monge and Lacroix. Although previous writers have suggested otherwise, the attitudes of the technological elites who directed French industrialization did not all originate in discipleship to a single prophet. To understand Technological Man as he emerged in nineteenth-century France, the investigative net must be cast more broadly, beyond the writings of Enfantin or the entrepreneurial adventures of the Pereires.

Perhaps it is unrealistic to expect an original, fully articulated social philosophy from men such as Lavallée and Dumas. Still, it is somewhat surprising that the curriculum dealt exclusively with technological subjects. Few engineers, after all, can carry out their projects in total isolation from social and economic instrumentalities. Yet the Ecole Centrale taught no statistics, no economics ("social arithmetic" or political economy), no law, no philosophy, no history. Despite his role in referring students to the school, Auguste Comte was never invited to lecture there on his new "sociology." The single course dealing in part with social institutions was Public Hygiene, a neglected stepchild of *la science industrielle* that made a fleeting appearance in the early 1830s.

At one point, Dumas did consider teaching his engineers about their social role. In fact, the course outlined in his "Program of a Course in Theoretical and Applied Political Economy, 1834–1835" would have offered more than a perfunctory introduction:

Above all, present the truths generally admitted by the most eminent thinkers. Show the rapprochement that exists among the doctrines of the principle economists. Show the applications of these principles that have been made and that can be made, and the difficulties that arise from misunderstanding them. The students at the Ecole Centrale are destined to produce: it is thus necessary to make them understand above all the importance of their role as producers in the social mechanism.

Lectures (*Leçons*)

1. Notions of value, exchanges, currency.
2. The classification of industries: the role of the scientist, the worker, and the entrepreneur. Functions of the general instruments of industry: land, labor, capital. Of intangibles.

3. Labor: notions about that instrument of production. About the principle of population.
4. Advantages that industries derive from the division of labor. Examples.
5. That in the interest of individuals and of society, there must be a free market in labor.
6. Land: Notions about that instrument of production, etc.
7. Capital: The notion of capital, its use in a private enterprise; in the work of a nation.
8. Capital in machines: advantages and difficulties for the worker, the entrepreneur, and the nation.
9. Capital in currency: The notion of currency. Ways of supplementing currency.
10. Credit: its use and its dangers.
11. Functions of banks; services rendered by banks and bankers.
12. Commercial paper, state notes, *assignats*.
13. Influence of the transportation system: A glance at the progress of canals and railroads.
14. Markets: how they can expand or decline.
15. That there must be free trade within nations and among the various nations.
16. Influence of customs duties on national production and the circulation of products.
17. How the profits resulting from production are divided and can be divided among workers, entrepreneurs, and capitalists. (Distribution)
18. (Consumption) How wealth is most usefully consumed by the individual and by the nation. Questions raised by private domestic and industrial consumption and by public consumption.
19. Impact of taxes on various branches of production.
20. Analysis of the national budget.
21. The results of public borrowing.
22. On the principle of property from an economic point of view.
23. On large and small-scale agriculture; on the division of land.
24. On the resources offered by the principle of association for the creation of products and the division of profits.
25. On treaties of commerce.
26. On colonies.
27. On the proposed methods for improving the lot of the working classes. Workhouses, hospitals and hospices, retirement funds, houses of refuge, aid societies, etc.
28. Special questions such as sugar, coal, iron, wool.
29. Diverse questions of current interest.
30. (Economic history). A glance at economic doctrines and the various sytems proposed.

31. Historical résumé of the great chemical and mechanical discoveries and the effects they produced.
32. The history of commerce.
33. (Industrial legislation). On the laws which have the most direct effect on the work of individuals and nations.
34. Laws relating to commerce.
35. (Administration). The administrative institutions of France.
36. A more intensive examination of those which function in the interest of agriculture, industry, and commerce: conciliation boards (*conseils de prudhommes*), chambers of arts and manufactures, higher councils.[24]

The format of thirty-six lectures suggests that Dumas intended the course to be given in the second or third year, where it would have been one of the shorter courses, longer than Steam Engines or Industrial Natural History, but only half as long as Industrial Physics, Industrial Chemistry, or Ferrous Metallurgy.

Such an outline can only begin to reveal the social philosophy Dumas sought to convey to his students. The boxes are tantalizingly empty: Just how *did* one divide up the profits among workers, entrepreneurs, and capitalists? What *were* the results of public borrowing? Yet the boxes in themselves do permit certain conclusions about Dumas' thinking. In the first place, political economy was a science not significantly different from the other sciences taught at the Ecole Centrale: it had both "theoretical" and "applied" aspects; it dealt with a small number of principles, such as the "principle of population" or the "principle of association," the understanding of which was indispensable for further effective action, the misunderstanding of which produced "difficulties" akin to the collapse of a bridge produced by the misapplication of stress formulas. Given the existence of such principles, one could proceed by deduction, the approach most congenial to the French mind, to the discovery of the likely outcomes of different actions. Society was a mechanism, of which the individual producer was a single part. The better the individual understood that mechanism, the better he could perform his designated function. Political economy thus promised to elucidate general patterns in the movement of the social mechanism that had the clarity and immutability of the laws governing the natural sciences. The Malthusian principle of population, with its comparison of arithmetic increase in food production and geometric increase in population, implied that demographic behavior was as measurable and predictable as a chemical reaction. In Dumas' suggestion that the thirty-first lecture deal with the history of the "great chemical and mechanical discoveries and the effects they produced" appears the interpenetration of the two sciences: technological changes become the driving force behind economic and social changes—as the next lecture, on the history of commerce, might have been intended to show.

Dumas knew, of course, that the laws of any science did not establish their authority without a struggle. In 1834 his bitter debate with Auguste Laurent over the nature of chemical radicals had just begun.[25] The authority of the propositions of political economy as taught in the lecture hall was not to be undermined by excessive attention to such struggles, however. Just as in his *Traité de chimie appliquée aux arts* he had avoided embroiling his readers in the debates over caloric or atomic theory, so in his program for the political economy course he insisted that the lecturer "show the rapprochement that exists among the doctrines of the principal economists." The heavy demands *la science industrielle* placed upon his young engineers' schedules left no time for savoring the nuances of scholarly controversy.

To represent science, *la science industrielle*, or political economy as realms in which all was rapprochement, order, unity, mechanism, in which tension, conflict, surprise, and the unexplainable appeared only as material for an occasional amusing anecdote, was to offer engineering students a view of reality as useful to the school's staff as it was distorted. (It has thus been the favored method of instruction in most engineering schools ever since.) One suspects that even lecture 31 would have depicted orderly marches from discoveries to benefits rather than confused skirmishes or lonely wanderings. Nor would lectures by Comte to his "intermediate class" of engineers have presented a less orderly, progress-oriented view of historical development: the law of the three stages of history was as fundamental to Comte as was the hierarchy of the sciences.[26]

That Dumas' course outline mentions specifically neither engineers nor scientists is not really so surprising. Although both received attention in the schemes of Saint-Simon and Comte, the writings of the more "orthodox," British-inspired economists such as J.-B. Say, upon whom Dumas depended heavily, concentrated on the functions of worker, capitalist, and, especially, the newly important entrepreneur. Besides, in these early years of French industrialization (and of the school's history), the role of civil engineer was rather broadly defined. *Centraux* were recruited from among the sons of entrepreneurs and owner-managers; they were taught by men who were both engineers and entrepreneurs (such as Walter de Saint-Ange, Payen, and Colladon); and it was expected that they would combine a number of roles in their own careers. Centrale spokesmen continued to be reluctant to assign engineers a single place in the hierarchy of *industriels*. If at one point they were the "leaders of the industrial army," at other times they seemed to be the staff officers rather than the commanders. In the concluding section of the *Rapport à présenter*, written after the school finally instituted a course on "industrial legislation" in 1856, the engineers were viewed as intermediaries whose appeal to the authority of political economy would play an important role in ending social conflict:

There are few men capable of answering scientifically the sophisms that stir up so much trouble in today's society and threaten to compromise its progress. The civil engineers of the Ecole Centrale, indispensable intermediaries between the great producers (*les grands producteurs*) and the working masses, can afford less than anyone to be ignorant of the precious observations and sage advice that constitute the science of Adam Smith and J.-B Say.[27]

For the period under investigation here, however, there is no evidence that political economy was taught to *Centraux*. Since J.-B. Say taught at the nearby Conservatoire des Arts et Métiers during this period, Dumas may have recommended that his students attend the lectures, but Say's audiences seem to have been fairly small.[28] Since Dumas never brought the proposal to the *Conseil des études*, why the course was never taught remains uncertain. Perhaps he calculated that the years of reconstruction after the cholera crisis were not the best time for experiments on the margins of *la science industrielle*, as the fate of Spiers's English course suggests. Or perhaps he never found a suitable instructor.

Dumas was a man of many projects and many responsibilities; the former not infrequently became casualties of the latter.

Politics had no place in the formation of the Centrale elite. Even Dumas' version of political economy had its dangers. Two of the most important groups in the school's clientele, the large merchants and the proprietors of industrial establishments, were sharply split on the issue of free trade (lecture 15.), one of the few on which Dumas' Program made a clear declaration of policy. Probably more would have opposed teaching such doctrines to their sons than would have supported the measure.[29] The backgrounds of the faculty members suggest that partisan politics could only have been seen as a divisive force. Perdonnet, dismissed from the Ecole Polytechnique for participating in a Republican conspiracy, and Péclet, whose Liberal views cost him his physics chair in Marseilles, sat in the councils of the school with Empaytaz, the Legitimist who resigned his commission rather than serve Louis-Philippe, and Achille Ferry, engineer of the Orleanist family domains.[30] Dumas, a man for almost all political seasons who held high posts in all governments from the July Monarchy through the Third Republic, served for twenty-five years with Olivier, the enemy of palaverers and parliamentary foolishness. As Olivier had mentioned in justifying the *externat*, the student body was equally diverse. The desire for harmony among parents, teachers, and students thus helped to dictate the exclusion of politics from the education of Centrale engineers.

Into the gap in moral formation left by the excision of politics, the Ecole Centrale placed a twofold ethic of work and of science. Olivier's declaration that work was the political dogma of the school has already been noted. Empaytaz, for whom the directorship of studies was "much needed port of calm and repose after a storm of political events," subscribed enthusiastically to the same doctrine. His obsession with work discipline emerges even through the restraints imposed by the obituary form in this account by Pothier, who attended the school during his administration:

He applied himself especially to the perfecting of graphic work. He loved to say that forcing the students to remain occupied in the classroom and the lecture hall was the best way to establish good order, and that the inspection of notebooks by the tutors was the best way to measure students' work. This tendency of his, perhaps a bit too exaggerated, suggests a certain confusion of work and knowledge; an understanding of something and the graphic reproduction of it do not always precisely correspond. . . . As a consequence of his policies, the keeping of notebooks acquired a great influence on examination grades, and they were kept with a perfection rarely observed before him.[31]

Gustave Eiffel, probably the most creative engineer, both technologically and artistically, that the Ecole has yet produced, found that his sense of balance could only be retained by continuing many of the *plaisirs* with which he had previously learned to renew himself, even at the expense of scholastic achievements.[32] Other students—the *boursiers* come to mind—may have felt less able to kick at the traces. How many would have benefited if the harness had been lighter from the start?

It is doubtful that the severity of the work discipline was dictated by the imperatives of technological change ("Learning complex scientific and technical subjects just *is* hard.") or the needs of the attempt to close the gap with Britain ("We have to try harder because we're only Number Two."). The directors of the Ecole Centrale clearly had additional motives. As the school's unpublished history put it, "Everything was calculated and organized to avoid leaving the students a moment of idleness, to break them to work routines (*les rompre au travail*), to firm up their moral character, to prepare them for the problems of practice and the difficulties of life."[33]

Much attention has been paid to the way the working classes were molded to a new life of labor in the early stages of industrialization: how they were forced to adopt new work rhythms, to develop new capacities of physical endurance, to accept the regimentations of the factory, the shop, and the work gang.[34] Although certain of the accompanying ideological pressures have been examined, especially political economy and English Methodism,[35] the center stage has been occupied by the economic whip, the institutional routine, and political *force majeure*. The entrepreneurial bourgeoisie, on the other hand, are more often portrayed as disciplined largely by ideology: the Protestant ethic, the sense of being a cultural minority, the drive to overcome domination by a neighboring nation or city, or a belief in the inevitability of technological progress.[36] The work load at the Ecole Centrale suggests that the institutional routines of such technical schools made them the factories of the bourgeoisie. True, a student who withdrew from the school rather than face another hour at the drawing board did not usually face starvation, but neither is it clear that this was the alternative for every refractory worker.[37] In any case, just as the ideologies of sobriety and discipline in the British Industrial Revolution performed functions other than merely increasing productivity, so clearly did productivity provide only part of the motivation for the work discipline at the Ecole Centrale. Order in the school and political stability in society were the rewards of a well-enforced program of student tasks.

In the view of the Centrale spokesmen, science helped in three ways to strengthen the moral order in which emerged their new technological elite: by the personal effectiveness that technical competence gave to the engineer, by the way its enterprises channeled students away from more dangerous careers, and by the general uplift produced by common participation in the acquisition and application of scientific knowledge.

In making the first point, they declared that technological advance created the need for a new sort of manager. In the past, the direction of an enterprise had required merely the ability to command men, the "sense of authority." But one could not command machines; only a knowledge of physics and mechanics could get the job done. If the director of an enterprise lacked the requisite technological knowledge, he could find himself at the mercy of both his employees and his competitors: "The study of machines has succeeded the study of men; he must now understand not a flexible, intelligent, and free being but a rigid, unconscious, and determinate mechanism."[38] Dumas claimed nevertheless that the study of men had not really been superseded: "Not that the industrialist, assuredly, can afford to disregard those moral qualities, that soundness of mind and rare good sense, which constitute the successful administrator."[39] It was simply that such qualities no longer sufficed.

Such was the substance of Centrale spokesmen's consideration of industrial enterprises as human organizations: the successful administrator needed little more than a sound mind, common sense, and the aura of authority brought by his knowledge of scientific technology. On the burning issues of the day closely connected with the engineer's managerial functions—the "organization of work," the "principle of association," the role of the State in the economy—or even mundane matters such as the implications of the *Code civil* or the *Code de commerce*, Centrale engineers were to be taught nothing. Only in the *Rapport à présenter* did there glimmer a dim awareness of the risk of intellectual and moral impoverishment this carried for the French technological elite. In observing that the course in hygiene had been short-lived and that no course in "industrial economy" had been taught to the school's first generation, the authors suggested that "it was in these courses . . . that the mind of the student might have been provoked to generalizations and preserved from the relative coldness of spirit (*sécheresse de l'esprit*) that an education too exclusively scientific and practical always risks developing."[40]

In their second argument for the moral value of science, the Centrale teachers pointed out that the technological careers opened up by their school constituted the best solution to the socially threatening excess of educated men produced by the traditional literary education:

Where leads a classical education taken alone? To nothing, unless it is to public functionaries' posts. The pressure that *l'Université* has exercised on the government through the medium of the families has been disastrous for France. It has contributed more than anything else to the enormous growth of the bureaucracy. . . . Each new bachelor's degree is a new threat to the budget, whereas each diploma of the Ecole Centrale is capital added to the wealth of the country. . . . The more the instruction the youth of the country receive is practical and applicable, the more that youth is sure of its future.[41]

The Ecole Centrale was thus one of the few places where a man without independent means could get an education without then becoming a placeseeker. In this way, then, the Ecole Centrale joined the debate over *enseignement secondaire* referred to above by presenting a Centrale education as a direct alternative to the classical *baccalauréat*. One could acquire both, of course, but, as was well known, Centrale did not require the degree from its candidates. In its emphasis upon the pernicious flooding of functionaries' posts, however, this argument also struck at the *Centraux's* great rivals, the *Polytechniciens*. The overwhelming majority of Polytechnique graduates became public functionaries in the military, education, or the technical *services publiques*. Centraux drew salaries from neither the army nor the bureaucracy but instead from the private sector, whose contribution to the wealth of the nation was becoming increasingly clear as France industrialized. Lavallée's notes for the "Industrie" section suggested that one should point out the "crowding of the careers to which leads the higher schools of the government." If they, too, taught a kind of science, it was nevertheless not *la science industrielle*.[42]

The Centrale spokesmen's third appeal to science stressed that scientific and technical education benefited society not only by reducing the state budget and strengthening industry, but also through its spiritual effects. In his contribution to the section on "Industrie," Walter de Saint-Ange listed these psychological transformations among the major "political advantages" of a Centrale education. The combination of hard work as a means with the knowledge that *la science industrielle* guaranteed the usefulness of the end produced men whose mentality made them pillars of the social order:

In directing the minds of our youth to a useful purpose, in showing them an honorable future secured by hard work, in spawning the praiseworthy ambition to be useful to their country in industrial careers, [a Centrale education] turns them away from the dangerous agitations that ceaselessly torment our society and debunks the deceptive utopias, showing their true value to minds that have now become more calm, more reflective, more discriminating. Such an education directs all their intellectual forces toward methods of attaining tranquillity, toward gradual and nondisruptive improvements.[43]

From a callow young man vulnerable to political fantasies, an Ecole Centrale education created a sober, industrious citizen who subjected all social commentary to the same critical scrutiny he gave to the blueprints for a steam engine.

Scientific knowledge had a crucial formative role to play not only for the engineering elite but for the population as a whole. Walter de Saint-Ange, who as a Saint-Cyr graduate had probably known personally neither Monge nor Lacroix, nevertheless subscribed to the central tenet of their faith:

It does not suffice that science exists; it must be propagated. It must be summoned to every place that it is currently applicable so that the country may profit from its blessings. It must branch out in all directions to carry everywhere its light and its continual ameliorative powers, to take everywhere its moral, civilizing force.[44]

The diffusion of science was especially important in one field that had received little attention from Lacroix and Monge, the social relations of industry:

Scientific knowledge brings enlightenment to both heads of factories and the working masses. It forms a link between them, establishes a more enlightened hierarchy based upon real knowledge. [Science permits] more powerful and rational methods of influence upon the newly intelligent masses (*masses devenues intelligentes*).[45]

The moral effect of science was thus crucial to the relations among strata in industrial society.[46] An environment permeated with science forged the crucial links in industry's "enlightened hierarchy"; the authority of science, not the personal prowess of the military commander or the dominance of the traditional master, cemented men together in a common endeavor. As Walter presented in detail in his section on the "material advantages" of a Centrale education, that common endeavor promised to bring a more plentiful life to everyone. The enlargement of the pie brought by scientific technology would vitiate arguments about the relative sizes of the pieces.

The great failure of the nineteenth century's first generation of engineering educators was that within the formal curriculum of engineering education, they did not examine what those "more powerful and rational methods of influence" might be. Early in the twentieth century, when systematic schemes for exerting that influence appeared within the engineering profession in the form of Taylor's Scientific Management and Fayol's principles of administrative leadership, the sense of common destiny upon which an effective and equitable system of industrial relations could be built had been eroded by two more generations of industrial and political conflict. By then it was not easy for those who attempted to put such schemes in practice to keep from appearing authoritarian, exploitative, or politically naive, despite the continuing appeal of productivist ideology. The social and political literature produced during the July Monarchy suggests that a different outcome was possible, if the serious consideration of such problems could have penetrated the drafting rooms of the Ecole Centrale. Even a critical examination of some version of Saint-Simonianism might have produced salutary results. As was shown by their behavior in the Revolution of 1848, where they provided the leadership cadres of the National Workshops, desperately attempting to turn that vast work-relief operation into an economically and politically viable enterprise, the young *Centraux* of that generation had no natural affinity for a "coldness of spirit."[47]

The propagation of science in which Walter de Saint-Ange placed such faith, as did Liberal, Saint-Simonian, and *Polytechnicien* "diffusers of useful knowledge," had social implications other than those he chose to stress in his "Industrie" note. As his colleagues had pointed out, the widespread arousal of ambitions could shatter the stability of both the political and the social order, but the first Revolution, they thought, had removed obstacles to that ambition, a removal they did not condemn. To meet this threat, they proffered the moralizing and civilizing influence of science, but that influence was inseparable from scientific knowledge itself. Yet the diffusion of such scientific and technical knowledge, by suggesting the possibility of economic success to anyone who listened, could excite even more ambitions than the political rhetoric of the Revolution. What made matters even worse was that the economic opportunities that the diffusion of scientific knowledge purported to open up were not entirely mythical. As the success stories cited by Charles Dupin in his lectures at the Conservatoire des Arts et Métiers, the record of the *écoles d'arts et métiers*, and the substantial response to the movement for the diffusion of useful knowledge were beginning to show, scientific and technical information that could serve as the key to economic ascent *was* intellectually accessible during the first half of the nineteenth century.

At the same time that Walter de Saint-Ange and his colleagues suggested that the propagation of science could satisfy widespread social ambitions, however, their appeal as school administrators was directed to the ambitions of a more restricted group. As was described, they sought to recruit the "heirs of the manufacturing domain," as well as other privileged strata. Moreover, they saw British civil engineers as "one of the richest and most honorable classes in society," and they hoped that their school would create such a class in France.[48] To the extent that (economically) useful knowledge was both accessible and successfully diffused, however, the advantages of a Centrale education seemed less clear—especially when substitute goods like lectures and journals were so much cheaper. Perhaps this is one of the reasons they sought to stretch their students' capacity for work: it was *this* aspect of their "special tempering," rather than their intellectual acquisitions, that would keep the *Centraux* ahead of lesser breeds.

The Culture of the Civil Engineer

Little originality informed the social vision of the men who shaped the educational policies of the Ecole Centrale. Their debts to others are clear enough, even if certain specific cases, such as Saint-Simon, remain somewhat problematic. If any conception can be seen as their creation, it is probably *la science industrielle*, but this construct specified rather little about the nature of society or the engineer's role in it. Even their institutional innovation was, first, modeled upon the Ecole Polytechnique of a generation earlier, and, second, clearly an idea whose time had come: in this case, the other simultaneous inventor was none other than Charles Dupin, who, in a speech in the Chamber of Deputies on 18 August 1830, suggested the creation of *une Ecole Polytechnique de l'industrie civile*, seemingly unaware that the Ecole Centrale had been established the year before.[49] Dumas, Péclet, Olivier, and the others revealed their originality largely in their scientific and technological endeavors, certainly not in social thought. This does not mean that their social and cultural pronouncements can be ignored, however. Even elites are formed by conventional attitudes as much as they are by philosophical innovations.

The construction of a coherent cultural education for their engineers would in any case have been even more difficult for these men than it was for the Ideologues, S.-F. Lacroix, or Gaspard Monge. The Centrale leaders were faced with two possible paths to the construction of an interrelated set of moral, aesthetic, intellectual, and practical standards—in other words, the creation of an educational culture. They could seek to unify all such standards, from classicism to mathematics to political economy to Catholicism, or they could propose a single set of scientistic or technological standards, as did Comte's early writings.[50] In either case the journey would not have been easy. The road would have had to be laid through the political thicket of new creeds, from De Maistrian ultramontanism to socialism,[51] the social thicket of bourgeois dominance and aristocratic opposition, and the economic thicket of dynamism and resistance to change. In the end, building such a road was one construction project the Ecole Centrale never attempted. Even when the limiting factors discussed above are considered, the narrowness of Centrale's technological culture remains surprising. After all, publications forming part of the movement for the diffusion of useful knowledge mentioned in chapter 2, such as the *Journal des connaissances usuelles*, managed to offer their readers articles on political economy, jurisprudence, and comparative education.[52] On the other hand, neither of these two forms of technological education, the school and the journal, took the most inclusive approach to its cultural mission. The *Journal*'s hostility to classical culture was no less strong than Centrale's. If Dupin spoke contemptuously in the *Journal* of "escapees from the *collèges*," ignorant of everything but a few Latin verses, Lavallée railed against the "false vocations given to youth by classical education."[53]

Nor did the Centrale professors or the *Journal* follow the Ideologues of the Revolutionary period in trying to use psychological theory as the basis for a unified cultural framework. Yet the educational theories of the Ideologues had held out the hope that "general" education and utilitarian education could be combined: that Latin could provide philosophical precision while it bestowed legal expertise, that *dessin* could provide useful skills for artisans while granting everyone perfected powers of observation.[54] The Ideologues simultaneously hoped that in the *écoles centrales*, the system of individually chosen *cours* would produce a fruitful marriage of specialism and generalism. Despite the fact that the various editions of S.-F. Lacroix's *Essais sur l'enseignement* had presented Restoration and July Monarchy readers with the same argument, the debates over *enseignement secondaire* stretching from Napoleon to 1848 had never succeeded in weaving the contrasting strands of culture—mathematical and classical, scientific and literary, "general" and utilitarian—into a single cloth. Nor was either strand strong enough to gain dominance, as the shelving of Dumas' 1847 plan demonstrated.

In Centrale's technical education, on the other hand, domination had brought exclusion, not unification. The victory of science was all but complete, yet theoretical science was granted legitimacy only by posing as the handmaiden of Utility. Even if the aggressiveness of Olivier's championing of Utility is contrasted with the more easygoing eclecticism of the *Journal des connaissances usuelles*, it is clear that both the *Journal* and the Ecole Centrale proclaimed the primacy of practice over theory even as they claimed to unite them. In reality, as demonstrated by Dumas' treatise or the distribution of articles in the *Journal*, pure science, applied science, and "positive knowledge" were often merely juxtaposed rather than truly integrated: in the *Traité*, the salts of yttrium followed Kuhlmann's contamination test; in the *Journal*, Francoeur's theories about sound adjoined recipes for making bread from chestnuts.

Part of the inadequacy of the culture of technological education may have arisen from a failure to understand the complexity of the problem. When Olivier claimed that the size of one's fortune and the size of one's knowledge were the criteria of class position in contemporary society, he presented too simple a picture. As the rest of his comments made even more clear, he conceived of education only as the acquisition of economically instrumental information. Greater knowledge brought class ascent only when applied to useful professional activities. If the level of contribution to the satisfaction of material needs formed the basis of social stratification, then, Laplace's achievement of esteem as a pure theoretician seemed bizarre, illegitimate, incomprehensible. The narrowness of Olivier's concept of Utility, indistinguishable from technological productivity, made it difficult for him to appreciate how powerfully appeals to other kinds of symbols could work to stratify society. Men cannot live by bread alone, nor did French society rank men only by their productivity as bakers. The myths successfully propounded to sanctify the classical *baccalauréat*, the continued prestige of Laplace's successors at the Polytechnique and the Ecole Normale,[55] the implicit hierarchies revealed in various reform schemes, and the changes in educational choices made by the Mulhouse elite all indicate that the technological culture espoused by Olivier and his colleagues could have become neither the sole arbiter of social status nor the sole repository of modern French values. The votaries of *la science industrielle* left too much cultural territory to their opponents. The Ecole Polytechnique knew better: not only did it pursue mathematics as a social cachet—and an assertion of the human spirit's adventuresomeness and passion for excellence—but it offered lectures on history and literature as well.[56]

The distortions built into a Centrale education reveal some of the limitations of any attempt to transmit culture by means of an elite *école*. The temptation to narrow the beam of inquiry so that it may shine more intensely on a few subjects is hard to resist, easy to justify. Yet even within that limited set of subjects—such as *la science industrielle*—the sheer bulk of the information to be mastered (increased all the more, in the case of *la science industrielle*, by the infinite particularities of "practice") threatens to crush the spirit of both teacher and student. And so one casts the politically, socially, and aesthetically limited but technologically vast curriculum in the form of a single "unified" mold and one proclaims that it stamps out generalists. But how many individuals can be made to fit that mold? Those whom personal difficulties with certain subjects or the demands of the institutional routine do not eliminate must confront the enforced versatility of the encyclopedic curriculum. This can be a wasteful way to train men. If France needed generalists, it was also true that most men realized their full potential as specialists—at least in the technological aspects of their daily professional work. Even if the Ecole Centrale had posed no barrier of financial cost, then, one suspects that only a limited number of individuals could have accommodated themselves to its education.

Where the *Centraux* needed a generalist's education was in the *non*technological aspects of their activities as managers, citizens, family members, bourgeois, and bearers of the divine spark. In theory, of course, they gained such an education in *enseignement secondaire*, but, as has been argued above, the cultural stalemate within that system masked (with declining success) its promotion of specialisms. In later generations the *collèges* and *lycées* found their cultural mission increasingly difficult to carry out. New "special education" tracks entered the secondary schools; new *baccalauréats* appeared; finally, in 1902, a "unifying" reform produced, in fact, four separate degrees.[57] The more *enseignement secondaire* divided, the more engineers insisted that the culture it gave to their future colleagues should be unified. By 1913 a Centrale teacher writing a history of the school could claim that "the general opinion is that we must restrict access to the higher technical schools to those candidates whose secondary education includes a full course of classical studies."[58] The generality of such a classical *culture générale* (the nineteenth century's refurbishing of *honnêteté*) was, however, as mythical as the unity of *la science industrielle*. The myths themselves had their uses, but they did little to solve the social and cultural problems brought by industrialization and political conflict.

Appendix:
Methods and
Classifications

The Samples

At the time the research for the present study was conducted, the archives of the Ecole Centrale des Arts et Manufactures contained an incomplete but extensive set of dossiers containing records on students who entered the school during the nineteenth century. Except for correspondence concerning requests for financial aid and occasional—not systematic—notations about personal crises such as death or withdrawal from the school, the dossiers contain information about the students up to the time at which they entered the school. Records of individual students' academic performance and comportment during their years at the Ecole Centrale were not generally available.

A run of dossiers permitting the study of social recruitment to an educational institution for such an early and extensive period is nevertheless a rare phenomenon.

The sample made from this set of dossiers consists of three parts. (1) For the purposes of the present study, the most important subgroup is that made up of all the available dossiers on students entering the school between the years 1829 and 1847. The dossiers of 1,319 students, about two thirds of all those who attended the Ecole Centrale during this period, have survived. Of these, 226 contained only the student's name (on the outside of the folder). Information from the remaining 1,093 cases was converted to machine-readable form. (2) Those who entered in 1881 ($N = 106$ usable). A midcentury rough check at the beginning of the Republican phase of the Third Republic. (3) Those who entered between 1910 and 1913, with the inclusion of 26 students from the years 1914–1917 whose dossiers happened to be included in the boxes for the earlier years ($N = 473$).

The series of dossiers contain certain gaps. Only two dossiers could be found for the class entering in 1831, the class that experienced the worst effects of the cholera epidemic that nearly closed the school in 1832. Out of more than 100 students who began studies in 1841, only a single dossier has been preserved. The median year of the 1829–1847 dossier sample was 1842, the same as the median for the estimate of total diplomas and certificates granted during the period given by Guillet's history.[1] Table A.1 gives the distribution of that sample by *promotion* and year of birth. Whereas nearly all the records for students entering in 1911 were made available, only half of the 1910–1912 dossiers could be found. In short, the possibility of systematic biases in the preservation of the dossiers cannot be dismissed, but the nature of the bias seems undiscoverable and possibly unimportant.

The completeness of the information contained in individual dossiers also varies widely. In the 1829–1847 group, the student's father's profession, secondary school, or place of birth often appears without the mention of one or both of the other key variables. Hence the valid totals involved will differ somewhat according to the variable or variables analyzed. As a general rule, unknowns have not been included in the tables, and to achieve the greatest possible commensurability for known figures, they have been removed, and percentages recalculated, for tables published elsewhere.

For obvious reasons, unless otherwise indicated, the tables refer only to French students.

In the case of the members of the Académie des Sciences, the two groups in the sample are part of a more extensive sample of all members of the physical sciences sections who were elected to the Académie (that is, the first class of the Institut National) up to 1930. When the individual scientist's dossier at the Archives of the Académie des Sciences did not reveal information about his social origins, a letter of inquiry was sent to the *Services des Actes civils* of his birthplace. This procedure was also followed in the case of the *secondaire* science teachers.

The two "generations" of the science teachers were extracted from the personnel dossiers in the Archives Nationales: F17/20,-000 series by a random sampling of teachers of all ranks—*profes-*

Table A.1
Sample distribution by date of birth and date of entry[a]

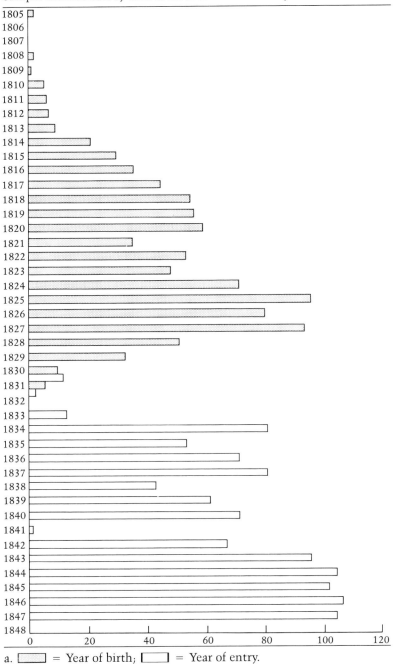

a. ▨ = Year of birth; ▭ = Year of entry.

seurs, chargés de cours, régents—in State secondary education whose title in the official annual from the years 1846 and 1881 carried either a specific mention of physics, chemistry, *sciences naturelles*—alone or in combination with mathematics—or the designation "*sciences*" without any mention of what kind of science was taught.[2] French archival procedures influenced the choice of years somewhat.[3] An individual dossier could not be delivered unless the subject of the dossier had been retired for fifty years. Since I found no easy way to determine the retirement date in advance, I postulated a mean career length of thirty-five years, took five years as a possible standard deviation, and arrived at the year 1881. Subtracting another thirty-five years placed the second sample date at 1846.

Social Origins: Problems of Definition and Method

The distribution of father's professions probably gives the best clue to the function of the Ecole Centrale in influencing patterns of social mobility during the nineteenth century. Measuring the changes in degree of democratization (that is, the extent to which students are represented according to the proportion of father's professional groups in the population as a whole) can only be an exercise in approximation, since the subject of profession appeared on official censuses only in 1851. Not until 1896 did such information become sufficiently detailed to lend itself to the more precise quantitative techniques.[4] Grouping the professions into a sufficiently small number of standardized categories, such as the ten used by the I.N.S.E.E., occurred much later. Not until 1946 did the French census bureau adopt its "code of social-professional categories," built around the criterion of "a certain social homogeneity." Persons belonging to the same category are "presumed susceptible of entertaining personal relations among themselves, often to have analogous behavior and opinion, and to consider themselves as belonging to the same category."[5] In the case of the nineteenth century, the historian must do the presuming. Groupings that seem historically meaningful, such as *haute bourgeoisie* and *classes populaires*, do not appear on census schedules.

The process of creating such broad social categories requires consideration of some of the fundamental problems in the study of social stratification. In the French case just as elsewhere, no single dimension of stratification adequately sums up the hierarchical relations in a given society. A number of scholars have argued that in modern industrialized societies hierarchies of class, status, and power tend to converge: technical elites become social elites.[6] Others, in trying to develop a unitary rank order for the second term in the trio, status, have argued that the relations can be expressed in a single formula. Thus Duncan claims that the prestige (status) of occupations has a correlation of 0.91 with a "combined measure" of education and income.[7] Treiman has even extended this approach across boundaries of time and nation; in a survey that includes fourteenth-century Nepal and Renaissance Florence as well as modern Europe and America, he finds "substantial uniformity in occupational evaluations throughout the world: the average intercorrelation between pairs of countries is 0.81."[8]

Whatever may be the global trend toward a single, uniformly determined rank order, the problem of status inconsistencies (that is, multiple criteria) and rival hierarchies (the "sectoral" question) cannot be so easily smoothed over by students of French history. The unevenness and tensions in that "process of convergence" make up an integral part—in some ways, the very heart—of the subject matter of social history. Ringer's study of German academic elites during the period of rapid industrialization illustrates this point brilliantly.[9] Even Treiman admits that the sociologist and the social historian ask different questions.[10]

In the case of a relatively long period of study, matters are further complicated by the fact that the categories themselves change over time. The lag between the historical birth of new strata and their appearance as a separate category on census schedules can be so long that it masks the phenomenon under study. For example, for decades, only independently established (*à leur compte*) engineers were counted as members of that profession on the census schedules.[11] Hence the 1891 census listed 12,490 "architects and civil engineers" under the rubric of Class VII, Liberal Professions.[12] The especially difficult *cadre moyen* classification was not officially established in the census codes until after World War II although a recognized category of positions corresponding to this designation existed in practice in the middle of the nineteenth century.[13]

The social significance of a particular profession can also change as shifts in patterns of social mobility transformed that group's relation to other strata. Chaline's study of marriage contracts in Rouen shows that by 1886 the shopkeepers (*boutiquiers*) had lost the role of "hinge category" they had played in 1819. In the earlier period, they had drawn spouses more or less equally from all strata; by the later date, the cosigners of their marriage contracts were largely artisans, workers (*ouvriers*), and day-laborers.[14]

Such considerations indicate the boundaries of the precision attainable in an analysis of the social origins of students at the Ecole Centrale. The nature of the documents also limits that precision. In the earlier period, a student's dossier might contain one or more of three types of documents indicating social origin: the birth certificate, a letter from a parent or official making reference to family background, or a notation on the dossier by school officials stating the father's profession at the time of entry.[15] Where more than one source was available, the computer program selected the notation on the birth certificate as most reliable and designated (rather arbitrarily) references in correspondence as least reliable. In any case, the dossiers usually contain only a single word or phrase concerning the father's profession. The present study thus lacks the detailed supporting information that gives both precision and authority to studies based on marriage contracts (with their frequent dowry notations), notarial archives (closed to researchers after midcentury), and registers of property-transfers after death (*mutations apres décès*).

In order to evaluate the historical significance of the information on social origins contained in the student dossiers without giving too short a shrift to the complexities of social stratification, chapter 3 presented the data in two ways. The first emphasized the horizontal divisions among social strata and the uniformities within the levels such as *haute bourgeoisie* and *classes populaires*; the second, comparing Shinn's results for the Ecole Polytechnique with those for the Ecole Centrale, gave somewhat greater attention to sectoral divisions. The tables are based on a scheme of seventy-nine categories that represents a modification of the *code socio-professionel* first proposed by Adéline Daumard in 1963.[16]

It is likely that none of my groupings of these seventy-nine categories into four broad strata, nor the names I have given those strata, will gain universal acceptance. An especially strong argument might be made in favor of labeling my "upper bourgeois" category simply "bourgeois" or "middle bourgeois" and my "middle bourgeois" category "petit bourgeois." My designation of "upper bourgeois" is based on a grouping suggested in the text of Daumard's 1963 article. But even one of the "typical cases" she cites as establishing the *haute bourgeoisie* category of *Professions libérales catégorie supérieure*, the notary (*notaire*), finds its claim to this high status shaken by the wide variations in the situation of notaries found by Theodore Zeldin.[17]

Terry Shinn's study of *Polytechniciens* excludes large merchants (*négociants*) and industrialists (*industriels*) from the *haute bourgeoisie*, placing them in the *moyenne bourgeoisie* along with middle-level functionaries.[18] He bases his classifications upon a sample of the 345 "most representative" (?) cases of *Polytechniciens* who become military officers and whose dossiers in the military archives at Vincennes happened to include information about *fortune* and annual revenue.[19] Although he notes that many of the richest *négociants* and *industriels* "enter into the economic sphere of the upper bourgeois," he concludes that because "an average fortune" for families of this designation "is situated at around 50,000 francs," a figure lower than the average for his *haut bourgeois* categories of *rentier* and *propriétaire*, and because a good part of this wealth lay in possessions other than land, large merchants and industrialists should be included in a "lower" category.[20] Although in general Shinn's system of categories is one of the most precise and sophisticated to be found anywhere in the literature, several considerations prompted me to follow Daumard on this matter of nomenclature: (1) On purely economic grounds, Shinn's own group of higher functionaires would qualify for upper-bourgeois status no more than the *négociants* and *industriels*. The average salary was only 6,000 francs, and the average wealth, "in both liquid and landed capital," was 47,000 francs, slightly less than that of the merchants and manufacturers group.[21] (2) It is likely, moreover, that the income produced by a given amount of wealth during this period was distinctly higher if that wealth lay in goods or capital equipment than if it lay mostly in land. Although this would have benefited the *négociants-industriels* category more than the others, even Shinn's upper-bourgeoisie groups seemed to have realized this. As early as 1820 57.6 percent of the wealth of Parisian members of the liberal professions, 43.4 percent of that of high functionaries, and 42.4 percent of that of proprietaires lay in sources other than real estate. By 1911 the corresponding percentages were 75.1, 71.3, and 57.2–and 77.9 for *négociants*.[22] The distinctions among these groups according to type of wealth thus diminished considerably throughout the century. (3) Shinn's system of classification lacks a notion of sector (or what Ringer calls socially horizontal segmentation)[23] as it has been used in this study. Despite important differences in outlook and life-style, commercial, financial, and manufacturing circles in Mulhouse, Lille, Lyons, or Paris were as conscious of the levels within the bourgeoisie, and of their membership in its *haute* category, as were the functionaries, generals, lawyers, and landowners who did not dirty their hands in "trade."[24]

As has been stated in chapter 3, Shinn's investigation of the fortunes of those *propriétaires* who sent their sons to the Polytechnique lends a valuable measure of precision to our understanding of this category into which 34 percent of the French population was placed in the census of 1851.[25] Clearly, Polytechnique (and most probably Centrale) fathers were not ordinary *propriétaires*, but some of the wealthiest in that category. Other evidence supports Shinn on this point. Dupeux assigns the class of *"propriétaires fonciers"* who profited from the Revolution to the upper levels of the bourgeoisie, but not a single such designation appeared in the dossiers.[26] The *"foncier"* may often have been assumed, however, since many entries appear specifying *propriétaires* of mines, industrial establishments, and commercial enterprises. The contrasting designation *cultivateur*, which more clearly indicates a peasant who worked the soil with his own hands, appears in the same set of documents and was listed under "popular classes." In his study of nineteenth-century Rouen, Chaline finds that those designating themselves as *"propriétaires"* mostly closely approximate the merchant-wholesalers (*négociants*) in the wealth they brought to the marriage contracts. The average wealth of those two categories in turn exceeded considerably the Chaline categories of "liberal professions" and "higher functionaries" (*fonctionnaires*). He finds nevertheless that all four of these categories "belong to the same social world."[27] In the 1891 census, all *propriétaires* except those *"cultivant exclusivement leurs terres"* ("working only their own land") were grouped with *rentiers* as "living on their income."[28]

The use of the *cadres moyens* category under "middle bourgeoisie" is technically an anachronism when referring to the early nineteenth century. The *cas typiques* in this category, *controleurs* and *géomètres*, would probably have been considered *employés* in the early nineteenth century since there seems to have been no conventionally accepted middle term between the *haut bourgeois fonctionnaire* and other salaried, nonmanual employees.[29] The designations found in the documents, however, clearly indicate that such a middle-level category was in the process of formation.[30] Such employees and officials had positions usually involving a significantly greater technical content (or formal education), responsibility, and salary than did such *cas typiques* of the "employees and lower cadres" group as *garçons de bureau* and *employés de commerce*.[31] At the same time, an examination of the description of the various positions in vocational guides cited above such as Charton, Massé, Jacquemart, and Bastien indicates that such positions had a clear ceiling, a break in the rungs of the career ladder limiting the responsibility and salary attainable by ordinary advancement.

The position of the *boutiquiers* in the social structure probably shifted considerably during the nineteenth century. Their inclusion in the middle bourgeoisie is based in part on Daumard's study of the Parisian bourgeoisie, which draws heavily upon the evidence of total wealth.[32] As was noted above, however, Chaline's study of Rouen finds the fewer shopkeepers were "marrying up" at the end of the century. Perhaps the drop in the percentage of boutiquiers' sons entering Centrale reflects a drop in the average resources at the command of this group, but the decline may also reflect a drop in their proportion of the entire population. The most recent general study of the lower bourgeoisie in France argues, moreover, that in 1848 shopkeepers typified to contemporaries not the *moyenne* but the *petite bourgeoisie*; this was Marx's view as well.[33] In any case, the percentages of shopkeepers' sons in the Polytechnique and Centrale classes were never especially significant.

The final category in the middle bourgeoisie includes the various specialties of *marchand* that indicated small-scale operations, retail trade, or the ownership of hotels and restaurants. The "sectoral factor" in this category undoubtedly complicates matters. A professor of *philosophie* in a Parisian *grand lycée* might have seen considerable social distance between himself and the owner of a small restaurant in a provincial town, but at the same time the owner of Maxim's would not be likely to consider an *alliance* with a *professeur de troisième* at the obscure Lycée de Roanne. The category assumes that students' fathers form a representative sample of the professions included, a sample grouped around the mean. A number of the "small merchants" in this subcategory (*quincaillier, libraire*) might also be classified as shopkeepers.

The "employees and lower cadres" category has been touched upon above in connection with the consideration of *cadres moyens*. Designations such as *garçon de bureau* (office boy) clearly indicate a level below the boundary of *cadres moyens*, but it is possible (though unlikely) that untitled *employé* designations such as *employé de commerce* and *employé des postes* would be classed in a higher category if more details about the father's position were available. As has been noted, the primary school teachers (*instituteurs*) form another case in which the status of the category itself changed over time. By the eve of World War I they had come to play the role of a "hinge category" similar to that played by *boutiquiers* during the July Monarchy.[34]

The "popular classes" category includes all those who worked with their hands or held low-level service positions: peasants (*cultivateurs, laboureurs*, and so forth), day-laborers (*journaliers, manoeuvres*), domestic servants, *concierges*, workers (*paveurs, appareilleurs, typographes*, and the aristocrats of the metallurgical trades—the *puddleurs*), and artisans (*tanneurs, tonneliers, forgerons, ébénistes*).[35] This latter grouping probably includes, of course, a number of *petits patrons* who employed their own apprentices or journeymen. Whatever may have been their precise relation to the means of production, however, most of the artisans united with other elements of the urban *classes populaires* in political actions from the Revolution of 1789 to the Commune. Unlike Shinn, then, I have grouped artisans in this category rather than with the *petite bourgeoisie* (see table A.2). Daumard's study of the estates of her *"façonnier"* class in 1847 places the median estate for this group in the 1,000–2,000-franc category, closer to *ouvriers* and domestics (200–500) than to *boutiquiers* and *employés de l'état*, a distinct group below *fonctionnaires* who averaged 10,000–20,000.[36] In any case, the French language itself did not draw clearcut, mutually exclusive distinctions between *artisan* and *ouvrier* until the twentieth century.[37]

Appendix

Table A.2
Social origins of *Polytechniciens* and *Centraux*, 1830–1847

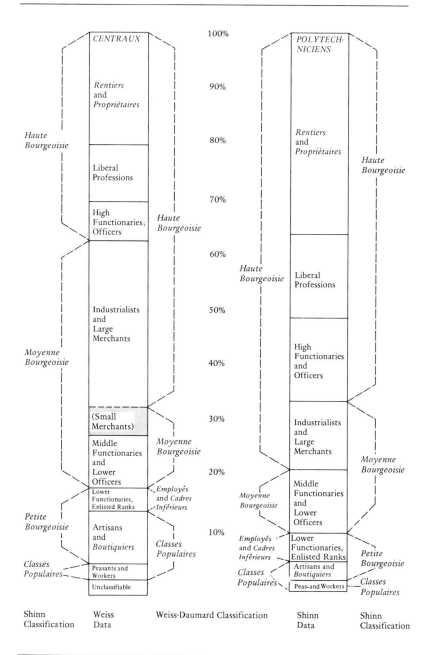

Notes

Chapter 1

1.
Anatole Mallet, "La Société des Ingénieurs Civils de France de 1848 à 1896," in Société des Ingénieurs Civils de France, *Annuaire de 1911* (Paris, 1911), 20–21.

2.
A discussion of the inadequacies of such an approach for historical research may be found in Fritz K. Ringer, *Education and Society in Modern Europe* (Bloomington, Ind., 1979), 17–18.

3.
Randall Collins, *The Credential Society: An Historical Sociology of Education and Stratification* (New York, 1979), 159.

4.
Unilinear scales can at least help to establish this fact, if not to explain it. William M. Evan, "The Engineering Profession: A Cross Cultural Analysis," in Robert Perrucci and Joel Gerstl, eds., *The Engineers and the Social System* (New York, 1969), 127, shows the wide variations in "prestige rankings" of the engineering profession for nine countries. France was not included.

5.
Ringer, *Education and Society*, 2.

6.
Göran Ahlström, "Higher Technical Education and the Engineering Profession in France and Germany during the 19th Century," *Economy and History* XXI:2, 57; Joseph Ben-David, "The Rise and Decline of France as a Scientific Center," *Minerva* VIII:2 (April 1970), 160–179.

7.
Rondo Cameron, *France and the Economic Development of Europe: 1800–1914* (Princeton, 1961), 54.

8.
Collins, *The Credential Society*, 1–21. Cf. Ivar Berg, *Education and Jobs: The Great Training Robbery* (Boston, 1971).

9.
Claude Fohlen, "The Industrial Revolution in France: 1700–1914," in Carlo Cipolla, ed., *The Emergence of Industrial Societies*, Part 1 (London, 1973), 20.

10.
Jean Pigeire, *La vie et l'oeuvre de Chaptal* (Paris, 1931), 380–381, 462–463.

11.
Jean-Antoine Chaptal, *Mes souvenirs sur Napoléon* (Paris, 1893), 41.

12.
Fohlen, "The Industrial Revolution in France," 20.

13.
Arthur C. Dunham, *The Industrial Revolution in France* (New York, 1955), 30, 46.

14.
David S. Landes, *The Unbound Prometheus* (Cambridge, England, 1969), 160.
15.
On pig iron and steam engines see John H. Clapham, *Economic Development of France and Germany, 1815–1914* (Cambridge, England, 1963), 59–60, 62. On coal see *Annuaire statistique de la France: Résumé rétrospectif* (Paris, 1966), 229.
16.
Dunham, *Industrial Revolution*, 129.
17.
Clapham, *France and Germany*, 62.
18.
Dunham, *Industrial Revolution*, 40–41.
19.
Clapham, *France and Germany*, 69.
20.
Ibid., 65. Of Dupin's many works on the subject the most influential was his *Forces productives et commerciales de la France* (Paris, 1827).
21.
Tihomir Markovitch, *Histoire quantitative de l'économie française*, Vol. 7, *L'industrie française de 1789 à 1964—conclusions générales* (Paris: ISE, 1966), 122.
22.
Their activities are discussed in detail in W. O. Henderson, *Britain and Industrial Europe* (Leicester, 1972), 10–75.
23.
Landes, *Unbound Prometheus*, 148. Cf. Antoine Léon, *La Révolution française et l'éducation technique* (Paris, 1968), 264, who gives the much higher figure of 15,000 for *ouvriers anglais*.
24.
The most recent such appraisals may be found in Maurice Lévy-Leboyer, "Innovation and Business Strategies in Nineteenth- and Twentieth-Century France," in Edward C. Carter II, Robert Forster, and Joseph N. Moody, *Enterprise and Entrepreneurs in Nineteenth- and Twentieth-Century France* (Baltimore, 1976), 107–108; Peter Lundgreen, *Bildung und Wirtschaftswachstum im Industrialisierungsprozess des 19 Jahrhunderts* (Berlin, 1973), 144–149; Ahlström, "Higher Technical Education," 62–63.
25.
On the levels of literacy corresponding to various occupational groups during the early stages of British industrialization, see Michael Sanderson, "Literacy and Social Mobility in the Industrial Revolution in England," *Past and Present* 56 (August 1972), 75–104.
26.
Brian Simon, *Studies in the History of Education, 1780–1870* (London, 1969), 32.
27.
Sidney Pollard, *The Genesis of Modern Management* (Harmondsworth, England, 1965), 129.

28.

Ibid., 137.

29.

The most thorough classificatory census of such institutions is still Nicholas Hans, *New Trends in Education in the Eighteenth Century* (London, 1951), 11–193. The case for Oxford and Cambridge is made in A. E. Musson and Eric Robinson, *Science and Technology in the Industrial Revolution* (Toronto, 1969), 166–178.

30.

Ibid., 10–189. The crucial role of subscription libraries in the life of an eminent engineer is described in William Pole, ed., *The Life of Sir William Fairbairn, Bart, Written Partly by Himself* (London, 1877), 74–83. The case of James Watt illustrates the limits of the institutional argument. Should he be classified as a student at Greenock Grammar School, an artisan who took up the career of instrument-making at the age of fifteen, or as an aide to the chemistry professor Joseph Black at the University of Glasgow?

31.

H. Hale Bellot, *University College, London, 1826–1926* (London, 1929), 38–39. The professor who had been appointed to teach such courses at the founding of the college in 1828 resigned before courses began. A chair was not created until 1841. Lectures on various subjects in engineering and the applied sciences had been given at Cambridge University since the end of the eighteenth century, but no engineering department was created until 1875. See T. J. N. Hilken, *Engineering at Cambridge University, 1783–1965* (Cambridge, England, 1967), 1–32.

32.

Karl-Heinz Manegold, "Das Verhältnis von Naturwissenschaft und Technik im Spiegel der Wissenschaftsorganisation im 19. Jahrhundert," *Technikgeschichte in Einzeldarstellungen* 11 (1969), 156–159; Charles C. Gillispie, "Science and Technology," in the *New Cambridge Modern History*, Vol. IX (Cambridge, England, 1957) 125; John T. Merz, *A History of European Thought in the Nineteenth Century*, Vol. I, Part I (New York, 1965) 166.

33.

Gerschenkron, *Economic Backwardness*, 7.

34.

Ibid., 12–16.

35.

Landes, *Unbound Prometheus*, 150.

36.

Idem. This is also the argument of Lundgreen, *Bildung und Wirtschaftswachstum, passim.*

37.

Wilhelm Von Dyck, "Historische Übersicht," in Wilhelm Lexis, *Die Technischen Hochschulen in Deutschen Reich*, Part I (Berlin, 1904), 8.

38.

See Taton, *Enseignement et diffusion*, 345–354, 386–391.

39.

Ibid., 385–508. Subterranean geometry was a response to the need for precise quarrying techniques in the Paris area.

40.
Walter Schöler, *Geschichte des naturwissenschaftlichen Unterrichts im 17. bis 19. Jahrhundert* (Berlin, 1970), 42.
41.
Von Dyck, "Historische Übersicht," 7.
42.
Raymond H. Merritt, *Engineering in American Society, 1850–1875* (Lexington, Ky., 1969), 3–6.
43.
As Edwin Layton points out in a review of Merritt in *Technology and Culture* 12:1 (January 1971), 121.
44.
Michael Argles, *South Kensington to Robbins* (London, 1964), 15–22.
45.
Landes, *Unbound Prometheus*, 150.
46.
A. Fourcy, *Histoire de l'Ecole Polytechnique* (Paris, 1828), 12; this work is the fundamental source for most later accounts.
47.
This motive is stressed, for example, in the account of Sylvestre-François Lacroix, *Essais sur l'enseignement en général et sur celui des mathématiques en particulier* (Paris, 1828), 29–30.
48.
Fourcroy's speech cited in Fourcy, *Histoire*, 21–22.
49.
Terry Shinn, *L'Ecole Polytechnique, 1794–1914: Savoir scientifique et pouvoir social* (Paris, 1980), 16. Shinn's work is the first full-length study by a professional historian.
50.
Charles C. Gillispie, *The Edge of Objectivity: An Essay in the History of Scientific Ideas* (Princeton, 1960), 177; Joseph Ben-David, *The Scientist's Role in Society* (Englewood Cliffs, N.J., 1971), 94–102; Everett Mendelsohn, "The Emergence of Science as a Profession in Nineteenth-Century Europe," in Karl Hill, ed., *The Management of Scientists* (Boston, 1964).
51.
The "cult of Vauban" within the Corps of Military Engineers, which he founded, is described in A. Blanchard, "Les ci-devant ingénieurs du Roi," *Revue internationale d'histoire militaire* 30 (1970), 100; Perronet's role in the Corps des Ponts et Chaussées, which he headed for half a century, is described in Jean Petot, *Histoire de l'administration des ponts et chaussées* (Paris, 1958), 185, *et passim*.
52.
René Taton, *L'oeuvre scientifique de Monge* (Paris, 1951), contains a brief biography, an analysis of Monge's scientific work, and a full bibliography.
53.
René Taton concludes in his *L'histoire de la géométrie descriptive* (Paris, 1954), 24, that although Monge did not invent the method of representation by two associated orthogonal projections, he still merits the title of creator of descriptive geometry "as a coherent and autonomous discipline."

54.
Shinn, *L'Ecole Polytechnique*, 16.

55.
F. De Dartein, "Notice sur le régime de l'ancienne Ecole des Ponts et Chaussées," *Annales des Ponts et Chaussées. Mémoires et documents* (1906), 73.

56.
C.-P. Marielle, *Répertoire de l'Ecole Impériale Polytechnique* (Paris, 1855), (a).

57.
Jean-Pierre Callot, *Histoire de l'Ecole Polytechnique* (Paris, 1958), 23.

58.
Ibid., 31.

59.
Ibid., 22; Petot, *Ponts et chaussées*, 431.

60.
Charles De Comberousse, *Histoire de l'Ecole Centrale des Arts et Manufactures* (Paris, 1879), 20.

61.
E. Mouchelet, *Notice historique sur l'Ecole Centrale des Arts et Manufactures* (Paris, 1913), 2.

62.
Manuel, *Saint-Simon*, 192. Claude Neuschwander, *L'Ecole Centrale des Arts et Manufactures* (Casablanca, 1960), 18, claims that the role played by the *Globe* in the founding of the school shows the Saint-Simonian influence. At the time the school was founded, however, the editor was not a Saint-Simonian. See Paul Gerbod, *Paul-Francois Dubois: Universitaire, journaliste, et homme politique* (Paris, 1967), 85–100.

63.
Mouchelet, *Notice historique*, 4.

64.
General J. B. Dumas, *La vie de Jean-Baptiste Dumas*. Typescript. Dumas Papers, A. Acad. Sci.

65.
Binet de Saint-Preuve's personal dossier at the Archives Nationales lists him as the director of the *"Ecole manufacturière et industrielle,"* which supposedly led to the founding of the Ecole Centrale, during the years from 1828 to 1831. A.N. F 17/23112.

66.
Rapport à présenter à M. le Ministre (hereafter: *Rapport à présenter*), unpublished history of the Ecole Centrale, Dumas Papers, A. Acad. Sci., Sheet 1-3.

67.
Ibid., 1-6. The full authorship of the history is uncertain, but the draft copy in the Dumas papers contains notations and corrections in Dumas' own hand. De Comberousse published a few extracts from the report in his *Histoire*. Since he was at this time a professor at the school and a member of the governing *Conseil*, he may also have contributed to it by assisting the 74-year-old Dumas.

68.
A.N. F 17/6770.

69.
Daniel Colladon, *Souvenirs et mémoires* (Geneva, 1893), 191.
70.
Francis Pothier, *Histoire de l'Ecole Centrale des Arts et Manufactures* (Paris, 1887), 15.
71.
Rapport à présenter, 1-3.
72.
Shinn, *L'Ecole Polytechnique*, 49, speaking of the longer period 1830–1880, gives an annual figure of 450 *admissibles* for 200 admitted.
73.
Ibid., 249, which states that candidates "almost never" succeeded on the first attempt.
74.
"Ordonnance du Roi portant organisation de l'Ecole Polytechnique" of 30 October 1832, reprinted in *Annuaire de l'Ecole Royale Polytechnique pour l'An 1837* (Paris, 1837), 41. Title III, Section 6, required that the student be less than twenty years old on January 1 of the year he took the entrance examination, given on August 1.
75.
Archives of the Ecole Centrale des Arts et Manufactures. Hereafter: A.ECAM Student dossiers. Precisely how many Polytechnique candidate students entered Centrale as a second choice is not certain; remarks indicating candidacy for the latter school were found in fifty-one of the July Monarchy dossiers, 4.7 percent of those containing any records at all inside the folder. Many others contained indirect evidence that the students may have tried for the Polytechnique, such as records of years spent in advanced mathematics courses or attendance at particular cramming schools. The Ecole Centrale's unpublished history states that during the period 1859–1871 one fourth of the students at the Ecole Centrale had been candidates for the Ecole Polytechnique, a figure that probably forms the upper limit.
76.
(Paris, 1830).
77.
Doré, *Nécessité*, 3.
78.
Ibid., 5.
79.
Ibid., 9–10. The base of the calculation appears to have been 1815–1830.
80.
Ibid., 13.
81.
Ibid., 18–22.
82.
Ibid., 24.
83.
Ibid., 25.

84.
Ibid., 26.

85.
Much the same view of the possible range of new careers is expressed in Yves-Delphis Bugnot, *De l'Ecole Polytechnique, dans ses rapports avec les services publics qu'elle alimente* (Paris, 1837). In 1853 Bugnot, an 1818 Polytechnique graduate, was director of studies at the Ecole Militaire de Saint-Cyr, but I have been unable to determine whether in 1837 he held the same kind of "insider" position that Doré did in 1830.

86.
Lévy-Leboyer, "Innovation and Business Strategies," 107.

87.
Rapport à présenter, 1-1.

88.
Ibid., 1-6.

89.
Ibid., 1-5.

90.
According to S. B. Donkin, "The Society of Civil Engineers (Smeatonians)," *Transactions of the Newcomen Society* XVII (1936/7), 51–60, in the Smeatonian Society, "Conversation, argument, and a social communication of ideas and knowledge in the particular walks of each member were, at the same time, the amusement and the business of meetings." Even the Institute of Civil Engineers, founded in 1818 and chartered ten years later, was closer in form to the "Lit and Phil" societies then in vogue than to a modern professional organization or a French *grand corps* of the Napoleonic variety. See the edition of Samuel Smiles's *Lives of the Engineers,* edited by Thomas P. Hughes (Cambridge, 1966), 398–399.

Chapter 2

1.
Augustin Sicard, *Les études classiques avant la Révolution* (Paris, 1887), 9.

2.
Frederick E. Farrington, *French Secondary Schools* (London, 1910), 4. Farrington's book is one of the best and least biased histories of the secondary schools in any language.

3.
Cited in Georges Snyders, *La pédagogie en France aux XVIIe et XVIIIe siècles* (Paris, 1965), 75.

4.
Ibid., 76. Clement Falcucci, *L'humanisme dans l'enseignement secondaire en France* (Toulouse, 1939), 14, states that the purpose of these "skillful *découpages* was to mask the dangers of paganism and present the Latin world as calm and well-arranged, in consonance with the harmonies of divine thought."

5.
See Falcucci, *L'humanisme,* 14.

6.
Sicard, *Les études classiques,* 23.

7.

Ibid., 17–24; Falcucci, *L'humanisme*, 14.

8.

The concept of *honnêteté* was multifaceted and anything but static. My intention here is only to sketch the broad outlines of what seem to have been the salient components of the ideology. A useful monograph on the subject is Maurice Magendie, *La politesse mondaine et les théories de l'honnêteté en France au XVIIe siècle* (Paris, 1925).

9.

The most psychologically perceptive description of the Jesuit system of individual competition and round-the clock "guidance" is given in Snyders, *La pédagogie*, 31–56.

10.

Sicard, *Les études classiques*, 368–378, 540.

11.

Snyders, *La pédagogie*, 140. See also Magendie, *La politesse, passim*.

12.

Falcucci, *L'humanisme*, 51.

13.

Sicard, *Les études classiques*, 541–542. Emphasis in original.

14.

The tensions caused by a pedagogical policy that combined isolation from an evil, corrupting world with an education in social polish are discussed in Snyders, *La pédagogie*, 35–56, 346–352, 142–145.

15.

G. Kersaint, *Antoine-François de Fourcroy, sa vie et son œuvre* (Paris, 1966).

16.

Archives parlementaires, 2d series, Vol. 3, 481. This set of volumes is cited hereafter as *Arch. parl.*

17.

See the table presented in *Recueil de lois et réglemens concernant l'instruction publique depuis l'Edit de Henri IV, en 1598, jusqu'à ce jour*, Vol. 2 (Paris, 1828), 304–309. This set of volumes is cited hereafter as *Recueil . . . 1598.*

18.

Ibid., Vol. 4, 2. The "little schools, primary schools where one learns reading, writing, and arithmetic," were also included in the decree on the University, but neither French scholars nor the most recent American students of the subject consider them to have been part of it. See Ringer, *Education and Society*, 114, and Joseph Moody, *French Education since Napoleon* (Syracuse, N.Y., 1978), 13–15.

19.

Arch. parl., Vol. 3, 569.

20.

Recueil . . . 1598, Vol. 5, 28–31.

21.

Ibid., Vol. 7, 66.

22.

Jean-Baptiste Piobetta, *Le baccalauréat de l'enseignement secondaire* (Paris, 1937), 304.

23.
Recueil . . . 1598, Vol. 7, 114–132. *Troisième* to *Rhétorique* were known collectively as the *humanités*.

24.
Ibid., Vol. 8, 81. Félix Ponteil, *Histoire de l'enseignement 1789–1965* (Paris, 1966), 173, gives a total of 88 class hours for literary studies and 31 for scientific studies in the full secondary program of 1809, and 122 for letters and 40 for sciences in 1830.

25.
Quoted in Georges Weill, *Histoire de l'enseignement secondaire* (Paris, 1902), 34.

26.
Ibid., 35. Cuvier's role in the University is discussed in Eugène Rendu, *Ambroise Rendu et l'Université de France* (Paris, 1861), 72–73.

27.
Weill, *L'enseignement secondaire*, 35.

28.
Ibid., 36.

29.
John W. Padberg, S.J., *Colleges in Controversy: The Jesuit Schools in France From Revival to Suppression, 1815–1880* (Cambridge, Mass., 1969), 75. Italics in original.

30.
Falcucci, *L'humanisme*, 116–144; Emile Durkheim, *L'évolution pédagogique en France*, Vol. 2 (Paris, 1938), 173–175.

31.
Weill, *L'enseignement secondaire*, 34.

32.
Ponteil, *Histoire de l'enseignement*, 172.

33.
Antoine A. Cournot, *Des institutions d'instruction publique en France* (Paris, 1864), 43–45; Edmond Goblot, *La barrière et le niveau* (Paris, 1925); Gordon Wright, *France in Modern Times* (New York, 1960), 215, 358.

34.
Both *peuple* and *classes populaires* were contemporary terms.

35.
Figures are from (Abel Villemain), *Rapport au Roi sur l'instruction secondaire* (Paris: M.I.P., 1843), 24 and 25, combining tables in such a way as to eliminate double enrollments. This does not mean that the remainder come from *enseignement primaire*; tutoring and other private arrangements were very common.

36.
Recueil . . . 1598, Vol. 2, 3, 18–24.

37.
On the cost of the uniform see Robert R. Palmer, *The School of the French Revolution: A Documentary History of the College of Louis-le-Grand and Its Director, Jean-François Champagne, 1762–1814* (Princeton, 1975), 208. Louis-le-Grand was the only important primary school never to close its doors during the Revolution. Cournot, *Des institutions*, 24, describes its role as the model for other schools.

38.
Padberg, *Colleges in Controversy*, 50.

39.
Edouard Charton, *Guide pour le choix d'un état ou dictionnaire des professions* (Paris, 1842), 324; *Annuaire statistique de la France: Résumé rétrospectif* (1966), 422.

40.
Calculated from Villemain, *Rapport*, 29.

41.
Ibid., 27.

42.
Ibid., 22–23, 36.

43.
Recueil . . . 1589, Vol. 2, 51.

44.
Ibid., Vol. 2, 8. Robert R. Palmer, "Free Secondary Education in France before and after the Revolution," *History of Education Quarterly* (Winter 1974), 437, finds that "46 percent of the pupils in the colleges paid nothing for instruction in 1789, and that the corresponding percentage fell to 16 per cent in 1842 and to 10 per cent in 1865." These figures do not refer to *bourses* for room and board, but only to tuition fees.

45. ·
Fourcroy's report in *Arch. parl.*, 2nd series, IX, 81.

46.
Padberg, *Colleges in Controversy*, 51.

47.
Recueil . . . 1598, Vol. 2, 51.

48.
Arch. parl., 2nd series, IX, 83.

49.
M. Gasc, *Observations sur le rapport au Roi et sur l'ordonnance du 26 mars 1829* (Paris, 1829), 17.

50.
Villemain, *Rapport*, 53. Dominique Julia and Paul Pressly, *"La population scolaire en 1789: Les extravagances statistiques du Ministre Villemain,"* *Annales: E.S.C.* (November–December 1975), 1516–1561, offer a detailed critique of Villemain's statistics that concludes that secondary education was indeed more accessible on the eve of the Revolution and that changes between the founding of the *lycée* system and the time of Villemain's report "noticeably altered the character of secondary education and accentuated even further its class character: *lycées* and *collèges* are henceforth peopled mainly by young bourgeois." Ringer, *Education and Society*, 316–317, using better population figures than those available to Villemain, estimates secondary enrollment at 1.2 percent of those eligible, versus the minister's 2.8 percent.

51.
For a discussion of the role of the educational system in transforming economic capital into cultural capital in contemporary France see Pierre Bourdieu, *Reproduction in Education, Society and Culture* (London, 1977), 30–31, 72–76, *et passim*.

52.

A more extensive discussion of the promotion of the *baccalauréat* as social cachet and professional credential can be found in John H. Weiss, "Bridges and Barriers: Narrowing Access and Changing Structures in the French Engineering Profession, 1800–1850," in Gerald L. Geison, ed., *The Professions in the Modern West* (forthcoming).

53.

Michel Chevalier, "Sur l'instruction secondaire," *Journal des économistes* (April 1843), 54.

54.

Edmond About, *Le progrès* (Paris, 1867), 350.

55.

Ernest Renan, *Souvenirs de jeunesse* (Paris: Hachette, 1880), 131. See also Francisque Sarcey, *Souvenirs de jeunesse* (Paris, 1885), 301, who says that there were "no skeptics" when exalted claims were made for the benefits of classical education. Ringer, *Education and Society*, 119, concludes that "during the Restoration and July Monarchy, the classicists were generally in control of French public secondary education."

56.

These changes may be followed in the official *Bulletin universitaire*, 1830: 148 (Arrêté of 11 September 1830 eliminating the teaching of *philosophie* in Latin); 1832: 78–79 (Arrêté of 28 September 1832 eliminating Latin responses to *philosophie* questions on the *baccalauréat*); 1840: 114–118 (Règlement of 25 August 1840 shifting sciences to later years, replacing *histoire naturelle* with ancient languages in the earlier years), 125–128 (Circular of 5 September 1840 in which Cousin moves *histoire naturelle* up to the *philosophie* year); 1841: 138–139 (Arrêté of 15 October 1841 in which Villemain attempts to get more physical sciences into the *philosophie* year); 1843: 13–17 (Arrêté of 20 February 1843 coordinating mathematics programs in the *collèges royaux* and *collèges communaux*, strengthening mathematics in the latter); 18–22 (Arrêté of 24 February 1843 on physics and chemistry programs in *philosophie*. Specified details of 15 October 1841 above); 1845: 125–130 (Arrêté of 2 September 1845. Salvandy appoints commission on mathematics teaching); 1847: 45–48 (Circular of 5 March 1847. Salvandy reintroduces mathematics teaching; 1847: 45–48 (Circular of 5 March 1847. Salvandy reintroduces mathematics teaching in early years, down to *Quatrième*.)

57.

Camille Leroy to Ministre de l'Instruction Publique, 6 September 1834. Archives Nationales (A.N.) F17/6894—the principal file on science in secondary education during this period.

58.

Professeurs de mathématiques to M.I.P., 2 May 1842. A.N. F17/6894.

59.

Inspecteur Général, Report of 1839. A.N. F17/6894.

60.

Moniteur universel (24 March 1837), 662–663.

61.

Ibid., 663.

62.
An account of the *écoles centrales'* attempts to realize the Ideologues' vision may be found in L. Pearce Williams, "Science, Education, and the French Revolution," *Isis* 44 (December 1953), 311–330.

63.
Lacroix's work is examined in detail in John H. Weiss, *Origins of a Technological Elite: Engineers, Education and Social Structure in Nineteenth-Century France* (PhD Dissertation, Harvard University, 1977), 87–118.

64.
François Picavet, *Les Ideologues* (Paris, 1891), 424.

65.
They were published in 1938 as *L'évolution pédagogique en France.* See Steven Lukes, *Emile Durkheim: His Life and Work* (New York, 1972), 379.

66.
Falcucci, *L'humanisme,* and Francisque Vial, *Trois siècles de l'enseignement secondaire* (Paris, 1936).

67.
Charles Dupin, *Essai historique sur les services et les travaux scientifiques de Gaspard Monge* (Paris, 1819), 43.

68.
See Maurice Crosland, ed., *Science in France in the Revolutionary Era* (Cambridge, Mass., 1969), and Alphonse Aulard, *Napoleon 1er et le monopole universitaire* (Paris, 1911), 103.

69.
Lacroix, *Essais,* 242–345.

70.
Ibid., 168–203.

71.
Louis Trénard, "L'enseignement secondaire sous la Monarchie de Juillet," *Revue d'histoire moderne et contemporaine 12* (April–June 1965), 81–133, and the bibliography in Paul Gerbod, *La condition universitaire en France au XIXe siècle* (Paris, 1965).

72.
Joseph N. Moody, "The French Catholic Press in the Educational Conflict of the 1840's," *French Historical Studies* VII:3 (Spring 1972), 394.

73.
Paul Gerbod, *Paul-François Dubois* (Paris, 1967), 32–53, and Gerbod, *Condition universitaire,* 78.

74.
Trénard, "L'enseignement secondaire," 121.

75.
Ibid., 119.

76.
Saint-Marc Girardin, *De l'instruction intermédiaire et son état dans le midi de l'Allemagne* (Paris, 1835) and *De l'instruction intermédiaire et de ses rapports avec l'instruction secondaire* (Paris, 1847); Philibert Pompée, *De l'éducation professionnelle en France* (1845), reprinted in his *Etudes sur l'éducation professionnelle en France* (Paris, 1863), 1–182; Victor Cousin, *Mémoires sur l'instruction*

secondaire dans le royaume de Prusse (Paris, 1837). The earliest full-length attempt to deal with the question within the Napoleonic structure is Charles Renouard, *Considérations sur les lacunes de l'éducation secondaire en France* (Paris, 1824).

77.
Cousin, *Mémoires*, 171.

78.
See, along with his other writings, *Du vrai, du beau, et du bien*, first published in 1819. Surprisingly, there exists no full study of the history of *culture générale*. Falcucci, *L'humanisme*, discusses the content of curricula without relating it to the social context of education. The sociologist Celestin Bouglé does little better in *The French Conception of 'Culture Générale' and Its Influence upon Instruction* (New York, 1937), a very brief treatment.

79.
Cousin, *Mémoires*, 172.

80.
Pompée, *De l'éducation*, 17–20.

81.
The report was printed—as "Rapport sur l'état actuel de l'enseignement scientifique dans les collèges, écoles intermédiares, et écoles primaires," 10 April 1847, in *Journal général de l'instruction publique* XVI (19 May 1847), 1–60. A longer MS version is contained in the Dumas papers, Archives of the Académie des Sciences (A. Acad. Sci.), Carton 16.

82.
Ibid., 9.

83.
Ibid., 24.

84.
Ibid., 10.

85.
Ibid., 11–12.

86.
Ibid., 21.

87.
Ibid., 36.

88.
Ibid., 60.

89.
Ibid., 42.

90.
Ibid., 41.

91.
The best discussion of the ill-fated "bifurcation" project is in Robert D. Anderson, *Education in France 1848–1870* (Oxford, 1975).

92.
Shinn, *L'Ecole Polytechnique*, 27.

93.
René Rémond, *La droite en France de 1815 à nos jours* (Paris, 1954), 80–81.

94.
Charles Gide and Charles Rist, *Histoire des doctrines économiques depuis les physiocrates jusqu'à nos jours* (Paris, 1926), 129.

95.
Say introduced the term "entrepreneur" into economic theory. Lewis H. Haney, *History of Economic Thought* (New York, 1936), 356.

96.
J. conn. u. (May 1825), 92. For a further discussion of this journal and its contents, see Weiss, *Origins of a Technological Elite*, 157–199.

97.
In 1815 Saint-Simon had joined J.-B. Say, the Comte de Lasteyrie, and others in establishing the Société de Paris pour l'Instruction Élémentaire. Frank Manuel, *The New World of Henri Saint-Simon* (Cambridge, Mass., 1956), 182.

98.
Ibid., 252.

99.
This was the way Saint-Simon addressed the workers on the title page of an 1821 pamphlet cited in Manuel, *New World*, 255.

100.
Georges Weill, "Les théories saint-simoniennes sur l'éducation," *Revue internationale de l'enseignement* XXXI (1897), 237–246.

101.
Manuel, *New World*, 288–294.

102.
For a discussion of the New Christianity, *Ibid.*, 348–363. The same author examines Comte's Religion of Humanity in his *The Prophets of Paris* (Cambridge, Mass., 1962), 248–296.

103.
Georges Weill, *L'école saint-simonienne* (Paris, 1896), 119, 170–176, 244–248, 276–278; Haney, *History*, 359.

104.
Alexander Gerschenkron, *Economic Backwardness in Historical Perspective* (Cambridge, Mass., 1962), 22–26; John C. Eckalbar, "The Saint-Simonians in Industry and Economic Development," *American Journal of Economics and Sociology* 38:1 (January 1979), 83–96.

105.
Barrie M. Ratcliffe, "The Economic Influence of the Saint-Simonians: Myth or Reality?" *Proc. An. Meeting W. Soc. Fr. Hist.* 5 (1977), 252–262; Cecil O. Smith, *French State and Private Engineers in the Age of Steam and Iron* (forthcoming). Cf. George Iggers, preface to the second edition of his translation of *The Doctrine of Saint-Simon: An Exposition; First Year, 1828–1829* (New York, 1972), vii, where he concludes that "the direct relevance of Saint-Simonianism on later thought and practice has, I now believe, been overstated."

106.
Michel Bouillé, *Enseignement technique et idéologique au XIXème siècle* (Doctorat de 3e cycle, Paris, 1972), 112–113. Bouillé's list of those who corresponded with the *Globe* during that newspaper's Saint-Simonian phase includes such clear non-Saint-Simonians as C. L. M. Navier and Frédéric Le Play.

107.
Raymond Oberlé, *L'enseignement à Mulhouse de 1798 à 1870* (Paris, 1961), 164–166.

108.
C. Rod Day, "The Making of Mechanical Engineers in France: The Ecoles d'Arts et Métiers, 1803–1914," *French Historical Studies* X:3 (Spring 1978), 443; Norman Graves, "Technical Education in France in the Nineteenth Century," *The Vocational Aspect* XVI (1964), 151.

109.
Antonin Monmartin, *Précis sur l'Ecole La Martinière* (Lyons, 1862), 4–8.

110.
Kenneth E. Carpenter, "European Industrial Exhibitions before 1851 and Their Publications," *Technology and Culture*, 13:3 (July 1972), 467.

111.
Bulletin de la Société pour l'encouragement de l'industrie nationale. Hereafter: *Bull. Soc. Enc.* (1825), 374.

112.
Association Polytechnique, *Histoire de l'Association polytechnique* (Paris, 1880), 197–199.

113.
Georges Duveau, *La pensée ouvrière sur l'éducation pendant la Seconde République et le Second Empire* (Paris, 1948), 62.

114.
Denis Poulot, *Le sublime*, 2nd ed. (Paris, 1872), 272.

Chapter 3

1.
Pothier, *Histoire*, 49.

2.
Pothier thanked Dubois in a letter of 3 December 1829, A. ECAM, Correspondence file.

3.
Pothier, *Histoire*, 49.

4.
A copy of the circular appears as an annex to chapter II in Pothier, *Histoire*, 448–449.

5.
Le cent cinquantième anniversaire de la Société d'encouragement pour l'industrie nationale et les problèmes actuels de l'économie française, 1801–1951 (Paris, 1951), 1–6. Dumas served for a number of years as president of the society.

6.
Bull. Soc. enc. (1830), 37–38.

7.
Pothier, *Histoire*, 50.

8.
Rapport à présenter, 6-2.

9.
Pothier, *Histoire*, 50; Colladon, *Souvenirs*, 12.

10.
Rapport à présenter, 2-6.

11.
Among many examples, see Dossiers Denis Rerolle, Emile Gruissillier, Pierre Frontin de Bellecombe.

12.
A. ECAM. Dossier Jean-Victor Duqueyroix.

13.
A. ECAM. Dossier Philippe Massabuau.

14.
A. ECAM. Dossiers Alfred Troupel, Gustave Pury, and Stanislas Michalowski.

15.
See the appendix, which describes the dossier sample and the methods used to classify the information.

16.
Adéline Daumard, "Les élèves de l'Ecole polytechnique de 1815 à 1848," *Rev. d'hist. mod. et contemp.* 3 (1958), 233. Bouillé, *Enseignement technique*, map 1, gives a Parisian figure of 21.1 percent for the period 1800–1880.

17.
In the case of the 1829–1847 group, for example, the Pearson correlation r is 0.9276, giving an R^2 of 0.8605.

18.
Pressly and Julia, "La population scolaire," 1535–1537. In discussing the case of Britain, Carlo Cipolla, *Literacy and Development in the West* (Harmondsworth, 1969), 68, observes that "by offering increased opportunities for the employment of children, the Industrial Revolution raised the opportunity cost of education and therefore affected negatively the consumption demand for it," but he suggests at the same time (p. 70) that greater state support for education prevented such a pattern on the Continent.

19.
Bouillé, *Enseignement technique*, 175.

20.
Jean Sutter, "L'évolution de la taille des polytechniciens (1801–1954)," *Population* 13:3 (July–September 1958), 392. The long-postulated association between prominence as a source of military officers and status as a frontier province, which would explain this aspect of the Polytechnique pattern, has been subjected to a detailed critical examination in William Serman, *Les origines des officiers français (1848–1870)* (Paris, 1979), 271–281.

21.
Daumard, "Les élèves de l'Ecole polytechnique," 233.

22.
Shinn, *L'Ecole Polytechnique*, 51. Shinn does not give more precise figures for this period, but in discussing the longer span, 1830–1880, he finds 19.7 percent from the Lycée Saint-Louis and 14.4 percent from the Lycée Louis-le-Grand, the next largest number.

23.
Throughout this period a "second-" or "third-category" *collège royal* had no teacher specializing in the teaching of the physical sciences. Most of them also had no class in *mathématiques spéciales*.

24.
Ringer, *Education and Society*, 116–117, 316, notes that these schools were not successful until the 1880s. As late as 1887 they enrolled only 21,000 pupils.

25.
The Senate's investigation prior to the State takeover confirmed this figure. See DeComberousse, *Histoire*, 155.

26.
Edouard Charton, *Guide pour le choix d'un état* (Paris, 1842), 51.

27.
Rapport à présenter, 2-7. Charton gives 7,000 francs for a lawyer's education, calculating three years in Paris at 2,000 francs and 1,000 francs for fees at the Faculty of Law. His figure for a four-year medical education is 10,070 francs.

28.
Charton, *Guide*, 254.

29.
Ibid., 324. Shinn, *L'Ecole Polytechnique*, 52, finds an "average annual urban proletarian income" of 1,400–1,800 francs for this period.

30.
A. ECAM. Dossier Hector Rigaud.

31.
A. ECAM. Dossier Jean Zetter.

32.
A. ECAM. Dossier François Delannoy.

33.
A. ECAM. Dossier Gustave Méraux.

34.
A. ECAM. Dossier Eugène Despeyroux.

35.
A. ECAM. Dossier Jules Duchamp.

36.
A. ECAM. Dossier Camille Polonceau.

37.
A. ECAM. Dossier Alfred de Bellefonds.

38.
Shinn, *L'Ecole Polytechnique*, 79.

39.
Daniel Massé, *Pour choisir une carrière* (Paris, 1908), 271.

40.
For an explanation of nomenclature and methods of classification, see the appendix.

41.
Adéline Daumard, "Une référence pour l'étude des sociétés urbaines en France aux XVIIIe et XIXe siècles: Projet de code socioprofessionel," *Rev. d'hist. mod. et*

contemp. X (July–September 1963), 200, notes the absence of this designation in the eighteenth century and its disappearance in the twentieth.

42.

Georges Dupeux, *La société française, 1789–1970* (Paris, 1972), 131.

43.

Shinn, *L'Ecole Polytechnique*, 66.

44.

A. ECAM. Dossier Charles Dollfus.

45.

See the table in Ringer, *Education and Society*, 347–348.

46.

This was also stated in the *Rapport à présenter*, 3-5.

47.

The titled nobility, at least fathers discernible as such, played a small role in all the samples; hence I have tended to group them together with the *haute bourgeoisie*. I do not wish to imply, however, that rivalries, both political and social, between the various nobilities, and between the nobility and the bourgeoisie, did not play an important part in the history of this period.

48.

See, for example, Guillet, *Cent ans*, 267–415, and the list of dynasties in Bouillé, *Enseignement technique*, 148–150. A parallel case in which a new profession began to produce its dynasties is provided by the *secondaire* teachers studied by Victor Karady, "Normaliens et autres enseignants à la Belle Epoque," *Rev. franc. de sociol.* XIII (1972), 35–58. The Ecole Normale Supérieure became the dominant institution in the secondary system only during the July Monarchy. By the end of the nineteenth century it testified to the inbreeding of the University: 31.4 percent of the *Normaliens* who taught between 1900 and 1914 were the offspring of *professeurs*.

49.

Shinn, *L'Ecole Polytechnique*, 49–50, 187. Shinn does not distinguish between large merchants (often wholesalers) and small merchants, nor does he employ upper and middle categories for his groups of liberal professions, as do Daumard and the present writer.

50.

Daumard, "Les élèves de l'Ecole polytechnique," 229, also concludes that artisans and shopkeepers considered the military and functionary posts available to *Polytechniciens* as "badly remunerated."

51.

Shinn, *L'Ecole Polytechnique*, 185.

52.

The contrast between the two schools is even more striking in view of the fact that the Centrale sample, as explained in the appendix, is heavily weighted toward the years immediately before World War I, whereas Shinn's sample is distributed uniformly between 1880 and 1914.

53.

Musson and Robinson, *Science and Technology*, 372–373; Gordon Roderick and Michael Stephens, *Scientific and Technical Education in Nineteenth-Century England* (New York, 1972), 98; Lord Hinton of Bankside, *Engineers and Engineering* (Oxford, 1967), 1–4; Institution of Civil Engineers, *The Education and Status of*

Civil Engineers in the United Kingdom and in Foreign Countries (London, 1870), viii–ix; W. J. Reader, *Professional Men: The Rise of the Professional Classes in Nineteenth-Century England* (London, 1966), 70–71.

54.
Antoine Léon, *La Révolution française et l'éducation technique* (Paris, 1968), 250–251.

55.
Ibid., 250.

56.
Idem.

57.
Ibid., 251.

58.
An F17 1144 doss. 13. *Extrait des registres des déliberations de la République*, cited in Léon, *Révolution française*, 251.

59.
Moniteur, 9 ventose An XI, cited in Léon, *Révolution française*, 252.

60.
Camille Ferdinand-Dreyfus, *Un philanthrope d'autrefois: La Rochefoucauld-Liancourt* (Paris, 1903), 376.

61.
Artz, *Technical Education*, 135–136.

62.
Léon, *Révolution française*, 252.

63.
C. Rod Day, "The Making of Mechanical Engineers in France: The Ecoles d'Arts et Métiers, 1803–1914," *French Historical Studies* X:3 (Spring 1978), 449.

64.
Norman J. Graves, "Technical Education in France in the Nineteenth Century," *The Vocational Aspect* XVI:34 (1964), 148–159; Antoine Léon, "Promesses et ambiguités de l'oeuvre d'enseignment technique en France de 1800 à 1815," *Rev. d'hist. mod. et contemp.* XVII (July–September 1970), 858.

65.
Correspondance Napoléon Ier, letter number 13643, Vol. XVI:411, cited in Léon, *Révolution française*, 254.

66.
Correspondance Napoléon Ier, letter number 1721m, 14 December 1810, Vol. XXI:318, cited in Léon, *Révolution française*, 254.

67.
Graves, "Technical Education," 151.

68.
Artz, *Technical Education*, 38.

69.
Ferdinand-Dreyfus, *Un philanthrope*, 377–413.

70.
Léon, "Promesses et ambiguités," 858.

71.
Le Brun (the inspector-general), "Notice sur les Ecoles Impériales d'Arts et Métiers," A.N. F17/14317.
72.
Prefect of the Gard to Minister of Commerce, 28 August 1840, A.N. F 12/1171, quoted in Day, "Making of Mechanical Engineers," 443. Between 1820 and 1847 the number of children enrolled in the *primaire* more than tripled, from 1.1 million to 3.5 million, while between 1829 and 1847 the number of communes without a primary school fell from 13,987 to 3,213. See Antoine Prost, *L'enseignement en France: 1800–1967* (Paris, 1968), 108.
73.
Day, "Making of Mechanical Engineers," 443.
74.
A contraction of the French *gars des arts*, "lads from the [schools of the industrial] arts."
75.
Ibid., 444: "By the 1860's about five thousand graduates were working in business, industry, and transport. They comprised about 40 percent of the trained engineers and middle-level technicians of France including almost all mechanical engineers and 20 per cent of the civil engineers."
76.
Ibid., 454.
77.
Ibid., 456, 451.
78.
Rondo Cameron, *France and the Economic Development of Europe: 1800–1914* (Princeton, 1961), 50–58; Ezra N. Suleiman, *Elites in French Society* (Princeton, 1978), 25, *et passim.*; Bouillé, *Enseignement technique*, 141–162.
79.
Maurice Lévy-Leboyer, "Innovation and Business Strategies in Nineteenth- and Twentieth-Century France," in Carter et al., *Enterprise and Entrepreneurs*, 107–111.
80.
Kindleberger, "Technical Education and the French Entrepreneur," in Carter et al., *Enterprise and Entrepreneurs*, 28–30; Edmonson, *From Mécanicien to Engineer*, V:1, 14, 18–19, 22–24.
81.
Terry Shinn, "Des Corps de l'Etat au secteur industriel: Génèse de la profession d'ingénieur, 1750–1920," *Rev. Franc. de sociol.* XIX (1978), 54–56; Christophe Charle, "Les milieux d'affaires dans la structure de la classe dominante vers 1900," *Actes de la Recherche en Sciences Sociales* 20/21 (March–April 1978), 87; Nicole Déléfortrie-Soubeyroux, *Les dirigeants de l'industrie française* (Paris, 1961), 199–263; and the sources cited in notes 78–80.
82.
Mouchelet, *Notice historique*, 22. The original name of the association indicated the school's role: Société Centrale des Ingénieurs Civils.
83.
Duveau, *La pensée ouvrière*, 269.

84.
Shinn, "Des Corps de l'Etat," 67.

85.
Of 5 non-*Centraux* identifiable in an incomplete list of 130 men who joined the Société in its first year published in Société des Ingénieurs Civils de France, *Annuaire de 1911* (Paris, 1911), 18–19, 3 had connections with the Union des Constructeurs. I have been able to identify only two presidents of the Société up to 1911 who were definitely not *Centraux*.

Chapter 4

1.
Lyon Playfair to Lord Taunton, 7 June 1867, printed as an appendix in Eric Ashby, *Technology and the Academics* (New York, 1963), 111–112.

2.
C. Lavollée, "L'Ecole Centrale des Arts et Manufactures," *Revue des deux mondes* (15 May 1872), 416.

3.
See the lists of "Ecole Centrale families" printed in Guillet, *Cent ans*, 219–413.

4.
The concept of a technological science is further developed in Edwin Layton, "Mirror-Image Twins: the Communities of Science and Technology in 19th Century America," *Technology and Culture* 12:4 (October 1971), 562–580.

5.
Ibid., 565.

6.
James Renwick, *Applications of the Science of Mechanics to Practical Purposes* (New York, 1842).

7.
The history of the use of the word *technology* in America shows that the role of science in technology was misconstrued from the very beginning. The first writer to popularize the use of the term was Jacob Bigelow, a Boston physician who first published his *Elements of Technology* in 1829. For Bigelow, technology concerned itself with the application of the sciences—"those departments of knowledge which are more speculative, or abstract, in their nature, and which are conversant with truths or with phenomena that are in existence at the time we contemplate them"—to the arts—"which have their origin in human ingenuity, which depend on the active, or formative processes of the human mind, and which without these, would not have existed." For Bigelow, known to the scientific world chiefly for his botanical survey *Florula Bostonienses*, "Discovery is the process of science; invention is the work of art." In this kind of terminological framework, there is little room for the idea of a technological science. When Bigelow asserts that "the application of philosophy to the arts may be said to have made the world what it is at the present day," the examples he gives—the printing press, gunpowder, and watchmaking—are not techniques that are usually considered as based on science. Bigelow's work goes on to describe all sorts of objects and techniques, from painting and sculpture to woodworking to the construction of the steam engine. It is perhaps not without significance that the book was drawn from lectures given at Harvard University during the years from 1819 to 1829. Bigelow's

claim that he was discussing the "applications of philosophy" may have been intended to lend his endeavors an academic legitimacy that would have been denied to something presented as a survey of arts and trades—or, for that matter, engineering. The claim of the Ecole Centrale's founders to have founded *la science industrielle* may have been motivated partly by the same need for legitimacy. Bigelow himself later helped to found the Massachusetts Institute of Technology.

8.

Layton, "Mirror-Image Twins," 566.

9.

Idem. Layton presents another more extensive study of the development of "technological science" by American millwrights in his "Millwrights and Engineers, Science, Social Roles, and the Evolution of the Turbine in America" in Wolfgang Krohn, Edwin T. Layton, Jr., and Peter Weingart, eds., *The Dynamics of Science and Technology* (Boston, 1978), 61–88.

10.

Bernard Forest de Belidor, *La Science des ingénieurs* (Paris, 1729), V–VI. Belidor had nevertheless been named a correspondant of the Académie des Sciences in 1722.

11.

James Kip Finch, "Engineering and Science: A Historical Review and Appraisal," *Technology and Culture* II:4 (Fall 1961), 324.

12.

Louis-Marie-Henri Navier, *Résumé des leçons sur l'application de la mécanique à l'établissement des constructions et des machines*, ed. Barré de Saint-Venant (3rd ed., Paris, 1864), lxxxiv.

13.

Stephen P. Timoshenko, *History of the Strength of Materials* (New York, 1953), 104–107.

14.

Ibid., 231. As Layton points out, "Scientists tend to explain their findings by reference to the most fundamental entities, such as atoms, ether, and forces. But these entities cannot always be observed directly. To be useful to a designer, however, a formulation must deal with measurable entities." Instead of grappling with molecular forces, "engineers were content with a simple macroscopic model—for example, viewing a beam as a bundle of fibers." Layton, "Mirror-Image Twins," 569.

15.

Timoshenko, *Strength of Materials*, 80–81.

16.

The case of the influence of the steam engine upon the origins of thermodynamics is the *locus classicus*. See Milton Kerker, "Sadi Carnot and the Steam Engine Engineers," *Isis* 51 (1960), 257–70, and his "Science and the Steam Engine," *Technology and Culture* II:4 (Fall 1961), 381–390. The influence of technological developments upon scientific advancement is perhaps most strongly emphasized in John D. Bernal, *Science and Industry in the Nineteenth Century* (Bloomington, Indiana, 1970).

17.

Alfred North Whitehead, *The Aims of Education* (London, 1932), 112 (emphasis added).

18.
Baron de Prony, "Notice biographique sur Navier," *Annales des ponts et chaussées* vii (1837), 108.

19.
Ibid., 109.

20.
Alexis De Tocqueville, *Democracy in America*, Bradley edition, Vol. II (New York, 1945), 45–47. By the end of the century, the language of the debate had become sharper, as in this passage from Henri Poincaré's *Science and Method* (New York, 1952), 16: "One has only to open one's eyes to see that the triumphs of industry, which have enriched so many practical men, would never have seen the light if only these practical men had existed, and if they had not been preceded by disinterested fools, who died poor, who never thought of the useful, and yet had a guide that was not their own caprice. What these fools did, as Mach has said, was to save their successors the trouble of thinking."

21.
Cf. Richard Shryock, "American Indifference to Basic Science during the Nineteenth Century," in Bernard Barber and Walter Hirsch, *The Sociology of Science* (New York, 1962), 98–110.

22.
Auguste Comte, *Cours de philosophie positive: Discours sur l'esprit positif*, new edition with introduction and commentary by Charles Le Verrier (Paris, 1949), xlviii.

23.
Ibid., 101.

24.
Ibid., 110. Emphasis in original.

25.
Ibid., 166–176.

26.
Ibid., 113.

27.
Ibid., 109.

28.
Ibid., 103.

29.
Guillet, *Cent ans*, 87, attributes to Dumas the invention of this maxim.

30.
Rapport à présenter, 2-2.

31.
Idem.

32.
Ibid., 2-3.

33.
"Cinquantenaire de l'Ecole" (Draft of a speech), 9. Dumas Papers, Carton 16.

34.
Cf. Alexis Bertrand, "Un réformateur de l'éducation: Auguste Comte," *La nouvelle revue* (15 January 1898), 286–304.

35.
"Cinquantenaire de l'Ecole," 9. Dumas Papers, Carton 16.

36.
Rapport à présenter, 3-5.

37.
The French word *technologie* has retained this meaning, i.e., a subject of study that describes and catelogs industrial processes.

38.
Rapport à présenter, 3-6.

39.
James Kip Finch, *The Story of Engineering* (Garden City, New York, 1960), 181.

40.
Rapport à présenter, 3-7.

41.
Idem.

42.
M. G. Muret, "Antoine de Chézy, histoire d'une formule d'hydraulique," *Annales des ponts et chaussées* 2 (1921), 165–269. First worked out by Chézy in the 1770s, it was originally written V (the average velocity of flow) equals $C\sqrt{DS}$, with D equaling the mean depth of the channel and S its slope.

43.
Coriolis acknowledges the importance of Bélanger's contribution in the "Avertissement" to his *Du calcul de l'effet des machines* (Paris, 1829), iii.

44.
See the report on Coriolis's memoir made to the Académie des Sciences on 8 June 1829 by Prony, Navier, and Girard, which became a preface to the published book, 2, where Coriolis's introduction of the term is acknowledged, and the definition of the term by Coriolis on 17.

45.
Ibid., "Rapport," 1.

46.
See, for example, his discussion of angular momentum, *ibid.*, 4–5.

47.
"Rapport de la commission chargée d'examiner l'état de l'enseignement de la mécanique." A.ECAM and P.V. Conseil des Etudes. 13 December 1837 and 30 April 1838.

48.
Cours de mécanique appliquée, 1840–1841. Compiled by the student Deligny; *ibid.*, 1841–1842. Compiled by the student Gouvy. A. ECAM.

49.
Rapport à présenter, 3-5.

50.
Idem.

51.
Eugène Péclet, *Traité de l'éclairage* (Paris, 1827), i.

52.
Ibid., ii.

53.
Ibid., iij.

54.
The longest French holdout against the wave theory was J. B. Biot, who taught at the Parisian Faculty of Sciences and the Collège de France. Péclet later made his debt to Fresnel and Young explicit in his *Traité de physique*.

55.
Péclet, *Traité de l'éclairage*, 1–2. In 1847 Péclet updated this figure to 80,000 leagues (198,220 miles) per second in his *Traité de physique*, II, 329. The correct figure, of course, is 186,000 miles per second.

56.
Péclet, *Traité de l'éclairage*, 5–11.

57.
T. K. Derry and Trevor Williams, *A Short History of Technology* (Oxford, 1960), 471, claim that "the Davy lamp because it was invented by the leading chemist of the day, is something of a landmark in the relations between science and technology, as also in the use of technology to serve humanitarian rather than purely economic purposes."

58.
He found that the eye position of the observer greatly affected the perception of the darkness of the two sets of shadow lines to be compared.

59.
See Péclet, *Traité de l'éclairage*, 212–283. Another characteristic of engineering science is that it often will accept methods of testing and experiment that are not the most rigorous possible, simply because they are the most practicable at the time. Péclet was not interested in the output of some ideal candle or wick but in the performance of devices that were actually available. On the other hand, although he acknowledged that there was variation among the specimens in his samples, he used no statistical presentation—not even something as simple as the means of a series of results—to help express the range of possible outcomes.

60.
Derry and Williams, *A Short History of Technology*, 630.

61.
Ibid., 610.

62.
A. ECAM. P.V. Conseil des Etudes. 2 May 1847. The first edition appeared in 1830, the fourth in 1847.

63.
Eugène Péclet, *Traité élémentaire de physique* I (Paris, 1847), i.

64.
Ibid., i–ii.

65.
One of the more useful general discussions of the process of scientific advance can be found in Imre Lakatos and Alan Musgrave, eds., *Criticism and the Growth of Knowledge: Proceedings of the International Colloquium in the Philosophy of Science, London, 1965* (Cambridge, England, 1970). The most influential discussion of the scientific method in nineteenth-century France was the *Introduction a l'étude de la médecine expérimentale*, published by Péclet's contemporary Claude Bernard, a close student of Auguste Comte, in 1865.

66.
Hence a gas expands 1/274 of its volume for each degree of increase in temperature above 0 degrees Celsius, but at 274 degrees below 0, if this law were completely general, the volume would be 0, and below that temperature it would reach a negative number. At the time Péclet wrote, temperatures low enough to test the generality of his law could not be attained.

67.
Péclet, *Traité de physique*, I, v.

68.
Idem.

69.
Ibid., vj.

70.
For information on the most recent experiments, Péclet drew largely upon the *Annales de physique et chimie* and the *Comptes-rendus de l'Académie des Sciences*. Foreign journals were only rarely mentioned.

71.
See, for example, his discussion of Galileo's finding that the weight of falling bodies does not determine their speed of fall in Péclet, *Traité de physique*, I, 32–33.

72.
Eugène Péclet, *Traité de la chaleur considérée dans ses applications*, 3 Vols. The third and last edition appeared posthumously in 1860.

73.
Péclet, *Traité de physique*, I, 203.

74.
Ibid., 361.

75.
Ibid., 362.

76.
Idem.

77.
The best discussion of the decline of the caloric theory in this period can be found in D. S. L. Cardwell, *From Watt to Clausius: The Rise of Thermodynamics in the Early Industrial Age* (Ithaca, N.Y., 1971), 89–294.

78.
For the assessment of their importance see Thomas Kuhn, "Energy Conservation as an Example of Simultaneous Discovery," in Marshall Clagett, ed., *Critical Problems in the History of Science* (Madison, 1969), 349.

79.
Charles C. Gillispie, *The Edge of Objectivity* (Princeton, 1960), 369.

80.
Marc Séguin, *De l'influence des chemins de fer et de l'art de les tracer et de les construire* (Paris, 1839), 380–397.

81.
Kuhn, "Energy Conservation," 326. Séguin also reported experiments along the same lines in *De l'influence des chemins de fer*, 381–393.

82.
Kuhn, "Energy Conservation," 324.

83.
Ibid., 326.

84.
Ibid., 341.

85.
Ibid., passim.

86.
Péclet, *Traité de physique*, II, 324–325.

87.
Leonard K. Nash, *The Atomic-Molecular Theory* (Cambridge, Mass., 1950), 114. The difficulty was caused by the need to reconcile the knowledge that equal volumes of gases contained equal numbers of molecules with the discrepancies between the atomic weights obtained by this process and the atomic weights determined by other means. Partly due to the influence of Berzelius' "dualistic" theory of molecules, scientists at the time did not accept the idea of a single molecule composed only of two or more atoms of the same element.

88.
Léon Velluz, *Histoire brève de la chimie* (Paris, 1966), 69.

89.
Maurice Daumas and Jean Jacques, "The Rebirth of Chemistry," in René Taton, ed., *Science in the Nineteenth Century* (New York, 1965), 275. Daumas discusses the polemical nature of this search for new universal principles in "L'école des chimistes français vers 1840," *Chymia* 1 (1948), 55–65.

90.
Jean-Baptiste Dumas, *Traité de chimie appliquée aux arts*, Vol. I (1828), II (1830), III (1832), IV (1834), V (1835), VI (1842), VII (1844), VIII (1846).

91.
Ibid., I, vi.

92.
Ibid., I, liv–lvi. L. Pearce Williams, *Michael Faraday* (New York, 1964), 227–257, and William C. Dampier, *A History of Science and Its Relations with Philosophy and Religion*, 4th ed. (Cambridge, England, 1966), 215–217, discuss the prevailing theories of electrochemical affinity.

93.
Ibid., I, lix. Especially troubling were the facts that chlorine was "positive with regard to oxygen but negative with regard to hydrogen" and that both chlorine and oxygen were "negative with regard to calcium." In consequence, in relation to calcium oxygen should have been more negative than chlorine, yet chlorine "chased" the oxygen from calcium oxide and took its place.

94.
Ibid., I, lx.

95.
Ibid., I, xxxiii.

96.
Ibid., I, xxxiv.

97.
Cf. Gerd Buchdahl, "Sources of Scepticism in Atomic Theory," *British J. for the History of Science* 10 (1960), 120–134.

98.
Dumas, *Traité de chimie*, V, 30–41.

99.
Jean-Baptiste Dumas, *Leçons sur la philosophie chimique professées au Collège de France en 1836*, 2d ed. (Paris, 1878), 315.

100.
Daumas and Jacques, "Rebirth of Chemistry," 282; Satish C. Kapoor, "Dumas and Organic Classification," *Ambix* 16 (1969), 1–66. Philip A. Anderson, *Case Studies in the Acceptance of Dalton's Atomic theory* (Thesis, Harvard University, 1970), 49–62, shows that Dumas "argued not against the idea of ultimate particles, but against the system of *Daltonian* atoms." His work had convinced him that since there was no way to determine the number of atoms (in his terminology, "chemical atoms") in a molecule ("physical atom"), the pursuit of experiments along these lines could lead to no useful results.

101.
Dumas, *Traité de chimie*, I, viii.

102.
Ibid., I, ix.

103.
Ibid., I, ix–x.

104.
Ibid., I, x.

105.
Ibid., II, 664.

106.
Ibid., VI, 415–431.

107.
Ibid., VIII, 662–676.

108.
John Ziman, *Public Knowledge* (Cambridge, England, 1968).

109.
In 1831 a young woman came to Dumas' office with a tearful request that he use his influence to dissuade her husband from continuing his research in chemical technology, which was threatening to bankrupt the family. Dumas agreed to help the woman, but, upon meeting the young Daguerre, became so interested in his work that he encouraged him to continue and offered him his support. Dumas, *Vie de J.-B. Dumas*, 172.

110.
Ibid., *passim*; Antoine Gilardin, *J. B. Dumas et ses oeuvres* (Paris, 1862).

Chapter 5

1.
Prospectus de l'Ecole Centrale des Arts et Manufactures (1830), 3, (1836), 3; *Rapport à présenter*, 2-2.

2.
Annuaire de l'Ecole Polytechnique (1837), 176–177, which also gives far greater details concerning the preparation in algebra than can be found in 1794 or in the early Centrale programs.

3.
Rapport à présenter, 2-2.

4.
Ibid., 2-3; Guillet, *Cent ans*, 63.

5.
Rapport à présenter, 2-2. De Comberousse *Histoire*, 39, reports that the minimum age was raised to sixteen in 1835. By contrast, entering students at the Polytechnique had to be between sixteen and twenty, with an upper age of twenty-five for those who were serving in the military.

6.
Rapport à présenter, 2-4, 2-6. The chemical specialty was originally divided into "mineral chemistry" and "organic chemistry and agriculture." Cf. De Comberousse, *Histoire*, 38, 62.

7.
At least according to De Comberousse, *Histoire*, 46, who uses the military term *appel* in his chart giving the allocation of the hours in the day. De Comberousse is not always accurate on such details—the same chart includes a course never taught—but the degree to which other aspects of student life were organized along military lines, especially in the later years, makes an *appel* not unlikely.

8.
I have capitalized when the name of a subject refers to a specific Centrale course.

9.
Pothier, *Histoire de l'Ecole Centrale*, 352.

10.
A. ECAM. P.V. Conseil des Etudes, 7 November 1838.

11.
De Comberousse, *Histoire*, 51.

12.
Idem.

13.
René Vallery-Radot, *La vie de Pasteur* (Paris, 1918), 25–26.

14.
Ibid., 26.

15.
Léon Velluz, a prominent industrial chemist and biographer of Priestley, Berthelot, and Lavoisier, told me that he had once considered a biography of Dumas but decided that it would inevitably lack dramatic tension: Dumas' life seemed free from struggle against obstacles or heroic conflicts. He simply rose quickly to the top and stayed there.

16.
Dumas, *Vie de J.-B. Dumas, passim*; Andre-Jean Tudesq, *Les grands notables en France*, I (Paris, 1964), 459–461.

17.
Pothier, *Histoire de l'Ecole Centrale*, 54.

18.
De Comberousse, *Histoire*, 50.

19.
Ibid., 51.

20.
Pothier, *Histoire de l'Ecole Centrale*, 54.

21.
Georges Barral, *Le Panthéon scientifique de la Tour Eiffel* (Paris, 1892), 260–263. Colladon was born in 1802, the son of a *régent de troisième* in a classical *collège* who was part of a large family of Protestant pastors, lawyers, and merchants. This active intellectual milieu of the polymath Genevan bourgeoisie was one of the most productive centers of European scientific activity in the late eighteenth and early nineteenth centuries. Colladon's parents and uncles were the close friends of most of its leading figures, such as the botanists Pyramus and Alphonse de Candolle. Geneva also served as a tributary for the larger, wealthier Parisian scientific community. Like his fellow Centrale professor Dumas (two years his senior), Colladon soon left Geneva to seek his fortune on the wider horizons of Parisian science. Working with his childhood friend Charles Sturm, who had gone to Paris as a tutor to the De Broglie family, he had won the Grand Prix of the Académie in 1827 for work on the compressibility of liquids and the speed of sound in water. He had worked closely with Ampère, Fourier, and Dumas' collaborator Constant Prévot. Colladon, *Souvenirs et mémoires*, 15–17, 141–142, 193.

22.
Ibid., 189–190.

23.
Ibid., 200. As Colladon explained in a letter to Pothier of 23 February 1887, which the latter reprints in his *Histoire*, 479, Colladon could not teach the full Mechanics course after 1832 because he spent his winters in Geneva. He gave a special steam engine course of 22 *leçons* in the summers.

24.
Idem.

25.
Ibid., 202–203.

26.
Ibid., 203.

27.
A ECAM. Registre de l'Ecole. 3 November 1836. Cf. De Comberousse, *Histoire*, 199.

28.
A ECAM. P.V. Conseil des Etudes. 10 December 1840. As was often the case, the *procès-verbaux* did not mention Regnault by name, but in this instance the mention of the course allows identification. I have been unable to establish that the supervision of note taking was also the rule at the Ecole Polytechnique, but it is not unlikely.

29.
Colladon, *Souvenirs et mémoires*, 197–198.

30.
Ibid., 198; Letter of Colladon to Francis Pothier of 29 November 1886, printed in Pothier, *Histoire de l'Ecole Centrale*, 473.

31.
Ibid., 472.

32.
Ibid., 473; Colladon, *Souvenirs et mémoires*, 199.

33.
When a French friend complained of this problem to Colladon, the latter pointed out that his mistake had been to present himself as an entrepreneur rather than a scientist. *Ibid.*, 220–221.

34.
American visitors to Britain during this period also made good use of their ability to reproduce machines from memory. The best-known case is that of Francis Lowell, one of the founders of the Massachusetts textile industry.

35.
A. ECAM. Registre de l'Ecole, 16 February 1834, 25 January 1835, 17 April 1835, 20 February 1839, 14 January 1843, and other instances.

36.
Ibid., 16 November 1830, 14 May 1831.

37.
Colladon, *Souvenirs et mémoires*, 200.

38.
Rapport à présenter, 3-1.

39.
Ibid., 4-2, 4-3; Guillet, *Cent ans*, 149; De Comberousse, *Histoire*, 74–78; Pothier, *Histoire de l'Ecole Centrale*, 123.

40.
Rapport à présenter, 4-1.

41.
A. ECAM. P.V. Conseil des Etudes. 3 March 1838.

42.
A ECAM. Registre de l'Ecole. 8 September 1832. (The later date was occasioned by the cholera epidemic's disruption of studies that year, to be described. Another list of projects is given in Pothier, *Histoire de l'Ecole Centrale*, 169.

43.
Rapport à présenter, 4-3, 4-4.

44.
See the *projets* for the refinery submitted by students Fontenay, Rochkoltz, Cayrol, Bineau, Psicha, Granié in the 1832 *concours*. A. ECAM.

45.
See the surviving *projets* for the 1838 and 1841 *concours*. A. ECAM.

46.
Projet file. A. ECAM. I would like to thank Prof. John Abel of the Cornell School of Engineering for assistance in the analysis of this case.

47.
Landes, *Unbound Prometheus*, 105.

48.
Péclet, *Traité de la chaleur*, I, 1.

49.
Pothier, *Histoire de l'Ecole Centrale*, 253.

50.
Rapport à présenter, 9-5.
51.
Colladon, *Souvenirs et mémoires*, 211.
52.
Ibid., 214–217. Colladon listed Championnière as coauthor when he published the results in volume XXXIX of the *Bibliothèque universelle de Genève*.
53.
Ibid., 220–222.
54.
"Travaux des Laboratoires de Recherche." A. ECAM. The *répétiteur* Emile Boistel signed fourteen of the reports filed between 1833 and 1835, but the rest were signed by students. In 1833 alone, 118 experiments were recorded.
55.
Pothier, *Histoire de l'Ecole Centrale*, 116, who also reports on 133 that Jules Péligot began his distinguished career as a chemist in the laboratories of the Ecole Centrale while he was a student there.
56.
Rapport à présenter, 4-3; A. ECAM. Registre de l'Ecole. 8 October 1831 and 1 December 1838.
57.
Rapport à présenter, 9-8.
58.
Ibid., 9-9.
59.
Prospectus de l'Ecole Centrale des Arts et Manufactures (1837), 4–5. Not until 1858 did physics and chemistry become part of the actual requirements for entry.
60.
Guillet, *Cent ans*, 94; A. ECAM Registre de l'Ecole. 14 November 1836.
61.
François Poncetton, *Eiffel: Le magicien du fer* (Paris, 1939), 69.
62.
Maurice Donnay, *Centrale* (Paris, 1931), 12–13.
63.
Pothier, *Histoire de l'Ecole Centrale*, 83.
64.
Mémoires de la Société des Ingénieurs Civils (hereafter: Mém. S.I.C.) (1878), 673–685.
65.
Rapport à présenter, 5-1.
66.
Ibid., 6-6, 6-7. In a letter to the *Journal des chemins de fer* 44 (4 February 1843), 378, Perdonnet pointed out that until 1843 the Ecole des Ponts et Chaussées had no course dealing with locomotives, hence "no course in which one could seriously consider the question of the design of railroads."
67.
Speech by J. B. Dumas at the funeral of Perdonnet, 4 October 1867, printed as an appendix in De Comberousse, *Histoire*, A.90–A.92.

68.
A. ECAM. P.V. Conseil des Etudes. 9 February 1838.
69.
Ibid., 2 December 1839.
70.
Idem.
71.
Ibid., 9 February 1838.
72.
Ibid., 2 December 1839; *Rapport à présenter*, 4-3, 4-4, 4-8.
73.
Ibid., 2 December 1839; *Rapport à présenter*, 4-3, 4-4, 4-7, 4-8.
74.
I have been unable to learn anything about the content of this course. The section of *procès-verbal* of 2 December 1839 that discusses the new regulations mentions that part of its subject matter was incorporated into Ferry's course on Ferrous Metallurgy and part into Walter's own more advanced course on Machine Construction. Pothier's biographical notice on him suggests, however, that he had no solution to the problem. He notes that Walter, a graduate of Saint-Cyr, who had left military service to manage a foundry for the Compagnie de Terre-Noire, then launched out on his own as an *ingénieur civil*. In Pothier's view this on-the-job education had left gaps: "Theoretical knowledge is best acquired while young; this knowledge may have been lacking in his explanations of the chemical phenomena in the making of pig iron and of theory of machines, but the students nevertheless respected this professor, always clear in his demonstrations and his practical descriptions." Pothier, *Histoire de l'Ecole Centrale*, 213. By 1846 Walter's Machine Construction course was devoid of any more general or theoretical explorations, at least according to the single course notebook retained in the school's archives, produced by a student named Maire.
75.
John Jewkes, David Sawers, and Richard Stillerman, *The Sources of Invention* (London, 1958), discussed European cases. Jacob Schmookler, *Invention and Economic Growth* (Cambridge, Mass., 1966); W. Paul Strassman, *Risk and Technological Innovation* (Cambridge, Mass., 1955); and Elting W. Morison, *Men, Machines and Modern Times* (Cambridge, Mass., 1971) are useful studies of the American case.
76.
Poncetton, *Eiffel*, 70.
77.
See the Ecole Centrale's *Annuaire* for 1903, 10–12; Pothier, *Histoire de l'Ecole Centrale*, 190; Neuschwander, *L'Ecole Centrale*, 220–225.

Chapter 6

1.
As in Viviane Isambert-Jamati, "La rigidité d'une institution: Structure sociale et systèmes de valeurs," *Rev. française de sociologie* VII (1966), 306–347, and the same author's *Crises de la société, crises de l'enseignement* (Paris, 1970).

2.
Charles De Comberousse, speech in "Cinquantenaire de l'Ecole Centrale," Dumas Papers, A. Acad. Sci. Carton 16 pp. 31–32.

3.
Ibid., 32.

4.
Idem.

5.
Ibid., 33.

6.
Ibid., 35.

7.
Ibid., 37–38.

8.
Ibid., 40.

9.
Ibid., 42.

10.
Idem.

11.
Ibid., 43.

12.
Ibid., 43–44.

13.
Ibid., 44.

14.
Ibid., 47.

15.
The principal instructor, Henri Milne-Edwards, was a specialist in invertebrate paleontology interested primarily in lacustrine fossils, especially corals. He spent considerable time opposing Darwin's theories. Taton, *Science in the Nineteenth Century*, 361, 371, 478.

16.
De Comberousse, speech at *Cinquantenaire*, 47.

17.
Ibid., 49.

18.
Annuaire statistique de la France: Résumé rétrospectif (1966), 170, 200.

19.
Paul Hohenberg, *Chemicals in Western Europe: 1850–1914* (Chicago, 1966); Clapham, *Economic Development of France and Germany*, 236–239.

20.
Théodore Olivier, *Mémoires de géométrie descriptive, théorique, et appliquée* (Paris, 1851), ii.

21.
Ibid., iii.

22.
Ibid., iv. Emphasis in the original.
23.
Idem.
24.
Ibid., v.
25.
The Doctrine of Saint-Simon: An Exposition, translated and annotated by Georg G. Iggers (New York, 1972), 29, where the views of Saint-Simon are contrasted with those of his followers.
26.
Olivier, *Mémoires*, xx.
27.
In discussing this delay in his *Turning Points in Western Technology* (New York, 1972), 148–149, D. S. L. Cardwell points to competition from the steam engine and gas lighting. The Ecole Centrale, in Colladon's course and in Péclet's course, was first in the world to develop these latter as subjects of engineering education. Did they perhaps become overcommitted to these first achievements, thus losing a certain flexibility of investigation: One would feel more confident that this was the case if the Ecole Centrale had not also been one of the first (in 1884) to establish a course in electrical engineering.
28.
Olivier, *Mémoires*, vi.
29.
Idem.
30.
Ibid., vii.
31.
Idem.
32.
L. Pearce Williams, in "Science, Education, and Napoleon I," reprinted in Leonard M. Marsak, *The Rise of Science in Relation to Society* (New York, 1964), 80–91, argues instead that Napoleon's reforms were directly detrimental to French scientific education.
33.
Olivier, *Mémoires*, viii, xiv.
34.
Ibid., xii.
35.
Ibid., xvi.
36.
Idem.
37.
C.-P. Marielle, *Répertoire de l'Ecole Impériale Polytechnique* (Paris, 1855), (a).
38.
The other members: the Vicomte de Caux, general in the *Génie militaire* and *conseiller d'Etat*; the Comte de Caramon, an artillery officer; Baron Héron de Villefosse, *inspecteur-divisionnaire des mines* and *maître des requêtes*; Paulimes

de Fontenneles, secretary-general at the Ministry of the Interior and a lieutenant-colonel in the *Génie*.

39.
Pinet, *Histoire de l'Ecole Polytechnique*, 427. This is also the interpretation of Shinn, *L'Ecole Polytechnique*, 33–34.

40.
Quoted in Fourcy, *Histoire de l'Ecole Polytechnique*, 351, and Pinet, *Histoire de l'Ecole Polytechnique*, 426.

41.
Ibid., 427. The Laplace commission also rejected proposals to introduce courses on political economy and on *technologie*.

42.
Quoted in Fourcy, *Histoire de l'Ecole Polytechnique*, 350.

43.
Ecole Polytechnique. *Livre du centenaire*, Vol. I (Paris, 1895), 51.

44.
Ibid., 52.

45.
Idem.

46.
Ibid., 53.

47.
Idem. Instruction was carried on by four part-time *maîtres*.

48.
Ibid., 56.

49.
Ibid., 58.

50.
Ibid., 60.

51.
Olivier, *Mémoires*, xii.

52.
Ibid., xviii. Olivier makes the same argument in his article, "Monge et l'Ecole Polytechnique," *Revue scientifique et industrielle*, Vol. 7, 3rd series (1850), 64–68.

53.
Pothier, *Histoire*, 206, gives a total of 75 students, but Thomas's assistants also included graduates of the school (as well as law students, veterinary students, and Saint-Cyriens). In his own *Histoire des Ateliers Nationaux* (Paris, 1848), Thomas gives the names of 41 *Centraux*, seven of them alumni.

54.
Ibid., 42–43. The episode is examined further in Weiss, *Origins of a Technological Elite*, 596–600.

Chapter 7

1.
Guillet, *Cent ans*, 232.

2.
The student dossiers did not systematically specify this fact. Because Paris experienced a large net in-migration during the years 1800–1848, it seems likely that a somewhat larger number of students than the 23.6 percent of the sample *born* in Paris lived there with their parents. On the other hand, some Parisian families may have preferred to send their sons to boarding institutions.

3.
Although I did not locate any list of correspondents, the dossiers contained a scattering of letters to and from such agents upon which the above statement was based. The case of Louis Pasteur illustrates how residence in Paris might be arranged using such correspondents. An officer of the Parisian *garde municipale*, Capt. Barbier, who regularly took his vacations at the Pasteurs' town of Arbois, came to know the family well. When the furthering of Louis's secondary education was being discussed between Joseph Pasteur and Romanet, principal of the local *collège*, Barbier offered himself as correspondent for the young student. When Joseph, a modest tanner, objected that the cost of such a Parisian education would be too great, Barbier replied that he knew a *pension* in the Latin Quarter run by a Franc-comtois named Barbet who gave reduced rates to his compatriots. And so in 1838 Pasteur arrived at Barbet's institution to prepare his *baccalauréat*. Vallery-Radot, *La vie de Pasteur.* 12–13.

4.
De Comberouse, *Histoire*, 65; A. ECAM. Registre de l'Ecole 31 October 1837.

5.
Rapport à présenter, 2-7.

6.
Ibid., 2-8.

7.
Olivier, *Mémoires*, xxii.

8.
Idem.

9.
See Table 3.1. Fifty-one of the 145 students in the first Centrale class were over twenty-one, for example.

10.
At the Polytechnique, after all, there were always windows, and tutors who might be persuaded to wink at unauthorized departures. *Polytechnicien* slang had a special term for such escapades: a *bélier*, which referred to the ram's-head wall ornament that helped students scale the walls. Callot, *L'Ecole Polytechnique*, 174, 191.

11.
Honoré de Balzac, *The Lily of the Valley* (Barrie edition), 16.

12.
Ibid., 17–18.

13.
Ibid., 19.

14.
Idem.

15.
The ushering to Charlemagne is described in the following sentence: "During the

recreation preceding the hour at which the ushers escorted us to the Charlemagne *lycée* the wealthy students went to breakfast at the porter's, a man called Doisy" (p. 16). Ushering is a convention: the emphasis of the sentence is upon the introduction of Doisy. In the later instance, the sentence quoted above describing the escorting of Felix to law lectures is followed by a sentence that lays dramatic emphasis on that act: "A young girl would have been guarded with less precaution than was suggested for the care of my person by my mother's fears" (p. 19). Customs, activities, and attitudes introduced as assumptions, as conventions, or as "background" elements may correspond more closely to the typical phenomena of the writer's social environment than do those passages clearly designed to contribute to the central dramatic or psychological structure. In his discussion of the use of literary sources for the study of class relations and criminality in the Paris of the July Monarchy, Louis Chevalier states that "the most important testimonies concerning [these changes in criminality] are those that the author could not avoid giving, and, even more generally, those that he was not conscious of giving." *Classes laborieuses et classes dangereuses à Paris pendant la première moitié du XIXe siècle* (Paris, 1958), 118.

16.
Guillet, *Cent ans*, 11; Fernand Lotte, *Dictionnaire biographique des personnages fictifs de la Comédie humaine* (Paris, 1952), 675.

17.
Colladon, *Souvenirs et mémoires*, 141.

18.
Poncetton, *Eiffel*, 75–76.

19.
Pinet, *Histoire de l'Ecole Polytechnique*, 208.

20.
Rapport à présenter, 2-8; see also 13-3.

21.
Poncetton, *Eiffel*, 69.

22.
Rapport à présenter, 13-4, which gives only this final index number.

23.
This would include students at the faculties of medicine, law, theology, arts, and sciences and the *élèves* of the specialized schools: Polytechnique, Centrale, Mines, Ponts et Chaussées, Commerce, Beaux-Arts, Normale Supérieure, and Alfort.

24.
As, for example, in the opening passages of *The Magic Skin*.

25.
Pinet, *Histoire de l'Ecole Polytechnique*, 208.

26.
I have found addresses for sixty-eight students clearly not living with their parents. Of these all but nine lived within the area bounded by the Seine, the Rue Saint-Martin, and the boulevard Saint-Martin and its southeastern continuations.

27.
Weill, *L'ecole saint-simonienne*, 26.

28.
As described in Gabriel Perreux, *Au temps des sociétés secrètes: La propagande républicaine au début de la Monarchie de Juillet* (Paris, 1931).

29.
Paul Raphael and Maurice Gontard, in *Hippolyte Fortoul* (Paris, 1975), show how their subject, when he was only a twenty-year-old law student, was readily received by the poet Béranger, by Saint-Simonions such as Pierre Leroux, Jean Reynaud, and Hippolyte Carnot, and by the historian Edgar Quinet. He quickly became an editorial collaborator and writer for half a dozen of the major periodicals of the time. Fortoul later served as minister of public instruction under Napoleon III, in which capacity he worked closely with J.-B. Dumas.

30.
A recent discussion of the encounter of a single young student with the messianisms of Paris is found in Barrie M. Ratcliffe's study of Gustave d'Eichthal, "Saint-Simonism and Messianism: The Case of Gustave d'Eichthal," *French Historical Studies* IX:3 (Spring 1976), 484–502. The *locus classicus* is Frank E. Manuel, *The Prophets of Paris* (Cambridge, Mass., 1962), 103–315.

31.
Poncetton, *Eiffel*, 72.

32.
Mouchelet, *Notice historique*, 11; Pinet, *Histoire de l'Ecole Polytechnique*, who refers to the "Ecole des Arts et Métiers."

33.
Callot, *Histoire de l'Ecole Polytechnique*, 79; Pinet, *Histoire de l'Ecole Polytechnique*, 208.

34.
Ibid., 208–209.

35.
Callot, *Histoire de l'Ecole Polytechnique*, 79.

36.
Pothier, *Histoire de l'Ecole Centrale*, 177.

37.
Rapport à présenter, 6-8.

38.
Calculated from Marielle, *Répertoire*, (c).

39.
Pothier, *Histoire de l'Ecole Centrale*, 125, 145, 174.

40.
A. ECAM. Dossiers Louis Meiner and Felix Vinchon.

41.
A. ECAM. Dossier Charles Debray.

42.
A. ECAM. Dossier Charles Bouchacourt.

43.
A. ECAM. P.V. Conseil des Etudes. 20 August 1839.

44.
The Ecole des Ponts et Chaussées also did not use such examiners.

45.
A. ECAM. Registre de l'Ecole. 18 March 1830.

46.
Ibid., 4 May 1831; 6 June 1831.

47.
Ibid., 18 April 1831; 25 April 1831.

48.
Pothier, *Histoire de l'Ecole Centrale*, 80. Of the other men identifiable as *chefs d'études* who later gained prominence Ludovic Martin served as vice-president of the S.I.C. and Marc Cayrol served as its treasurer.

49.
Cagnol, Guepin, Loustau, Bouisset, Du Pan (de Geneva), de Fontenay, Granie, Vassenot, Verdavainne, and Eurard. An eleventh, Bineau d'Aligny, had been chosen by them as an *aide-préparateur*.

50.
At least according to Colladon, *Souvenirs et mémoires*, 196, the only source that mentions him.

51.
A. ECAM. Registre de l'Ecole. 2 November 1830. Also mentioned in Pothier, *Histoire de l'Ecole Centrale*, 137.

52.
Rapport à présenter, 3-1.

53.
Quoted *Ibid.*, 3-7.

54.
This was all related by Loustau to Pothier, who reports it in his *Histoire de l'Ecole Centrale*, 66.

55.
Ibid., 67.

56.
Prospectus de l'Ecole Centrale des Arts et Manufactures (1830), 2.

57.
Idem. The French *brave* can mean courageous, but in this context Lafayette probably meant something closer to "worthy."

58.
Ibid., 5.

59.
See Table 7.1. In 1830 126 of the approximately 460 candidates were admitted to the Polytechnique; in 1831, 142 of 480 were admitted. Callot, *Histoire de l'Ecole Polytechnique*, 348, and Marielle, *Répertoire*, (c).

60.
Colladon's memoirs show how a faculty member at the Ecole Centrale could suddenly find himself in the middle of a panicked population. During a meeting in his workshop with the stockholders of the steamboat *La Seine*, one of his draftsmen asked to leave work because he felt sick. Colladon instead put the man to bed in an adjoining room and did what he could to make him comfortable. When he returned to the meeting, he found that all the shareholders had fled. When the theaters of Paris were ordered to stay open despite the epidemic, Colladon and

two Swiss friends found themselves at the Théatre des Variétés with only eight others in the audience. Colladon, *Souvenirs et mémoires*, 207–208.

61.

Pothier, *Histoire de l'Ecole Centrale*, 83.

62.

Printed as an appendix in Pothier, *Histoire de l'Ecole Centrale*, 83.

63.

Ibid., 88.

64.

Colladon, *Souvenirs et mémoires*, 209–210. According to Pothier the school had proved more costly than his father-in-law had predicted. Of the 100,000 francs Lavallée had pledged to use in founding the school, he had planned to use 70,000 for "initial expenditures" *(frais de premier établissement)*, leaving the rest in reserve. Instead, between 1 November 1828 and 15 December 1832, the initial expenditures totalled 162,814 francs. Lavallée estimated at that time that 45,000 francs more would be needed to accommodate the optimal number of 350 students. On 1 January 1832 the lease on the Hôtel de Juigné, for which Lavallée was personally responsible, had seventeen years more to run, at 14,000 francs a year. Lavallée's parents, anxious not to jeopardize the dowry of their remaining unmarried daughter, refused to lend him any support. Lavallée's wife's parents, who lived in Paris on the income from buildings and plantations in Louisiana, apparently gave him considerable aid, as did Lallemand. See Pothier, *Histoire de l'Ecole Centrale*, 84–86.

65.

Rapport à présenter, 4-5; Pothier, *Histoire de l'Ecole Centrale*, 85–90; Colladon, *Souvenirs et mémoires*, 210. Colladon and Perdonnet argued that the Raucourt proposal would leave them all "subject to the caprices of the shareholders" while burdened with administrative responsibilities that would restrict their freedom to teach. Pothier claims that the Chamber of Deputies would not have given the required consent for such a limited liability company, especially in view of the separation between direction and ownership. Raucourt apparently did not consider the more easily accepted *commandite* option, perhaps because he had no one to propose as *commanditaire*.

66.

Rapport à présenter, 4-5, 4-6; Pothier, *Histoire de l'Ecole Centrale*, 93.

67.

A. ECAM. Registre de l'Ecole. 3 August 1832.

68.

Jesse Pitts et al., *In Search of France* (Cambridge, Mass., 1963), 254–259; Michel Crozier, *The Bureaucratic Phenomenon* (University of Chicago, 1964), and the criticism of Viviane Isambert-Jamati, "L'autorité dans l'éducation française," *European J. of Sociol.* 6:2 (1965), 149–166.

69.

Pothier, *Histoire de l'Ecole Centrale*, 86, asserts that "the school was saved by the confidence and courage of the students who remained at the school during the epidemic and by the professors who were willing to give their lectures without payment."

70.

A. ECAM. P.V. Conseil des Etudes. 3 August 1833. Olivier presented the petition on behalf of the students, recommending that the council "accept it and ask Raucourt what his plans were." Raucourt replied that the pressure of his other work made it impossible for him to continue to teach at the Ecole Centrale.

71.
They are so designated in the school's Musée des Centraux. See also *Rapport à présenter*, 4-7; Neuschwander, *L'Ecole Centrale*, 45.

72.
Eugene Péligot, speech at Olivier's funeral, 10 August 1853, reprinted in De Comberousse, *Histoire*, A.60–A.61.

73.
Pothier, *Histoire de l'Ecole Centrale*, 94.

74.
Letter of 20 April 1884 written on the occasion of Dumas' death, quoted in Colladon, *Souvenirs et mémoires*, 190.

75.
The most important work in this series was the *Théorie géométrique des engrenages*, published in 1842.

76.
Edward Shorter and Charles Tilly, *Strikes in France: 1830–1968* (Cambridge, England, 1974), 107–108.

77.
In a speech in the Chamber of Deputies on 18 May 1835 Arago claimed that at this time the students did *not* want the *externat*: "They realized, first, that they would have had difficulty reconciling the rigorous demands of their course work with the social distractions of their evenings. They foresaw that if they became involved in the burning political questions of the moment, they would not have minds free to profit from their very difficult courses. They replied to the investigator that they now lived on excellent terms with each other despite the differences in their political opinions: Could this remain true if they returned to class each day with new political impressions gained in the social intercourse of the previous evening? Would there not be grave disorders?" Quoted in Pinet, *Histoire de l'Ecole Polytechnique*, 178.

78.
Ibid., 183–186.

79.
Ibid., 193.

80.
Ibid., 198, 214–215.

81.
Ibid., 226.

82.
Ibid., 220.

83.
Ibid., 223.

84.
Ibid., 221, in which the incorrect spelling Quenneville is given.

85.
Ibid., 190.

86.
Annuaire de l'Ecole Polytechnique (1837), 82–93. Average annual total enrollment at the school from 1832 to 1839 was 297.

87.
A. ECAM. P.V. Conseil des Etudes. 21 November 1832, 3 January 1833; *Rapport à présenter*, 3-1.

88.
Copies of the letters were not available. Nor does Pothier give any quotations from them that he might have found in Olivier's private register. However, the *registre Correspondance*, which listed communications between Olivier (and Lavallée, who dealt with all financial matters) and parents, recorded at least eight such letters each month. Copies also occasionally found their way into students' dossiers. See Dossiers J.-L.-T. Jacquelart, Pierre Witz, Louis Vuillemin, Celestin Bouvet, Jean-Baptiste Chavassier, Louis Luc, Auguste Martenot, Jules de Miniac, Jacques Perry, Alexandre Praileur.

89.
A. ECAM. Registre de l'Ecole. 6 November 1834.

90.
Olivier to Dumas. 14 February 1834. Dumas Papers. Carton 16.

91.
Pothier, *Histoire de l'Ecole Centrale*, 126–127.

92.
Olivier to Dumas. 13 November 1833. Dumas Papers. Carton 16.

93.
A. ECAM. P.V. Conseil des Etudes. 12 September 1833, 14 October 1833, 4 May 1834, 2 September 1835, 16 September 1836; Olivier to Dumas, 20 October 1834 and 30 September 1835. Dumas Papers. Carton 16.

94.
Order of 3 February 1835. Quoted in Pothier, *Histoire de l'Ecole Centrale*, 135.

95.
Ibid., 135–136.

96.
A. ECAM. P.V. Conseil des Etudes. 3 December 1834, 7 January 1835. For the petition asking for the course see *Ibid.*, 3 January 1834.

97.
Ibid., 28 January 1835.

98.
Ibid., February 1836.

99.
Quoted in Pothier, *Histoire de l'Ecole Centrale*, 127.

100.
Colladon reports on the Olivier-Dumas tension in *Souvenirs et mémoires*, 190.

101.
Dumas had a large private laboratory, but he also made increasing use of the laboratories at the Ecole Centrale, which he later equipped using funds from the Ministry of Education.

102.
Olivier to Dumas, 2 February 1834, 11 December 1834, and 21 March 1836. Du-

mas Papers. Carton 16. The petition of 2 February 1834 is also mentioned in Pothier, *Histoire de l'Ecole Centrale*, 127.

103.

Ibid., 129–130, and Dossier Roberto Valeriani.

104.

Olivier, *Mémoires de géométrie descriptive*, xxii.

105.

Prospectus de l'Ecole Centrale des Arts et Manufactures (1829), 2. The rules of the Ecole Polytechnique stated that no student could bring *any* book or newspaper into the school without the permission of the authorities. See Shinn, *L'Ecole Polytechnique*, 54.

106.

Pothier, *Histoire de l'Ecole Centrale*, 124. A. ECAM. Registre de l'Ecole. 20 November 1833.

107.

A. ECAM. P.V. Conseil des Etudes. 1 June 1834.

108.

Ibid., 30 March 1834; *Prospectus de l'Ecole Centrale des Arts et Manufactures* (1835), 104.

109.

William L. Langer, *Political and Social Upheaval* (New York, 1969), 9–10, 80, 109–110, discusses briefly the importance of the Polish events for Western Europe.

110.

Their dossiers indicate that they were not merely refugees from the revolution but active participants. Dossiers Eloy Bontemps, Paul Kaczanowski, Clutoine Mirecki, Albert Lutowski, Jean Metrebski, Edouard Dworzaczek, Louis Gobbiowski, Napoleon Dzialynski, Antoine Wolski, Vincent Stawecki, Charles Chobrzynski, Joseph Laznieski, Felix Szlubowski, Stanislas Mickaniewski, and Jean Piotrowski.

111.

Lafayette to Lavallée, 29 December 1832, cited in Pothier *Histoire de l'Ecole Centrale*, 106. They thus accounted for thirteen of the fifteen full scholarships awarded in 1834, the year of their greatest influx. In at least one earlier case, Lavallée showed himself willing to accept a student for political reasons. In a letter of 24 January 1831, he informed Michel Alcan, an apprentice bookbinder who had distinguished himself in the *journées* of July, that he was being accepted in order to second the kind intentions of the Commission of National Recompense. (Alcan had asked the commission "to be allowed to continue his industrial studies.") Lavallée and his colleagues did this despite their awareness that Alcan had "not yet mastered all the subjects generally required for admission." Alcan went on to distinguished career as textile engineer and entrepreneur. *M.S.I.C.* (1878), 673–685.

112.

A. ECAM. P.V. Conseil des Etudes. 17 November 1834.

113.

See Pinet, *Histoire de l'Ecole Polytechnique*, 190. Only eleven students from "Russia" were admitted during the entire period 1805–1833. Marielle's *Répertoire* lists no Polish or Russian names for the 1834–1838 period.

114.

A. ECAM. P.V. Conseil des Etudes. 6 November 1835.

115.
Dossier Eloy Bontemps. The name suggests French parentage, but the dossier clearly indicates that he was a "Polish refugee."
116.
A. ECAM. P.V. Conseil des Etudes. 20 February 1836.
117.
Dossiers Jean Piotrowski and Stanislas Mickaniewski.
118.
M.S.I.C. (1871), 443–462.
119.
Pothier, *Histoire de l'Ecole Centrale*, 56.
120.
Ibid., 185.
121.
Ecole Polytechnique, *Livre du centenaire*, I, 175.
122.
A. ECAM. Registre de l'Ecole. 18 January 1834.
123.
René Taton, "L'Ecole du Génie de Mézières," in Taton, *Enseignement et diffusion*, 602.
124.
Pothier, *Histoire de l'Ecole Centrale*, 61, says that Achille Ferry "received from his father the tradition of Monge's teaching; this great innovator in the applications of science inspired a profound admiration in him."
125.
Marielle, *Repertoire*, 152.
126.
Pothier, *Histoire de l'Ecole Centrale*, 318–319; Dossier William Priestley.
127.
Annuaire de l'Ecole Polytechnique (1837), 166–167. It is unlikely that the faculty members whose schooling is unknown include graduates of either of the two schools. Cf. the discussion in Shinn, *L'Ecole Polytechnique*, 45–48, who sees the recruitment from the inside that became marked after 1830 as producing "sclerosis" in curriculum and teaching methods.
128.
A. ECAM. P.V. Conseil des Etudes. 3 March 1834.
129.
J.-B. Dumas, Speech at the Funeral of Auguste Perdonnet, 4 October 1867, reprinted in De Comberousse, *Histoire*, A.91–A.92.
130.
Lavallée to Dumas, 13 May 1836. Dumas Papers. Carton 16.
131.
Ministère to Lavallée (draft) 2 June 1836. A.N.17/6770. Although for purposes of administrative control and the collection of the *rétribution universitaire* the school still came under the jurisdiction of the Ministry of Public Instruction, the funds were contributed by the Ministry of Commerce. Legally, the Ecole Centrale was an institution of *enseignement secondaire* and Lavallée was classed as a *chef*

d'institution. In 1838 this anomaly was eliminated with the transfer of administrative oversight to the Ministry of Commerce.

132.
A. ECAM. P.V. Conseil des Etudes. 18 November 1836. After Olivier's resignation, Lavallée was asked to sit in on the sessions of the *Conseil des études*, which, in effect, made him a member; hence he could serve on a committee designated by the *Conseil des études* without seeming subordinate to it.

133.
See table 7.3.

134.
A. ECAM. P.V. Conseil des Etudes. 29 April 1837; Dossier Gabriel Guyot; Pothier, *Histoire de l'Ecole Centrale,* 157–158. Since only the most serious cases reached the *Conseil des études,* where insertion of a student's name in the record was the most severe penalty short of suspension, Pothier's account, based on *Conseil d'ordre* records no longer available, becomes especially valuable here.

135.
A. ECAM. P.V. Conseil des Etudes. 5 August 1838, 10 August 1838, 6 November 1838, 27 November 1838, 16 January 1839, 28 January 1839.

136.
Ibid., 26 January 1839. Pothier repeats this argument as his own in *Histoire de l'Ecole Centrale,* 171.

137.
A. ECAM. P.V. Conseil des Etudes. 18 March 1839.

138.
Ibid., 2 April 1839. See also 27 April 1839, 29 April 1839.

139.
Ibid., 3 April 1839; Dossiers Jean Marsillon and François Guy (the two ringleaders). Pothier, *Histoire de l'Ecole Centrale,* 173–174, incorrectly claims that Olivier was able to save all the students.

140.
A. ECAM. P.V. Conseil des Etudes. 2 December 1839; *Rapport à présenter,* 5-1.

141.
Pothier, *Histoire de l'Ecole Centrale,* 315.

142.
Ibid., 194.

143.
Ibid., 195.

144.
Poncetton, *Eiffel,* 71.

145.
Pothier, *Histoire de l'Ecole Centrale,* 200–202.

146.
Rapport à présenter, 7-8. This was divided among Lavallée and the members of the *Conseil des études.*

147.
Pinet, *Histoire de l'Ecole Polytechnique,* 231.

148.
Ibid., 236.

149.
A. ECAM. P.V. Conseil des Etudes. 21 April 1844; Dossier Antoine-Frédéric Suisse.

150.
A. ECAM. P.V. Conseil des Etudes. 22 November 1840; Mouchelet, *Notice historique*, 10–11.

151.
A. ECAM. Dossier Constant Robaut and Adolphe Cauvet; Lavallée to Augustin Robaut, 30 June 1845, Correspondence File; Registre de l'Ecole, 11 May 1845, 18 November 1846, 19 November 1847.

152.
Association des Elèves de l'Ecole Centrale des Arts et Manufactures, *Règlement*, Title I, Article 3. A. ECAM.

153.
Ibid., Title II, Articles 7–8.

154.
Ibid., Title II, Article 6. Mouchelet, *Notice historique*, 21, describes the faculty's opposition.

155.
In 1848 a letter from a group of Centrale civil engineers in Marseilles congratulating their Parisian colleagues on their success in finally establishing a professional society claimed that attempts to set up such a group had been an "almost annual affair." *Compte-Rendu des travaux de la Société Centrale des Ingénieurs Civils* (March–May, 1848), 10.

156.
"Projet de note pour Monsieur le Ministre de l'Instruction Publique," Dumas Papers, Carton 16. A. Acad. Sci.

157.
See Association *Règlement*, Title I, Article 4.

158.
Mouchelet, *Notice historique*, 22; A–C. Benoit-Duportail, "Notice rétrospective sur la Société des Ingénieurs Civils de France de 1848 à 1886," in Société des Ingénieurs Civils de France, *Annuaire de 1911* (Paris, 1911), 7–8.

159.
These are minimum figures. The *Annuaire de 1911*, 18–19, gives a list of those admitted in 1848, but a comparison with the *Compte-Rendu* and other sources shows that it is incomplete, or though it may list all those who were officially admitted, it does not include all those who played a role in the meetings.

160.
Emile Thomas, *Histoire des Ateliers Nationaux* (Paris, 1848), *passim*.

161.
The consternation of Lavallée, Dumas, Olivier, and the others is evident in all the archival documentation from the period of the Revolution. See, e.g., P.V., Conseil des Etudes. 15 April 1848.

162.
Charles de Freycinet, *Souvenirs, 1848–1878* (Paris, 1892), 1–34; Léon Lalanne (the *Polytechnicien* and Ponts et Chaussées engineer who succeeded Thomas), "Lettres sur les Ateliers Nationaux," *Le National* (14, 15, 19 and 23 July; 6, 16, and 26 August 1848).

Chapter 8

1.

Lavallée, Dumas, Olivier, and Péclet to Ministre de l'Instruction Publique, 2 March 1846. A. N. F17/6770.

2.

Lavallée, in "Industrie" (no date but probably 1847), Dumas Papers, Carton 16. Lavallée opened his contribution to this section with the note "Development of industry more advanced in England and in Germany than in France." The inclusion of Germany at this date is somewhat puzzling, and would seem to undermine the importance of his two principal causes since it is doubtful that either the progress of science or the reduction of social barriers was more advanced in Germany than in France.

3.

Olivier, *Mémoires*, xxii, written in July, 1947.

4.

Joseph Bélanger, in "Prospectus," the preliminary draft of the note for the Ministries of Public Instruction and of Commerce, Industry, and Agriculture (no date but references to the scholarship question indicate mid-1840s). Dumas Papers, Carton 16.

5.

Lavallée, "Industrie."

6.

Olivier, *Mémoires*, xxii.

7.

Lavallée, "Industrie."

8.

J.-B. Dumas, "Draft of a speech on the Ecole Centrale and the scholarships" (January, 1848), Dumas Papers, Carton 16.

9.

Oliver, *Mémoires*, xxii–xxiii.

10.

A ECAM. Dossiers Philippe Picher de Grandchamp, Jean-René Bourgougnon, and Jules Gastelier.

11.

Letter to Minister of Commerce from "directors" of the Ecole Centrale (Draft, 1859), Dumas Papers. In *Les fonctions sociales de l'enseignement scientifique* (Paris, 1971), Monique de Saint Martin argues that scientific education serves at present to "reproduce" the upper strata from which it recruits.

12.

Cf. Roger Priouret, *Les origines du patronat français* (Paris, 1963), 13–56; Lhomme, *La grande bourgeoisie*, 13–29. Jean-Hervé Donnard compares Balzac's portrayal of the accumulation of wealth with other evidence in *La vie économique et les classes sociales dans l'oeuvre de Balzac* (Paris, 1961), especially 249–270.

13.

Rapport à présenter, 1-6.

14.

Belanger in "Prospectus." Dumas later drew a closely similar portrait of French industrial leadership in the years before the establishment of the Ecole Centrale:

"Let us remember how [at that time] most industrial enterprises were founded or directed. Former military men bored with inactivity; bankers looking for somewhere to place their capital; businessmen and owners of factories who had prepared at a faculty of law; and a few rare engineers from the Polytechnique: such were the hands in which rested the apparent direction of manufacturing enterprises governed in reality by a few intelligent but uneducated foremen. So many causes of failure, ruin, and disaster!" Dumas' speech at Cinquantenaire, 19.

15.

Ibid., 18.

16.

In his own section of the "Industrie" memorandum, Walter de Saint-Ange stated the case clearly: "England acquired its superiority in mechanical engineering only by long labor marked by countless fruitless projects and massive sacrifices of capital, which she has doubtless regained, but only because for so many years she was the workshop of the entire world. We French, on the contrary, guided by applied science, can arrive more promptly and more surely at the same goal . . . and soon place ourselves at the same level as our precursors."

17.

Dumas, "Projet de note pour M. le Ministre de l'Instruction Publique" (1852), Dumas Papers, Carton 16. In his fiftieth-anniversary speech, Dumas said that many thought that "because the British could do without theoreticians, so could the French." The difference between the 1829 industrial exposition, which revealed the "most shocking inequality" between the two countries, and its 1878 counterpart, where fifty years of Centrale education were reflected in "achievements that filled him with pride," was for him the best evidence that they had been mistaken. That apprenticeship training for engineers during this period could in fact produce excellent results has now been clearly demonstrated by the work of James Edmonson, *From Mécanicien to Engineer*.

18.

Belanger dealt at greatest length with this question in his "Prospectus."

19.

Rapport à présenter, 2-1. Even Perdonnet made this argument in his own contribution to the "Industrie" memorandum of 1847.

20.

Bruce Sinclair, *Philadelphia's Philosopher Mechanics: A History of the Franklin Institute, 1824–1865* (Baltimore, 1974), 324, finds a similar element of mystification in the parallel claims made by the leaders of that body: "These men gave new meaning to the old idea that technical progress would flow from a union of theory and practice. When Dr. Jones spoke of teaching the principles of science to workingmen, he gave no clear indication of the way a craftsman would actually use them. Neither Dr. Jones nor anyone else who invoked that catch-phrase knew precisely how the union was to be effected. Nor could they, since it represented an ideological stance rather than a program."

21.

De Comberousse, "Cinquantenaire," 42–43.

22.

Rondo E. Cameron, *France and the Economic Development of Europe: 1800–1914* (Princeton, N.J., 1961), 45–61; Charles Kindleberger, "Technical Education and the French Entrepreneur," in Edward C. Carter II, Robert Forster, and Joseph N.

Moody, Eds., *Enterprise and Entrepreneurs in Nineteenth- and Twentieth-Century France* (Baltimore, 1976), 19.

23.

Manuel, *New World*, 401, claims that the "neologism" *industriel* began to appear in Saint-Simon's works in 1817. Neither Saint-Simon nor the Centrale teachers used it consistently in this inclusive sense, however. Orthodox political economists such as Charles Dupin also used the term in the wider sense. Somewhere in the middle of the century, the usage narrowed to its current denotation of "industrialist."

24.

"Program of a Course in Theoretical and Applied Political Economy, 1834–1835" (handwritten draft). Dumas Papers. Carton 16.

25.

Daumas and Jacques, "The Rebirth of Chemistry," 288–289.

26.

Walter M. Simon. *European Positivism in the Nineteenth Century* (Ithaca, 1963), 4–7.

27.

Rapport à présenter, 16-6.

28.

In the same years that Dupin drew 2,000 to his lectures at the Conservatoire, Say drew 50. Robert Fox, "Education for a New Age: the Conservatoire des Arts et Métiers, 1815–1830," in D. S. L. Cardwell, *Artisan to Graduate* (Manchester, 1974), 31.

29.

Priouret, *Origines du patronat français*, 52–81; Bertrand Gille, *Recherches sur la formation de la grand entreprise capitaliste (1815–1848)* (Paris, 1959).

30.

Pothier, *Histoire de l'Ecole Centrale*, 61, 74, 194, 252.

31.

Ibid., 315–316.

32.

Poncetton, *Eiffel*, 71–75.

33.

Rapport à présenter, 6-8.

34.

For example, in Lewis Mumford, *Technics and Civilization* (New York, 1962), 151–211; Sidney Pollard, *The Genesis of Modern Management* (Baltimore, 1968), 189–243; Neil Smelser, *Social Change in the Industrial Revolution* (Chicago, 1959), 180–308; E. P. Thompson, *The Making of the English Working Class* (New York, 1963), 189–313, and his "Time, Work-Discipline, and Industrial Capitalism," *Past and Present* 38 (December 1967), 56–97.

35.

As in Thompson, *Making*, 350–400, *et passim*; and Reinhard Bendix, *Work and Authority in Industry: Ideologies of Management in the Course of Industrialization* (New York, 1963), 46–85.

36.

Landes, *Unbound Prometheus*, 21–24; Bendix, *Work and Authority*, 99–116, *et passim;* David McClelland, *The Achieving Society* (New York, 1961), 259–300, 336–390; Max Weber, *The Protestant Ethic and the Spirit of Capitalism* (New York, 1958), 155–184.

37.
We can only guess at how a student at Centrale was affected when he left in rebellion or fatigue or was dismissed as a poor student: Such men seem to leave memoirs even less often than successful engineers. Nor do they begin strike movements, and thus appear collectively in historical documents. What is most bitterly paralyzing about scholastic failure is that it seems an individual failure.

38.
Dumas, "Cinquantenaire" speech, 21.

39.
Ibid., 22.

40.
Rapport à presenter, 4-5.

41.
Lavallée, "Industrie."

42.
Lavallée, "Industrie."

43.
Walter de Saint-Ange, "Industrie."

44.
Idem.

45.
Idem.

46.
A perceptive study of the attempt to give "moral effect" to the stylized instruction in mathematics and the physical sciences offered to "worthy" (i.e., upwardly mobile) artisans at the British Mechanics' Institutes so much admired by Liberals like Charles Dupin has been published by Steven Shapin and Barry Barnes, "Science, Nature, and Control: Interpreting Mechanics' Institutes," *Social Studies of Science* 7:1 (February 1977), 31–74.

47.
First-hand accounts of the role of the *Centraux* in running National Workshops are Emile Thomas, *Histoire des Ateliers Nationaux* (Paris, 1848), the most extensive source; Edward Delessard, *Souvenirs de 1848: L'Ecole Centrale aux Ateliers Nationaux* (Paris, 1900), a tribute to the idealism of the class of 1848 by one of its members; and Edouard Jaime, *Souvenirs de 1848 à 1871* (Versailles, 1872), by a playwright who served as Thomas's principal lieutenant.

48.
Prospectus de l'Ecole Centrale des Arts et Manufactures (1831), 2.

49.
Pothier, *Histoire de l'Ecole Centrale,* 69.

50.
Joseph Ben-David, *The Scientist's Role in Society* (Englewood Cliffs, N.J., 1971), 78, defines the "scientistic" movement as "a group of people who believe in science (even though they may not understand it) as a valid way to truth and to effective mastery over nature as well as to the solution of problems of the individual and society."

51.

In his introduction to *The Doctrine of Saint-Simon: An Exposition*, Georg Iggers discusses the elaborate but short-lived efforts of Enfantin's group of Saint-Simonians to make a kind of "totalitarian" synthesis between the thought of De Maistre, scientism, and socialist industrialism. Given the differing political philosophies of the Centrale teachers, a more pluralistic form of educational culture might have emerged. For the argument that this occurred with the disciples of Comte, see John Eros, "The Positivist Generation of French Republicanism," *Sociological Review* III (1955), 255–277.

52.

The *Journal des connaissances usuelles*, with contributors such as Charles Dupin, Anselme Payen, and Clément Désormes, was only the most distinguished of a large number of such publications that flourished in this period. Information about other examples may be found in Eugéne Hatin, *Bibliographie historique et critique de la presse périodique française* (Paris, 1866).

53.

Lavallée, "Industrie."

54.

L. Pearce Williams, "Science, Education, and the French Revolution," *Isis* 44 (December 1953), 311–330.

55.

Mayer, "Science," 271–273, and Maurice Caullery, *La science française depuis le XVIIe siècle* (Paris, 1933).

56.

Ecole Polytechnique, *Livre du centenaire*, I, 86–87.

57.

The most recent studies are C. Rod Day, "Technical and Professional Education in France: The Rise and Fall of l'Enseignement Secondaire Spécial, 1865–1902," *Journal of Social History* (Winter, 1972–1973), 177–201 and Viviane Isambert-Jamati, "Une réforme des lycés et collèges: Essai d'analyse sociologique de la réforme de 1902," *Année sociologique* (1971), 9–60.

58.

Mouchelet, *Notice historique*, 46.

Appendix

1.

Guillet, *Cent ans*, 186.

2.

Almanach de l'Université royale, 1846; *Annuaire de l'Instruction Publique*, 1881.

3.

And also the size of the sample: researchers are allowed to consult only twelve dossiers a week.

4.

France. Ministère de Commerce. Service du Recensement. *Résultats statistiques du recensement général de la population effectué le 24 mars 1901*. Vol. 1 (Paris, 1904), 7–8. See also Michel Huber, et al., *La population de la France* (4th ed.), 9, and J. C. Toutain, "La population de la France de 1700 à 1959," *Cahiers de l'I.S.E.A.* 133.3 (January 1963), especially the discussion of pre-1851 sources, 103–104.

5.
Institut National de la Statistique et des Etudes Economiques, *Code des catégories socio-professionnelles* (Paris, 1954), I, 1–2.

6.
The argument that technical change causes this convergence is concisely stated in Pierre Naville, "Technical Elites and Social Elites," *Sociology of Education* 37:1 (Fall 1963), 27–29.

7.
Otis Dudley Duncan, "A Socioeconomic Index for all Occupations," in Albert J. Reiss, *Occupation and Social Status* (New York, 1961), 124.

8.
Donald J. Treiman, "The Validity of the 'Standard International Occupational Prestige Scale' for Historical Data," paper delivered at the Conference on International Comparisons of Social Mobility in Past Societies, Princeton, New Jersey, Institute for Advanced Studies, 15–17 June 1972, p. 4.

9.
Fritz K. Ringer, *The Decline of the German Mandarins* (Cambridge, Mass., 1969).

10.
Treiman, "Validity," 7.

11.
Georges Dupeux, *La société française, 1789–1970* (Paris, 1972), 31.

12.
Statistique Générale de la France. *Résultats statistiques du dénombrement de 1891* (Paris, 1894), 305.

13.
This can be seen perhaps most clearly in Massé, *Pour choisir une carrière*, and Paul Jacquemart, *Professions et métiers; Guide pratique à l'usage dés familles et de la jeunesse pour le choix d'une carrière*, 2 Vols. (Paris, 1892). In his *Du choix d'une carrière* (Paris, 1902), 275–276, the noted Third Republic historian and sometime diplomat Gabriel Hanotaux describes three levels of administrative employment that correspond closely to the current *cadres supérieurs, moyens,* and *inférieurs* classification, but he calls these *directeurs, agents,* and *employés.*

14.
Jean-Pierre Chaline, "Les contrats de mariage à Rouen au XIXe siècle," *Rev. d'hist. écon. et sociale* 2 (1970), 258–261. The importance of marriages in helping Parisian shopkeepers to avoid *déclassement* is also examined in the detailed intergenerational study of *boutiquier* families by Jean Le Yaouanq, "La mobilité sociale dans le milieu boutiquier parisien au XIXe siècle," *Le mouvement social* 108 (July 1979), 89–112.

15.
The number of cases showing a change between the son's birth and his entry to Centrale was too small to be of much value, consisting often of predictable shifts such as promotion from *lieutenant* to *capitaine* or from medical student to physician.

16.
Adéline Daumard, "Une référence pour l'étude des sociétés urbaines en France aux XVIIIe et XIXe siècles: Projet de code socio-professionel," *Rev. d'hist. mod. et*

contemp. X (July–September 1963). The suggested code seems to have survived criticism pretty much intact. See Jean-Yves Tirat, "Problèmes de méthode en histoire sociale," *Rev. d'hist. mod. et contemp.* X (July–September 1963), 211–218; Adéline Daumard and François Furet, " 'Problèmes de méthode en histoire sociale: Réflexions sur une note critique," *Rev. d'hist. mod. et contemp.* XI (October–December 1964), 291–298; J. Dupàquier, "Problémes de la codification socio-professionnelle," in *L'histoire sociale: Sources et méthodes,* 157–182. Each distinguishable entry within the seventy-nine categories has also received a code number. For example, cases entered on the computer tape having one or more of the "social-professional category" fields (one for each of the three possible types of source) filled with a 74 (proprietors of industrial establishments) also have three "social-professional individual code" fields. Sample entries: 151 (*propriétaire de forges*), 156 (*constructeur de navires*), 158 (*filateur*). By this method, I hope to correct and modify my categories and redesignate individual cases most easily. Technically, of course, the "social-professional category" field is unnecessary, but the limits of the SPSS RECODE program militated against separately collapsing categories for each run to produce the seventy-nine groups, which would usually require further RECODE collapses to produce usable tables.

17.

Theodore F. Zeldin, *France: 1848–1945* (Oxford, 1973), 43–52.

18.

Shinn, *L'Ecole Polytechnique,* 70–71. *Négociant,* literally translated, means "wholesaler," but, as Daumard points out in "Projet de code," 205, in the nineteenth century, merchants with large businesses used the term whether they sold to retailers or directly to the public.

19.

Shinn, *L'Ecole Polytechnique,* 251.

20.

Ibid., 70–71. On p. 252, however, Shinn presents without demurral a summary of Georges Dupeux's study of the Loir-et-Cher that describes *industriels* as one of "several social groups within the upper bourgeoisie."

21.

When one compares the distribution of estate sizes within the high functionaries group with that within the *négociant-industriel* group for the years 1820, 1847, and 1911, the latter group, in fact, appears slightly wealthier. Adéline Daumard, *Les fortunes françaises au XIXe siècle* (Paris, 1973), 220–221.

22.

Ibid., 230–235.

23.

Ringer, *Education and Society,* 29–30.

24.

See, for example, Jean Lhomme, *La grande bourgeoisie au pouvoir (1830–1880)* (Paris, 1960), *passim;* Adéline Daumard, *La bourgeoisie parisienne de 1815 à 1848* (Paris, 1963), 25–27.

25.

Shinn, *L'Ecole Polytechnique,* 66.

26.

Dupeux, *La société française,* 131.

27.
Chaline, "Les contrats de mariage," 268.

28.
Statistique Générale de la France. *Résultats Statistiques du dénombrement de 1891* (Paris, 1894), 311.

29.
See Daumard, "Projet de code," 206. The *cas typiques* are taken from those used in post-World War II census codes. See also "L'enquête par sondage sur l'emploi," *Bull mensuel de statistique. Supp.* (April–June 1953), 5, which gives as *"fonctionnaires moyens et assimilés"*: *controleurs des finances, commis et adjoints techniques des administrations générales, des P.T.T.*

30.
Daumard, "Projet de code," 206, also seems to suggest this possibility.

31.
In addition to the career guides already cited, Paul Bastien, *Les carrières commerciales, industrielles, et agricoles* (Paris, 1906), 53–76, has an informative discussion of the *employés de commerce* category.

32.
Daumard, *La bourgeoisie parisienne*, 74–76.

33.
Christian Baudelot, Rober Establet, and Jacques Malemort, *La petite bourgeoisie en France* (Paris, 1974), 21–28.

34.
Francois Goguel, "Six Authors in Search of a National Character" in Stanley Hoffman et al., *In Search of France* (Cambridge, Mass., 1963), 369. For the sons of *instituteurs* who became *professeurs*, see Gérard Vincent, "Les professeurs du second degré au début du XXe siècle; essai sur la mobilité sociale et la mobilité géographique," *Le mouvement social* 56 (April–June 1966).

35.
In the case of the artisans, identification and assignment to sectors were based in part on classifications in Charles Tilly, *The Vendee* (New York, 1964), 348–349.

36.
Daumard, *La bourgeoisie parisienne*, 118–119. It is probably a mistake to try to summarize the complexity of Daumard's categorizations by some statement about whether or not she views artisans as part of the *classes populaires*. At times she seems to include them as part of the *petite bourgeoisie*, at others (see p. 255) she discusses how the *petite bourgeoisie* is "closely tied" to *"milieux populaires et artisanaux,"* but nevertheless distinct from it.

37.
Compare the entries of *artisan* and *ouvrier* in Littré's *Dictionnaire of 1881* and Robert's *Dictionnaire alphabétique et analogique de la langue française* of 1960. Robert refers to a legal definition of *artisan* established in law of 26 July 1925. Cf. George Rudé, *The Crowd in History* (London, 1964), 195–196.

Bibliography

I. Primary Sources

A. Archival

Archives of the Ecole Centrale des Arts et Manufactures

When I first visited the Ecole Centrale, the school had just begun to transform a room where records were stored into an organized archive. For the purposes of the present study, the student dossiers described in chapter 3, the bound registers containing "orders of the day," and the *procès-verbaux* of the *Conseil des études* were the archives most valuable holdings. When I last visited the school at its new location in Chatenay-Malabry, the documentation was easily available and in good order. The *Registre*, and even the *procès-verbaux*, gave highly summarized accounts: only rarely can one detect the full conflict of views behind a controversial decision. The school's correspondence files are a rather spotty collection, not cross referenced to letters in individual dossiers. There are also prospectuses, textbooks, course attendance rosters, a few student notebooks, a selection of student projects, research laboratory reports, and works by members of the faculty and other contemporary scientists and engineers. In essence, in the years before the Ecole Centrale became a state-supported institution it was a place where crucial decisions emerged informally in daily contacts between Lavallée and key faculty members such as Dumas, Péclet, Olivier, Perdonnet, and Walter de Saint-Ange. This would also explain the rarity of letters between Dumas and the others in the Dumas Papers. The more formal written communications became common only when relations became strained, as between Dumas and Olivier.

Archives of the Académie des Sciences

The papers of Jean-Baptiste Dumas constitute one of the largest collections. Cartons 16, 17, 18, and 28 were of relevance to the present study. A number of individual documents should be mentioned in view of their importance to the present study and self-contained nature: (1) The draft-printed but unpublished history of the school mentioned above. The annotations indicate that Dumas, the only surviving founder at the time it was written, was the principal author. The *Rapport à présenter* seems to have been written as a justification and explanation of the school at a time when it was especially vulnerable, having been without a director for three years. (2) A typescript biography of Dumas by his grandson, General Jean-Baptiste Dumas. (3) A handwritten draft of Dumas' 1847 report on scientific education in the secondary schools, somewhat longer than the published version.

For the statistical study of careers and social origins referred to in chapter 3, the cartons of the following members of the Académie were consulted: Emile-Hilaire Amagat, Jacques Babinet, Antoine-Jérôme Balard, Antoine Becquerel, Henri Becquerel, Auguste Béhal, Paul Bert, Daniel Berthelot, Marcelin Berthelot, Claude Berthollet, Gabriel-Emile Bertrand, Emile Bourquelot, Edmond Bouty, Henry Braconnot, Edouard Branly, Marcel Brillouin, Charles Cagniard de la Tour, Auguste Cahours, Jean Chaptal, Michel-Eugène Chevreul, Alfred Cornu, Aimé Cotton, Pierre Curie, Gaston Darboux, Jean-Pierre-Joseph Darcet, Henri Debray, Pierre-Paul Dehérain, Stéphane-Michel Delépine, Paul Desains, Cesar Despretz, Nicolas Deyeux, Alfred Ditte, Jean-Marie-Constant Duhamel, Pierre-Louis Dulong, René-Joachim-Henri Dutrochet, Charles Fabry, Hippolyte Fizeau, Joseph Fourier, Edmond Fremy, Augustin Fresnel, Charles Friedel, Emile-Justin-Armand Gautier, Louis Gay-Lussac, Désiré Gernez, Pierre-Simon Girard, Octave Gréard, Edouard Grimaux, Albin Haller, Emile Haüy, Louis-Felix Henneguy, Jules Jamin, Emile Jungfleisch, Paul Langevin, Albert-Auguste Cochon de Lapparent, Henry Le Chatelier, Georges Lemoine, Louis Liard, Gabriel Lippmann, Louis-Alexandre Mangin, Eleuthère Mascart, Artheme-Camille Matignon, Henri Milne-Edwards, Henri Moissan, Charles Moureu, Jules Pelouze, Jean Perrin, Denis Poisson, Alfred Potier, Claude Pouillet, Joseph Proust, Edouard Quenu, Henri-Victor Regnault, Pierre-Jean Robiquet, Alexis-Marie de Rochon, Henri-Etienne Sainte-Claire-Deville, Felix Savart, Paul Schutzenberger, Georges Simon Sekullaz, Jacques Thénard, Louis-Joseph Troost, Georges Urbain, Nicolas-Louis Vauquelin, Paul Villard, Jules Violle, Charles-Adolphe Wurtz. Information about the origins of these and the remaining scientists in the sample, as well as the *secondaire* teachers, was obtained by mailing questionnaires to the Services des Actes Civils of the individuals' birthplaces.

Archives Nationales

The sample of science teachers was based largely upon the personal dossiers contained in the F17 20,000 series. The dossiers of certain professors, such as Binet de St. Preuve, also proved useful for the discussion in the main text. I shall not list the individual dossier numbers.

The following files were also useful for the present study:

F14 9999. Contains a list of father's professions for scholarship students at the *écoles d'arts et métiers* in the 1830s. Officers as well as enlisted ranks appear on the list.

F17* 1852–1853. *Procès-verbaux* of the *Conseil royal de l'Université*, 1847–1848.

F17* 78486. Report on an experiment in mathematics teaching at Louis-le-Grand.

F17 1559. Lists of books authorized for use in teaching, 1802–1850.

F17 2476. Library holdings of the *collèges royaux* and *collèges communaux*

F17 2782–2807. Reports on particular textbooks, 1837–1870.

F17 6770. Ecole Centrale des Arts et Manufactures. Correspondence and other documents, 1829–1848. *Ecoles commerciales*: Various documents, 1821–1829.

F17 6808–6816. Inspectors-general; reports from 1810 to 1873.

F17 6833. Report on secondary education of 1 January 1848 by Salvandy. Other reports on secondary education for period after 1848.

F17 6876–6877. Officially prescribed programs of study, 1840–1899. Also includes requirements for admission to Ecole Centrale and other special schools.

F17 6888. Library holdings of secondary schools.

F17 6894. The final resting place for most of the surviving documents dealing with science teaching in the secondary schools from 1809 to 1848.

F17 6931–6951. Alphabetical listing of former functionaries in the educational system, but does not systematically record place of birth or date of receiving *grades*.

F17 7040–7094. Information relating to the *concours d'agrégation*, 1821–1866.

F17 8701–8710. *Enseignement spécial*, general documents, 1847–1887. Includes early plans.

F17 9096. Miscellaneous, but includes reports of inspectors-general, including the report of 1868 by Charles Glachant that reads as if it were written by Michel Crozier to prove his points about the powerlessness of the center in a French bureaucracy: opposition from the teachers killed *bifurcation*.

F17 11706–11708. *Ecoles professionnelles*, 1843–1886.

Archives of the Académie de Paris

Very little here of relevance to the present study. Files A.J. 16 64 and 65 give information on candidates and performances in the scientific *agrégations.* There is also an unnumbered file of examination questions with answers for the baccalaureat examinations, 1843–1844.

B. Official Publications

Association Polytechnique. (and Association Philotechnique) *Distribution des prix aux élèves réunis des deux Associations Polytechnique et Philotechnique.* Paris: Paul Dupont. 1857, 1858, 1859. A useful series of pamphlets containing personal accounts of the early history of the two associations.

———. *Documents pour servir à l'histoire des cours de l' Association Polytechnique.* Paris: Dupont, 1862.

———. *Rapport sur la fusion des Associations Polytechnique et Philotechnique.* Paris: Dupont, 1857. A circular by Perdonnet prefaces this report, written by Lionnet, first president of the Association Philotechnique.

Bureau de la Statistique Générale. *Résultats Généraux du Dénombrement de 1872.* Nancy: Berger-Levrault, 1874. Used for the statistical investigations in Chapter 7.

———. *Résultats généraux du dénombrement de 1876.* Paris: Imprimerie Nationale, 1878.

Chambre des Députés. *Enquête sur l'enseignement secondaire.* 6 Vols. Paris: Chambre des Députés, 1899. Crucial source for many aspects of the history of secondary education as well as for the background to the reforms of 1902.

Dion, A. *Recueil complet sur la législation de l'enseignement secondaire.* Paris: Berger-Levrault, 1922.

Dumas, Jean-Baptiste. *Rapports addressés à M. le Ministre de l'Instruction Publique. 20 June 1846.* Paris: Ministère de l'Instruction Publique, 1846. Dumas' report on degree programs in the Faculty of Sciences, in which he advocates the introduction of more applied sciences and the creation of a *licence ès sciences mécaniques.*

———. *Rapport sur l'enseignement scientifique dans les collèges, les écoles intermédiaires, et les écoles primaires.* Paris: *Ministère de l'Instruction Publique,* 1847. Discussed in text. This report was also published in the *Journal général de l'instruction publique,* 19 May 1847.

Ecole Centrale des Arts et Manufactures. *Prospectus de l'Ecole Centrale des Arts et Manufactures.* Paris: Bachelier, 1829–1848. One of the most important sources for changes in the school's curriculum and self-image.

Ecole Polytechnique. *Annuaire de l'Ecole Royale Polytechnique.* Paris: Bachelier, 1818–1848.

Galeron, B. *Code spécial des établissements particuliers d'instruction secondaire.* Paris: Mesnage, 1846. Details the inspections, certification standards, and taxes imposed on private institutions.

Gobron, Louis. *Législation et jurisprudence de l'enseignement public et de l'enseignement privé en France.* Paris: L. Larose, 1896. Standard reference guide.

Institut National de la Statistique et des Etudes Economiques. *Annuaire statistique de la France: 1954.* Paris: P.U.F., 1955.

————. *Annuaire statistique de la France: 1966. Résumé rétrospectif.* Paris: INSEE, 1966. Invaluable.

————. Code des catégories socio-professionnelles. Paris: Imprimerie Nationale, 1954, 1962, 1969 editions. Used to prepare the appendix.

————. *Recensement général de la population; état civil et activité professionnelle de la population présente.* Paris: P.U.F., 1949. Introduction used for the preparation of the appendix.

————. *Recensement général de la population. Population légale.* Paris: P.U.F., 1956. Introduction used for the appendix.

Institution of Civil Engineers. *The Education and Status of Civil Engineers in the United Kingdom and Foreign Countries.* London: I.C.E., 1870. Compiled from questionnaires and their own investigations. Contains some statistics about the Ecole Centrale unavailable elsewhere.

Jourdain, Charles. *Rapport sur l'organisation et les progrès de l'instruction publique.* Paris: Imprimerie Nationale, 1867. Prepared for the Exposition of 1867 by Jourdain, a *chef de division* at the Ministère de l'Instruction Publique. Relies heavily, but not entirely, on Villemain for the pre-1843 period.

Ministère de l'Agriculture, du Commerce, et des Travaux Publics. *Enquête sur l'enseignement professionnel.* 2 Vols. Paris: Imprimerie Impériale, 1864–1865. Indispensable.

Ministère de l'Instruction Publique (This heading includes all designations for the State's educational administration.)

————. *Almanach de l'Université royale de France et des divers établissements d'instruction publique.* Paris: Hachette. Issued annually, 1816–1848. Introductions contain information on basic institutional structure and fundamental regulations.

————. *Annuaire de l'instruction publique et des beaux-arts.* Paris: Delalain Frères, 1881, 1883. Used for compilation of *secondaire* science teachers.

————. *Programmes officiels de l'enseignement spécial dans les lycées et collèges de l'Université, publiés conformément à l'arrêté du 17 septembre 1849.* Paris: Delalain, 1849.

————. *Recueil de lois et règlements concernant l'instruction publique, depuis l'édit de Henri IV en 1598 jusqu'à ce jour.* 8 Vols. Paris: Bachelier. 1814–1828. Indispensable.

Ministère de la Guerre. *Rapport sur l'enseignement de l'Ecole polytechnique adressé à M. le Ministre de la Guerre par la Commission mixte nommé en exécution de la loi du 3/8/1850.* Paris: Imprimerie Nationale, 1850.

Ministère du Commerce, de l'Industrie, des Postes et des Télégraphes. *Résultats statistiques du dénombrement de 1891.* Paris: Imprimerie Nationale, 1894. Consulted for the appendix.

✓ ———. *Résultats statistiques du dénombrement de 1896.* Paris: Imprimerie Nationale, 1899 [!].

———. *Résultats statistiques du recensement général de la population.* Vol. I. Introduction. Paris: Imprimerie Nationale, 1904.

———. *Resultats statistiques du recensement général de la population.* Vol. 1. Introduction. Paris: Imprimerie Nationale, 1913.

Rendu, Ambroise. *Code universitaire.* Paris: Hachette, 1835. The publication which helped to establish Rendu as the "living law" of the *Université.*

Villemain, Abel. *Rapport au Roi sur l'instruction secondaire.* Paris: Dupont, 1843. The first official statistical investigation since the *Université* was established although Kilian's earlier work made some of the same points.

Wissemans, Albert, ed. *Code de l'enseignement secondaire.* Paris: Hachette, 1906. Standard reference. More useful than Gobron for *enseignement secondaire.*

C. Periodicals

Annales des ponts et chaussées. 1841–1850. Two separate series were published, the *Mémoires et documents* and the *Lois, ordonnances, etc.* By the time the *Annales* began publication, the *Ponts* engineers were clearly interested in railroads although the project reports in the first series show that their previous passion for canal building did not immediately die away. In any case, this publication is an indispensable source for watching the growth of the organization that by the end of the nineteenth century, according to Jean-Claude Thoenig, "held all the infrastructures of the country in its hand."

Bulletin de la Société d'encouragement pour l'industrie nationale. 1802–1850. The official organ of the patronage society whose role in French industrialization has yet to be fully assessed.

Bulletin universitaire. 1828–1848. Serves as a more regularly published continuation of the *Recueil . . . 1598* mentioned above.

Journal de l'Ecole Polytechnique. 1799–1848. Really more a publication series for scientific memoris than a journal. Widespread circulation throughout *l'Université.*

Journal des chemins de fer. 1842–1850. A good source for following both the rivalry and the cooperation between *Ponts* and *ingénieurs civils.* Perdonnet had access to its columns, as did spokesmen for dissatisfied *conducteurs* seeking entry to the corps.

Journal des économistes. 1842–1849. The main organ for the propagation of the French version of political economy and its transformation in the 1840s. In addition to full-length articles on both theoretical and practical problems in economics, it also carried shorter notes and notices in which it commented upon matters such as the reform of the Ecole Polytechnique.

Journal du génie civil, des sciences, et des arts à l'usage des ingénieurs, constructeurs de vaisseaux, des ponts et chaussées, des mines et mécaniciens; des architectes, des sculpteurs des peintres, des entrepeneurs de maçonnerie; de charpente, de serrurerie, de peinture et de tous les artistes qui contribuent par leurs connaissances aux constructions civiles. 1828–1831, 1846–1847 (numbered continuously despite the fifteen-year gap). A curious journal. On the one hand, a typical technological compendium of the day, fat volumes of articles on the full range of subjects relevant to the professions described above. One can get a good idea of the occupational and intellectual matrix in which the French *ingénieur civil* appeared. On the other, the personal creation of its *"directeur de l'administration,"* Alexandre Corréard, whose brother Charles was a graduate of the Polytechnique. Despite the breadth of professions in the intended readership, all but one of the thirty-three men listed as collaborators were either engineers, *savants*, or architects.

Journal des connaissances usuelles et pratiques. 1825–1837.

Société des Ingénieurs Civils (Originally, Société Centrale des Ingénieurs Civils). *Mémoires et compte-rendu des travaux de la Societé des Ingénieurs Civils.* 1848–1919 (First issues carry only title *Compte rendu des travaux de la Société des Ingénieurs Civils*). After 1849 this journal confined itself largely to the nuts and bolts of its daily professional activities. Education appeared again as a subject only during World War I. This journal is still an important source of biographical information through the detailed *notices nécrologiques* that it regularly carried.

D. Contemporary Works and Published Document Collections

Arago, François, *Gaspard Monge.* Paris: Firmin-Didot, 1853. The last important biography of Monge by one of his students. A portrait of an inspiring teacher and scientist. Arago attributes to Monge the creation of the *chefs de brigade* system.

―――. *Oeuvres complètes.* Vol. XII: *Mélanges.* Paris: Gide, 1859. Contains an important essay on the organization of the Ecole Polytechnique and several general essays on education.

Archives parlementaires: Recueil complet dès débats législatifs et politiques des chambres françaises de 1800 à 1860. Paris: Librairie Administrative de Paul Dupont, 1864–1867. Indispensable for the Napoleonic period.

Audiganne, A. "L'enseignement industriel en France." *Revue des deux mondes* (June 1851), 860–893. Men no longer follow in the paths of their fathers. The burgeoning of ambition makes industrial education an urgent matter of *salut social.* Yet if you ask "a manufacturer, a merchant, or a worker who has found financial success what he dreams of for his son, you will always get the same answer: a so-called liberal profession or a place in the government." Considers first institutions of industrial education abroad, especially in Britain, where, in his view, "artisans and workers" are taught applied sciences and useful arts. Suggests, nevertheless, that much industrial training in Britain is descended from education for the poor, and that, like in France, this gives it a social stigma. But we must now realize that industrial education has a much larger clientele. Surveys French institutions, starting with the Conservatoire. Defends the *écoles d'arts et métiers*, which he says are designed to train "skilled workers," against charges that their graduates do not pursue industrial careers. Claims that more than half of the graduates become *ajusteurs, forgerons, menuisiers, fondeurs,* or *mécaniciens.* Among the remainder "a certain number" become *conducteurs* in the *ponts et chaussées.*

Bastiat, Frédéric. *Baccalauréat et socialisme.* Paris: Guillaumin, 1850. Bastiat the political economist has become Bastiat the deputy to the National Assembly. He attacks the State monopoly on the granting of degrees. The result of the *baccalauréat* system is that "the youth of the nation calculate with mathematical precision what they have to learn and what they can ignore." Classical studies suffer just as much as scientific studies.

Bastide, Jules. *De l'éducation publique en France.* Paris: Hetzel, 1847.

Bigelow, Jacob. *Elements of Technology.* Boston: Hilliard, 1829. Demonstrates an ambiguity about technology as the application of science that closely resembles that of many French commentators.

Bobin, A. *Questions importantes concernant les juenes gens que l'on destine à l'Ecole Polytechnique.* Paris: Bachelier, 1842. Bobin was the official in charge of the *Bureau des écoles militaires* at the Ministère de la Guerre. He claims that members of the *Université* are about to open a campaign to require the *baccalauréat-ès-lettres* for all those admitted to the Polytechnique, a move he violently rejects.

Brothier, Léon. *Du parti social, exposition des principes économiques et politiques devant servir de base à ce nouveau parti.* Paris: "Chez les principaux libraires," 1839. The current parties are seized with lassitude and ignorant of the social question. So a new party must be formed in order to create "a general impulse that impassions the masses for industry just as before they were impassioned for war." Brothier then sets out a complete social, economic, and political program for this party he hopes to form. The educational sections are a hodgepodge of ideas about education for an industrial society, some of which resemble those of the engineers or Dumas, some of which seem original, as well as bizarre. To reduce the opportunity cost of education to child factory workers, he would have only two hours a day of primary instruction, supplemented by lessons from a teacher who would tour the factories and workshops. The *écoles d'arts et métiers* suffer from being totally isolated from the primary system. They also have another problem: they mix together theory and practice. This produces men who are "not practicians enough to be workers but not theoreticians enough to be engineers." He proposes a four-stage hierarchy of institutions linked by generous scholarships. In the second level, the *collèges d'arrondissement,* "bourgeois education [literary studies] and popular education will come into contact." He is hostile to classical studies, however, because "the dead languages will

Bibliography

never help one become a good engineer or a good farmer." One of the most pressing problems is giving a suitable education to *ingénieurs civils*. Lately the title has become much abused, with all sorts of men claiming the title, the use of which should be regulated by law, just as are those of *avocat* and *médecin*. He proposes to educate the new *ingénieurs civils* in four separate "higher industrial schools" roughly on the level of the Polytechnique.

Bugnot, Yves Delphis E. *De l'Ecole Polytechnique, dans ses rapports avec les services publics qu'elle alimente.* Paris: Goeury, 1837.

Busset, F-C. *De l'enseignement des mathématiques dans les collèges.* Paris: Chamerot, 1843. Busset was a topographical engineer. This is an excellent example of the attempt by engineers and their allies, still very much on the defensive, to make mathematics an accessible subject and, in this case, to argue both for its utility and its cultural value. Pays tribute to both Lacroix and Lazare Carnot.

Cahour, Arsène. *Des études classiques et des études professionnelles.* Paris: Poussielgue Rusand, 1852. One of the rare works by a Jesuit on the question of industrial education. Admits that classical studies are not for everyone, but essentially comes to their defense: "Literary education, as we have seen, cultivates all the faculties of the soul; mathematics neglects two, sensitivity and imagination . . . and even one's intelligence acquires less breadth in the hands of geometers and algebraists than in the hands of *littérateurs philosophes*."

Charton, Edouard. *Guide pour le choix d'un état ou dictionnaire des professions.* Paris: Lenormant, 1842. Indispensable source for the study of early-nineteenth-century social stratification. In the preface, he explains why his guide is now needed, giving a rather overdrawn picture of the possibilities for upward mobility: "Under the Old Regime . . . birth, law, custom, paternal authority, public and private experience—all this combined to limit each man's horizon and to restrict uncertainty. The greatest problem arose not from an excess of choices but rather in the frustration of aptitudes caught in circles which held them captive. In such a situation a dictionary of professions might have aroused a certain curiosity as a tableau of society or a study of mores, but it would have been of little use as a practical guide. Today, the freedom of choice is unlimited. Each citizen has the right to aspire to everything; and his right is not absolutely a fiction: all that is needed to exercise it is the attainment of a level of instruction which is within the means of a considerable minority, if not yet of the majority. At this point of departure, which is like a second birth, the chances are almost the same for everyone, and, lacking wealth, one can succeed by study, talent, and perseverance." In 1848 Charton assisted Hippolyte Carnot in his attempted reforms at the Ministère de l'Instruction Publique.

Chasles, Michel. *Aperçu historique sur l'origine et le développement des méthodes en géométrie.* Brussels: Hayez, 1837. A prominent geometer pays tribute to Monge and to Lacroix's *Essais sur l'enseignement*.

Chevalier, Michel. *Cours d'économie politique.* Vol. I (1842), Vol. II (1844), Paris: Capelle. Lectures at the Collège de France by the former Saint-Simonian and later advisor to Napoleon III. A course in political economy at the Ecole Centrale might have drawn its material from these lectures although Chevalier's enthusiasm for state entrepreneurship might not have captivated men like Dumas. Chevalier regularly recommended students for admission to Centrale.

————. "Politique industrielle et système de la Méditerranée." Paris, 1832. One of a series of "Religion saint-simonienne" articles that appeared in *Le Globe*. Chevalier advocates using the army to educate and industrialize the country. Each regiment would became an école d'arts et métiers, men would flock to it, and the army would be transformed into a *corps industriel* that could only assure peace.

————. "Sur l'instruction secondaire." *Journal des économistes*. (April, 1843), 23–57. Chevalier's analysis of Villemain's report to the King.

Cochut, A. "De l'instruction publique en France," *Revue des deux mondes* (15 September 1838) 838–855. Defense of classical studies and nonutilitarian education against Girardin's attacks.

Comte, Auguste. *Cours de philosophie positive: Discours sur l'esprit positif*. New edition with introduction and commentary by Charles Le Verrier. Paris: Hachette, 1949. Comte's attempt to discover the interrelationships of the sciences and how they could best be studied.

Condorcet, Caritat de. *Esquisse d'un tableau historique des progrès de l'esprit humain, suivie de fragments de l'histoire de la quatrième epoque, et d'un fragment sur l'Atlantide*. Paris: Au bureau de la Bibliotheque choisie, 1829. Condorcet's testimony to the faith of the Enlightenment, published in the same year as Comte's *Cours* and the founding of the Ecole Centrale. In this Restoration edition, the fragment of the *Atlantide* contains the following passage: "One may still fear a kind of rivalry that exists among the sciences. It can only be in the interest of truth that they unite because all are more or less interdependent. One can break the chain nowhere without harming the broken portions." And this included the social sciences. Comte's emphasis, on the other hand, was on the hierarchical organization of the sciences and the precise order in which they should be studied.

Considérant, Victor. *Théorie de l' éducation naturelle et attrayante*. Paris: Ecole Societaire, 1844. A *Polytechnicien*'s Fourierist utopia in the guise of a treatise on education.

✓ Cournot, Antoine. *Des institutions d' instruction publique en France*. Paris: Hachette, 1864. Work of major importance by a mathematician and inspector-general of public instruction.

Cousin, Victor. *Défense de l'Université et de la philosophie*. Paris: Joubert, 1844. Principally relates to the attacks on the *Université* stemming from the "freedom of teaching" issue.

————. *Du vrai, du beau, et du bien*. Paris: Didier, 1873. A sampling of the thought of the July Monarchy's official academic philosopher, one of the key figures in the transformation of *honnêteté* into *culture générale*. Cousin's lectures on the subject began in 1819.

————. *Mémoires sur l'instruction secondaire dans le royaume de Prusse*. Paris: Levrault, 1837. Cousin likes the *Realschulen*, but only for the lesser breeds unable or unwilling to appreciate the sublimities of the Good, the True, and the Beautiful.

Crosland, Maurice, ed. *Science in France in the Revolutionary Era, As Described by Thomas Bugge.* Cambridge, Mass.: MIT Press, 1969. The diary of a Danish astronomer who visited Paris at the turn of the century. Principally of interest to historians of science, especially astronomy and mathematics. Relevant to the present study for its portraits of the *savants* as teachers.

Documents pour servir à l'histoire de l'Association polytechnique. Paris: Chaix, 1862. Lists of *cours du soir* for the Association Philotechnique and the Association Polytechnique for the years 1849–1852.

Dumas, Jean Baptiste. *Eloges et discours.* Vol. II. Paris: Bachelier, 1883. Dumas wrote no memoirs, nor was he much given to public statements of his social views. Hence one must resort to manuscript sources or to funeral orations and academic *éloges* such as these. Good examples of Dumas' style, as smooth as it was florid, but not much hard information. The speech in memory of Victor Regnault makes no mention of his teaching at the Ecole Centrale.

————. "L'enseignement supérieur agricole à l'Ecole Centrale des Arts et Manufactures." *Revue scientifique* 11:25 (4 May 1872), 1053–1058. Dumas' only published statement for his support of agricultural education at the Ecole Centrale, a subject he had taken up privately with his colleagues since 1848. As the Dumas Papers show, most of them opposed him on this. Only when the school was weakened by its lack of a permanent director in the years after the Franco-Prussian War was Dumas able temporarily to force an agricultural specialty into the curriculum.

Dupin, Charles. *Défense des corps des ponts et chaussées et de l'Ecole polytechnique.* Paris: Panckoucke, 1850.

————. *Discours d'inauguration de l'amphithéatre du Conservatoire des Arts et Métiers.* Paris: Bachelier, 1821.

————. *Essai historique sur les services et les travaux scientifiques de Gaspard Monge.* Paris: Bachelier, 1819. The admiring biography of devoted disciple.

————. *Forces productives et commerciales de la France.* Paris: Bachelier, 1827.

————. *Harmonies des intérêts industriels et des intérêts sociaux.* Paris: Bachelier, 1833.

————. *Historique de l'enseignement et de son influence sur le sort du peuple de 1819 à 1839.* Paris: Bachelier, 1839. An important source for the history of technical education and the diffusion of useful knowledge during this period.

————. *Lettre à M. Berryer au suject des écoles publiques* Paris: N.P. (no publisher), 1842. Berryer had proposed to cut 84,000 francs from the budget of the *écoles d'arts et métiers.* Dupin points to the high positions many of the graduates have obtained, even if there were also some disappointments.

————. *La morale, l'enseignement, et l'industrie.* Paris: Firmin Didot,

1838. Speech at the opening of his course at the Conservatoire des Arts et Métiers. Dupin apparently never bothered to collect all these speeches and papers and publish them in one or two volumes. A modern biography of Dupin would be of considerable value.

—————. *Rapport fait au nom de la commission chargé de donner son avis sur la prise en considération de la proposition relative aux emplois d'ingénieur des ponts et chaussées.* Paris: N.P., 1850.

—————. *Tableau comparé de l' instruction populaire avec l'industrie des départements.* Paris: N.P., 1827. A lecture given in the second (not just the opening) session of his course.

Essai d'un project de loi sur l'instruction secondaire. Anonymous pamphlet, probably written in 1837. The *classes populaires* have received their means of education in the law on primary education of 1833. Now *la classe bourgeoise*, which includes today all that remains of *l'ancienne noblesse*, demands more funds for secondary education.

Etat de l'enseignement scientifique dans les collèges royaux et particuliers depuis 1815 jusqu'en 1847. Paris: Fain et Thunot, 1847. Reply of the *proviseurs* of the *collèges royaux* and the "*directeurs*" of "*collèges particuliers*" in Paris to the Dumas report. They challenge the authoritativeness of the Dumas Commission, claiming that only one member had taught "long and seriously" in a *collège royal*. (Both Pouillet and Milne-Edwards had taught in Parisian *collèges*, however, so the question is what they meant by *sérieusement*). The Commission only visited three of the seven *collèges* in Paris, but it is in the other four that "the teaching of mathematics has seen its greatest development." They then proceed to a detailed account of their accomplishments in science teaching. The authorship of the document makes it difficult to judge its impact. Would Dumas have considered the *collèges particuliers* outside the *Université*, and thus his opponents, whereas the *collèges royaux* were institutions whose interests (at least in theory) he wanted to protect? This seems likely.

Flachat, Eugène, and E. Vuigner. *Discours.* Neuilly: Guiraudet, 1861. A joint speech by the two prominent engineers in which they justify the request of the Société des Ingénieurs Civils to be recognized as *d'utilité publique*. They point to the work in the Association Polytechnique of their honorary president, Perdonnet, and to the fact that their own *science professionnelle* is nowhere carried higher than at the Ecole Centrale des Arts et Manufactures, from which the majority of their members came. In short, the message is that the Société exists to diffuse useful knowledge, carry on activities for the advancement of engineering science, and protect the profession of *ingénieur civil* from lowered standards which might result from the intrusion of non-members if the S.I.C. did not exist.

Flachat, Eugène, and Stéphane Flachat. *Vues politiques et pratiques du les travaux publics en France.* Paris: N.P., 1832. Their only "political view" is an advocacy of State financing of canals and railroads. The rest of the 336 pages are devoted to detailed schemes for the execution of such projects.

✓ Flachat-Mony, Stéphane. *L'industrie: Exposition de 1834.* Paris: Terre,

1834. His report on the Paris industrial exposition of 1834. In his historical introduction, he states that the real rapprochement between science and industry came when the *savants* responded to the needs of the nation in arms. It was then that the *fabricants* began to appreciate the value of theories and to respect the men who could "penetrate by science into the secrets of industrial processes." Claims that the 10,000 candidates who were unsuccessful in gaining entry to the Polytechnique between 1794 and 1834 nevertheless benefited from "the solid instruction that they must have received in order to prepare for the examinations." He thus suggests that the Polytechnique has a kind of multiplier effect that helps to spread science among men who may make use of it in many ways, such as assisting in industrialization. On the other hand, in discussing the backgrounds of the men who made the exhibits, he gives evidence for the considerable accomplishments of men who received no elaborate, formal scientific education. Even products such as mathematical and optical instruments, "for which a substantial knowledge of geometry and physics is indispensable," were made by men who could barely write well enough to make out the exhibition cards.

Fourcy, A. *Histoire de l'Ecole Polytechnique.* Paris: Ecole Polytechnique, 1828. A full-scale, detailed account by the school's librarian. Even the special influence of Monge can be detected, although as *persona non grata* during the Restoration he receives less attention than other accounts would indicate he deserves.

Gasc, M. *Observations sur le rapport au Roi et sur l'ordonnance du 26 mars 1829.* Paris: N.P., 1829.

Gasc, P. E. *Pétitions adressées à la Chambre des Députés.* Paris: Appert, 1845.

Girardin, Emile de. *De l'instruction publique en France.* Paris: Michel Levy, 1840. Description of the system, various suggestions for reform, from greater State support of the primary system to the establishment of a Faculty of Administrative Sciences. Of special interest for the historical study of social stratification is the second section of the book, the *Guides des familles,* a survey of each of the professions for the benefit of parents, which serves as a useful check against Charton's guide.

Girardin, Saint-Marc. *De l'instruction intermédiaire et son état dans le midi de l'Allemagne.* Paris: Levrault, 1835. After a favorable report on the *Realschulen,* Girardin addresses himself to the question of the "unity of civilization" that education should promote and technological culture seems to threaten. He agrees that there should be a common principle, but asserts that it can be found neither in literary nor scientific studies but in religion.

———. *De l'instruction intermédiaire et de ses rapports avec l'instruction secondaire.* Paris: Delalain, 1847. One of the best sources for evidence that local pressures were demanding the industrial and commercial courses be offered in the secondary schools in cities such as Nancy and Metz.

Guiraud, A. *Lettre à MM. les Pairs sur le project de loi relatif à l'organisation de l'instruction secondaire.* Paris: N.P., 1844. Principally concerned with the "freedom of teaching" issue. He claims that he is far less concerned with where his son learns his mathematics and science than with where he learns his philosophy.

Hole, James. *An Essay on the History and Management of Literary, Scientific and Mechanics' Institutions.* London: Longman, 1853 (republished by Frank Cass, 1970). The earliest full-scale history of the Mechanics' Institutions by the secretary of the Yorkshire Union of Mechanics' Institutions. Dupin does not appear in it.

Iggers, Georg, ed. *The Doctrine of Saint-Simon: An Exposition. First Year 1828–1829.* 2nd ed. New York: Schocken, 1972. The transformation of Saint-Simon's thought as it was expounded to the faithful by Enfantin and Bazard. In his introduction, Iggers makes the case for his view that the "totalitarian" themes in their thought came not from scientism, as Hayek argues, but from the new authoritarianism of De Maistre, Bonald, and the early Lamennais. Or perhaps all yearned to recapture the "spirit of association" (one of the more persistent catchphrases of the entire period) that they had experienced, directly or vicariously, in Napoleon's armies? Iggers notes in his introduction, in any case, that the doctrine remained a "totalitarian fantasy," concluding that "the direct relevance of Saint-Simonianism on later thought and practice has, I now believe, been overstated in the original Introduction." It certainly proved less relevant than expected for my study of engineers.

Jomard, Edmé-François. *Souvenirs sur Gaspard Monge et ses rapports avec Napoléon.* Paris: Thunot, 1853. Deals with the years 1792–1816, especially the Egypt expedition.

Kilian, C. *Tableau historique de l'instruction secondaire en France depuis les temps les plus reculés jusqu'à nos jours.* Paris: Delalain, 1841. The first full-scale history of secondary education to be attempted in the nineteenth century. Hostile to the Bourbons and to the "complete separation of letters and sciences." Best source for statistics about the situation in 1789. Points out that in 1841 in the *collège* at Mulhouse, Latin was an optional course, but that drawing, French, German, history, geography, mathematics, physics, and chemistry—these last *dans leurs rapports avec l'industrie*—were obligatory.

Lacroix, Sylvestre-François. *Essais sur l'enseignement en général et sur celui des mathématiques en particulier.* Paris: Bachelier. 1805 and 1828 editions consulted.

————. *Notice historique sur la vie et les ouvrages de Condorcet.* Paris: Sajou, 1813.

Lamartine, Alphonse de. *Sur l'enseignement.* Paris: Duverger, 1837. A version of Lamartine's reply to Arago of 24 March 1837.

Landais, Napoleon. *De l'éducation et de l'instruction en France.* Paris: N.P., 1837. A rambling, 507-page hodgepodge of ideas. His proposals either copy Condorcet or partake of the utopian (such as the idea that all children from five to twelve should be raised at the expense of the State).

Laprade, Victor de. *Le baccalauréat et les études classiques.* Paris: Didier, 1869. One of the first book-length attacks on the *baccalauréat.* Takes the line that the *baccalauréat* actually harms classical studies by warping their study according to the needs of the examinations.

Bibliography

Laurent, Alphonse, *Réfléxions générals sur le mode d'enseignement suivi à l'Ecole Polytechnique.* Paris: Dupont, 1849. Laurent was a *Polytechnicien* graduated in 1832 and a captain in the *Génie militaire* at the time of writing. He argues that *Polytechniciens* are not taught to give enough attention to detail and that they do not know how to deal with practical problems.

Olivier, Theodore. *Mémoires de géométrie descriptive, théorique, et appliquée.* Paris: Carilion-Goeury et Dalmont, 1851.

―――. "Monge et l'Ecole Polytechnique." *Revue scientifique et industrielle.* Vol. 7. 3rd Series (1850), 64–68. Repeats the arguments in the 1847 preface of the work above. The timing is not without interest, however: the years 1849–1850 were those of the most concerted attack on the Polytechnique made during the entire century.

Pâris de Meyzieu. "Ecole militaire," in Diderot and D'Alembert, eds. *Encyclopédie.* Vol. 5 (1755).

Passy, Frédéric. *De l'instruction secondaire en France, de ses défauts, de leurs causes, et des moyens d'y rémédier.* Paris: N.P., 1840.

Perdonnet, Auguste. *Conférences publiques et gratuites faites dans le grand amphithéâtre de l'Ecole de Médecine.* Paris: Simon Racon, 1864. A lecture at the Association Polytechnique.

―――. *Des grandes inventions au point de vue des services publiques qu'elles ont rendues ou qu'elles peuvent rendre.* Paris: N.P., 1862. Republished in 1867. One of Perdonnet's more popular sets of lectures.

―――. *De l'utilité de l'instruction pour le peuple.* Paris: Hachette, 1867.

Pompée, Philibert. *Etudes sur l'éducation professionnelle en France.* Paris: Pagnerre, 1863. Republishes a number of previous essays dating from 1832. One of the best-informed and comprehensive sets of accounts of the struggle for technological education. Ranks with Cournot as one of the first books to consult on this period.

Quatrefages, Armand de. "De l'enseignement scientifique en France." *Revue des deux mondes* 22 (15 May 1848), 489–507.

Rendu, Ambroise. *De l'instruction secondaire, et spécialement des écoles secondaires ecclésiastiques.* Paris: Hachette, 1842. Deals mainly with the "freedom of teaching" question.

Renouard, Charles. *Considérations sur les lacunes de l'éducation secondaire en France.* Paris: Chez A. A. Renouard, 1824. The earliest full-length attempt to deal with the question of the need for modern studies in secondary education. The opening shot in a long debate.

Saint-Simon, Henri de. *Oeuvres complétes de Saint-Simon et d'Enfantin.* 47 Vols. Paris: Dentu and Leroux, 1865–1878.

Séguin, Marc. *De l'influence des chemins de fer et de l'art de les tracer et de les construire.* Paris: Carilian-Goeury, 1839. Séguin, who first patented the multitubular locomotive boiler while building the first French railroads, was one of the half-dozen leading *ingénieurs civils* of the period. In the introduction to this important technological treatise, he states that "a few years ago, when the construction of works of public utility was monopolized by government engineers, one could speak the

language of science with all its abstractions and complex formulas."
Now that this monopoly is broken, the writer addresses a group not
trained in special mathematical studies. He now has a different duty:
"Their reasoning and their explications must no longer cover merely
theoretical principles or mathematical solutions that they accept *a
priori*; technical writers must now penetrate to the elements of the science, simplify them, summarize them, and make them accessible to all
classes of readers."

Smiles, Samuel. *Self-Help*. Boston: Estes and Lauriat, N.D. First English
edition is 1859. The important question, which bears further research, is
whether this work—or others like it—had any influence in France.

Tabareau, Hippolyte. *Discours prononcé à la séance solenelle de rentrée
des facultés et de l'inauguration de l'école des sciences appliquées.*
Lyons: Vintrinier, 1855.

———. *Discours prononcé par M. Tabareau dans la séance d'inauguration
de l'ecole théorique des arts et métiers dite La Martinière.* Lyons:Perrin,
1826. Tabareau, *Polytechnicien* and physicist, was the leading force behind the school's curriculum. In this speech he addresses himself to *ouvriers* and *chefs d'atelier*, inviting them to make use of the school. He
reports, with considerable exaggeration, that "chairs of industrial chemistry have been established in all the cities of England."

———. *Exposé de la méthode Tabareau.* Lyons: Perrin, 1863. One of
many examples of Tabareau's attempts to popularize his method of
mathematics teaching, which, as is clear from A.N. F17 6894, came to
the attention of the Ministère de l'Instruction Publique, but was not
adopted. In his preface Tabareau suggests that his method of more rapid
mathematics teaching will result in students' having more time to study
literature.

Tabareau, Hippolyte, and Elisée Dévillac. *Rapport présenté à l'Académie
royale des sciences, belles-lettres, et arts de la ville de Lyon sur l'Ecole
La Martiniére.* Lyons: Barret, 1832. In the course of a defense of La Martinière, Tabareau and his colleague attack the education at the *ecoles
d'arts et métiers* as insufficiently theoretical.

Thomas, Alexandre. *Note à consulter sur l'état présent de l'Université.*
Paris: Comptoir des Imprimeurs-Unis, 1848. An attack on Salvandy.

Thomas, Emile. *Histoire des Ateliers Nationaux.* Paris: Michel Lévy,
1848. Indispensable.

Tocqueville, Alexis de. *Democracy in America.* New York: Vintage,
1945. The Phillips Bradley edition.

E. Textbooks and Scientific Treatises

Coriolis, Gaspard. *Du calcul de l'effet des machines.* Paris: N.P., 1829.
This edition contains the report to the Académie des Sciences made by
Prony, Girard, and Navier, which explains the importance of the work.

Dumas, Jean-Baptiste. *Leçons sur la philosophie chimique professées au Collège de France en 1836.* Paris: Gauthier-Villars. 2nd ed., 1878.

———. *Traité de chimie appliquée aux arts.* 8 Vols. Paris: Bachelier, 1828–1846.

Lacroix, Sylvestre-François. *Manuel d'arpentage ou instruction élémentaire sur cet art et sur celui de lever les plans.* First edition is Paris: Hachette, 1826.

———. *Traité élémentaire d'arithmétique.* 1st ed. is Paris: N.P., 1798.

———. *Traité élémentaire du calcul différentiel et du calcul intégral.* 1st ed. is Paris: Courcier, 1802.

Monge, Gaspard. *Géométrie descriptive.* Paris: Gauthier-Villars, 1922. Reprint of the 1820 edition published by Monge's student Barnabé Brisson.

Navier, Claude L.-M.-H. *Résumé des leçons de mécanique données à l'Ecole Polytechnique.* Paris: Carilian-Goeury, 1841.

———. *Résumé des leçons données à l'Ecole des Ponts et Chaussées sur l'application de la mécanique a l'etablissement des constructions et des machines.* Paris: Dunod, 1864. This 3rd ed. includes a very useful 307-page history of mechanics by Navier's colleague Barré de Saint-Venant, as well as a biographical notice on Navier by Prony.

Péclet, Eugène. *Traité élémentaire de physique.* 4th edit. 2 Vols. Paris: Hachette, 1847.

———. *Traité de l'éclairage.* Paris: Malher, 1827.

———. *Traité de la chaleur.* 3rd ed. 3 Vols. Paris: Masson, 1860.

Poncelet, Jean-Victor. *Traité de mécanique industrielle.* Paris: Bachelier, 1832. Poncelet's counterpart of Dumas' *Traité de chimie appliquée aux arts.* An accessible work: it uses no mathematics more complicated than quadratic equations.

F. Memoirs, Novels, and Published Correspondence

About, Edmond. *Le progrès.* Paris: Hachette, 1867.

Advielle, Victor, ed. *Journal professionnel d'un maître de pension de Paris au XVIIIe siècle.* Pont-l'Eveque: Delhais, 1868. Accounts of life in preparatory institutions of the eighteenth century are rare. This journal of the directors of the most important Parisian private *pension*, covering the period from 28 December 1773 to 5 September 1783, is thus of special value although its entries are hardly detailed. The very first entry, however, gives something of the flavor of the place: "Day of the Innocents, 28 December 1773. On this day the *Pensionnaires* [students] refused to begin study at three o'clock in the afternoon, on the pretext that it was a holiday. M. Berthaud forced them to begin and made them pay a forfeit for having disobeyed."

Balzac, Honoré. *The Lily of the Valley.* Philadelphia: Barrie, 1897.

Bibliography

————. *Louis Lambert.* Paris: Albin Michel, 1951.

————. *The Magic Skin.* Philadelphia: Barrie, 1897.

————. *A Start in Life.* Philadelphia: Barrie, 1897.

————. *The Village Curé.* Philadelphia: Barrie, 1897. One of the earliest portraits of an *ingénieur civil* in French literature is given here in the character Grégoire Gérard. Gérard writes a letter of despair attacking the engineering schools and the Ponts et Chaussées administration that summarizes most of the important complaints made about the profession. The Polytechnique and the Ecole des Ponts et Chaussées are great "factories of incapacity," as is evidenced by the collapse of the bridge over the Seine designed by Navier, one of their most revered professors. He questions whether such schools are really needed at all: "Did Vauban graduate from any other school than the great school called vocation? Who was Riquet's teacher? When men of genius, prompted by vocation, rise as they do above their social surroundings, they are almost always completedly equipped; in such cases, man is not simply a being created for a special purpose, he has the gift of universality." The *Corps des Ponts et Chaussées* is a stifling bureaucracy, sending highly trained young men off to the provinces to atrophy intellectually and morally in the performance of routine tasks that could be done by an illiterate. "Not a stone is laid in France until half a score of Parisian scribblers have made foolish and utterly useless reports." At the end of the novel Gérard, originally a *Ponts* engineer, is hired by the heroine, Madame Graslin, to carry out drainage and irrigation projects on her estates. After several years of work under his direction (acting now as an *ingénieur civil*), the estate is transformed and "science, the inheritor of Moses' staff, has caused abundance, prosperity, and happiness for a whole canton to gush forth."

Bertrand, Joseph. "Souvenirs académiques: Auguste Comte et l'Ecole Polytechnique." *Revue des deux mondes* (1 December 1896), 528–548.

Chaptal, Jean-Antoine. *Mes souvenirs sur Napoléon.* Paris: Plon, 1893. This addition also contains a fragmentary autobiography edited by his great-grandson.

————. *De l'industrie française.* 2 Vols. Paris: N.P., 1819. Surprisingly little about technical education or the importance of engineers.

Colladon, Daniel. *Souvenirs et mémoires.* Geneva: Aubert-Schuchardt, 1893. Poorly organized, repetitive, devoted chiefly to an account of his activities as a civil engineer in various Alpine projects, but, still, one of the best sources available for the early history of the Ecole Centrale and the French civil engineering profession.

Cournot, Antoine-Auguste. *Souvenirs d'A. Cournot.* Paris: Hachette, 1913. The early chapters were useful for the present study. The later chapters give a rare insider's glimpse of the heights of the educational administration. He claims that in the early 1850s three men ran the show: Fortoul, Le Verrier, and J. B. Dumas.

Cousin, Victor. "Huit mois au Ministère de l'Instruction Publique." *Revue des deux mondes* (15 September 1841), 371–396. Cousin's account of his policies during his tenure in the Ministry. Principally deals with primary education and the higher primary schools, which he sees as the solution to the need for "intermediate instruction." Claims that he did much to improve scientific education.

Delessard, Ernest. *Souvenirs de 1848. L'Ecole Centrale aux Ateliers Nationaux.* Paris: E. Bernard, 1900. The only account of the National Workshops by a *Centralien* other than Thomas himself. Confirms the *Pont-Centraux* hostility and, like Thomas, accuses the *Ponts* of refusing to provide the workshops with project plans.

Freycinet, Charles de. *Souvenirs: 1848–1878.* Paris: Delagrave, 1912. Contains a brief account of de Freycinet's activities during the Revolution of 1848, when Polytechnique students were used as aides-de-camp by the Provisional Government.

Grimaux, Edouard, ed. *Charles Gerhardt: Sa vie, son oeuvre, sa correspondance—1816–1856.* Paris: Masson, 1900. Shows the other side of Dumas, who seems to have used his influence to curtail the career of the talented chemist Gerhardt when the latter challenged his work.

Guernon-Ranville, Comte de. *Journal d'un ministre.* Caen: Blanc-Hardel, 1873. Guernon-Ranville, a rather vapid lawyer, served as Polignac's education minister. Not a very useful account.

Henry, Charles, ed. *Correspondance inédite de Condorcet et de Turgot.* Paris: Charavay Frères, 1883. A useful corrective to the favorable studies of the Ponts et Chaussées engineers by Petot and De Dartein. Condorcet has nothing but contempt for the corps. On 10 October 1775 he writes to Turgot, "Trudaine will give you no help in this affair; he will not permit the least doubt to be cast upon his *premiers commis*. Perronet, who is at the head of the whole group, is a most vain and ignorant character, who instituted the *corps des ponts et chaussées* and who would let the whole kingdom collapse before he would let his *bel établissement* suffer the least slight."

Jaime, Edouard. *Souvenirs de 1848 à 1871.* Versailles: Laurent, 1872. Account of the days of the National Workshops by one of Thomas's lieutenants. Unlike Delessard and most of the others, Jaime was not even a graduate of or a student at the Ecole Centrale. He was born in 1804, the son of a maker of bronze statues who was "ruined by the events of 1815." He himself studied drawing, then began to write plays. He had written more than 100 by 1848. When the Revolution broke out, he offered his services to the *mairie* of the fifth *arrondissement*, then linked up with Thomas. By a mixture of bravado and cleverness, he proceeded to maintain order and to enlist even the most rebellious workers on his side.

Laffitte, Jacques. *Mémoires de Laffitte.* Paris: Firmin-Didot, N.D. Not much use for the present study.

Le Canu, L. R. *Souvenirs de M. Thénard.* Paris: Donday-Dupré, 1857. Portrait of another rags-to-riches chemist who remained far rougher at the edges than did Dumas.

Michelet, Jules. *Nos fils.* Paris: Calmann-Lévy, 1902. First published in 1869, the last of Michelet's series on education. This edition includes a study by Octave Gréard, literary historian and long-time vice-rector of the Academy of Paris. Speaks of the Ecole Centrale as offering France something new, "the solidarity of the [industrial] arts."

Palmer, Robert R. *The School of the French Revolution:A Documentary History of the College of Louis-le-Grand and Its Director, Jean-François Champagne, 1762–1814.* Princeton: Princeton Univ. Press, 1975. Indispensable.

Pole, William, ed. *The Life of Sir William Fairbairn, Bart., Written Partly by Himself.* London: Longmans, Green, 1877. Perhaps the most detailed autobiography of an early-nineteenth-century civil engineer.

Rémusat, Charles de. *Mémoires de ma vie.* 4 Vols. Paris: Plon, N.D. Volume 4 contains portraits of Villemain and Salvandy and useful discussions of the political fate of their educational policies.

Reybaud, Louis. *Jérome Paturot à la recherche d'une position sociale.* Paris: Calmann-Lévy, 1843. Might be read as a novelist's response to Charton's *Guide du choix d'un état.* Paturot is a *naïf* in bourgeois society, constantly failing at various careers and schemes as he savors downward mobility in July Monarchy society.

Sarcey, Francisque. *Souvenirs de jeunesse.* Paris: Ollendorff, 1885.

Sengler, A. *Souvenirs d'académie: Séances littéraires et dramatiques, données dans les collèges de la Compagnie de Jésus en France de 1815 à 1878.* A Jesuit's attempt to show the *collèges'* contribution to French nineteenth-century culture through demonstrating the survival of the traditions of the *honnête homme.*

Taton, René, ed. "Auguste Laurent et J.-B. Dumas d'après une correspondance inédite." *Revue d'histoire des sciences* VI:4 (October-December 1953), 329–347.

Tocqueville, Alexis de. *The Recollections of Alexis de Tocqueville.* New York: Meridian, 1959.

II. Secondary Sources

A. Bibliographical Aids and Reference Works

Albert, Colette. *Bibliographie annuelle de l'histoire de France du cinquième siècle à 1939.* Paris: N.P., 1955 + .

Bastien, Paul. *Les carrières commerciales, industrielles et agricoles.* Paris: Albert Foutemoing, 1906. Used as an occasional supplement to the more detailed and comprehensive guides by Massé and Jacquemart.

Ben-David, Joseph. "Professions in the Class System of Present-Day Societies." *Current Sociology* XII:3 (1963–1964), 247–329. Surprisingly few historical studies.

Caron, Pierre and Marc Jaryc. *Répertoire des peri diques de la langue française.* Paris: Fédération des Sociétés françaises, '935.

———. Supplement, 1939.

Compayré, Gabriel. "The Educational Journals of France." *Educational Review* (February 1900), 121–142. Unique.

Dantès, Alfred. *Dictionnaire biographique et bibliographique, alphabétique et méthodique.* Paris: Boyer, 1875. Regularly lists some of the more obscure members of the French scientific scene.

Eells, Walter C. *American Dissertations on Foreign Education.* Washington, DC: Committee on International Relations of the National Education Association, 1959.

Ferguson, Eugene S. *Bibliography of the History of Technology.* Cambridge, Mass.: MIT Press, 1968. An essential guide, well annotated, which should be supplemented by the additions regularly published in *Technology and Culture.*

Glenn, Norval, and David Weiner. *Social Stratification: A Research Bibliography.* Berkeley, Calif: The Glendessary Press, 1970. Very little on France, and not much on Europe as a whole.

Hanotaux, Gabriel. *Du choix d'une carrière.* Paris: Flammarion, 1902. Of some interest given its author's eminence. For Hanotaux there are really only two *grandes écoles* to which all families aspire: Polytechnique and Normale Supérieure. The Ecole Centrale—which he mentions next—"despite the services it renders, is not quite at the same level." The Ecole de Saint-Cyr is a good place for *braves garçons* with "muscles and mustaches."

Hatin, Eugène. *Bibliographie historique et critique de la presse périodique française.* Paris: Editions Anthropos, 1965. A reprint of the 1866 edition. Extremely useful. The number of journals devoted to the diffusion of useful knowledge that appeared after about 1815 is remarkable.

Index biographique des membres et correspondants de l'Académie des Sciences Paris: Gauthier-Villars, 1968. Gives birth and election dates, posts held with the Académie, and official positions held, if scientific.

Jacquemart, Paul. *Professions et métiers.* 2 Vols. Paris: Armand Colin, 1892. The most detailed guide, of great value for the historical study of social stratification. Especially good on bureaucratic career ladders.

Jolly, Jean. *Dictionnaire des parlementaires français.* 3 Vols. Paris: P.U.F., 1963.

Marielle, M. C. P. *Répertoire de l'Ecole Impériale Polytechnique.* Paris: Mallet-Bachelier, 1855. Marielle was archivist of the school. A comprehensive guide to the date of entry and leaving and first career assignment or current position of all *Polytechniciens.* Adds a few biographical details on notable graduates.

Massé, Daniel. *Pour choisir une carriére.* Paris: Larousse, 1908. Not as detailed as Jacquemart, but far more useful than Hanotaux or Bastien.

Palmer, Robert R. "Some Recent Work on Higher Education." *Comparative Studies in Society and History,* 13:1 (January 1971), 108–115.

Paul, Harry W. "La science française de la seconde partie du XIXe siècle vue par les auteurs anglais et américains." *Revue d'hist. des sciences* XXVII:2 (1974), 147–163.

Pfautz, Harold W. "The Current Literature on Social Stratification: Critique and Bibliography." *Am. J. Sociol* LVIII:4 (January, 1953), 391–418.

Rudé, George. *Debate on Europe, 1815–1850.* New York: Harper and Row, 1972. A useful quick introduction but, given Rudé's talents, something of a disappointment in interpretive quality. Very little on education or the professions.

Taton, René. "Sur quelques ouvrages récents concernant l'histoire de la science française." *Revue d'hist. des sciences* XXVI:1 (1973), 69–90.

B. General Studies

Ashworth, William. "Backwardness, Discontinuity, and Industrial Development." *Economic Hist. Review* XXIII:1 (April 1970), 163–169. A critical examination of Gerschenkron's ideas, which observes, "The concept of economic backwardness has proved to be an important aid in studying the moderately retarded. Its usefulness for understanding the more numerous countries which lag much farther behind has not been adequately tested but appears likely to be very much less."

Bertier de Sauvigny, Guillaume de. *The Bourbon Restoration.* Philadelphia: Univ. of Penn. Press, 1966. Bertier's defense of the Restoration as one of the happier periods in modern history.

Cameron, Rondo E. *France and the Economic Development of Europe, 1800–1914.* Princeton: Princeton Univ. Press, 1961. The chapters on the work of French engineers and entrepreneurs abroad were especially useful.

Caron, François. *Histoire de l'exploitation d'un grand réseau: La compagnie du chemin de fer du nord: 1846–1937.* Paris: Mouton, 1973. An excellent study in economic history, but nothing about the recruitment of engineers.

Charle, Christophe. "Les milieux d'affaires dans la structure de la classe dominante vers 1900." *Actes de la recherche en sciences sociales* 20/21 (March–April 1978). 83–96.

Chéruel, A. *Dictionnaire historique des institutions, moeurs, et coutumes de la France.* 2 Vols. Paris: Hachett, 1855. A useful reference work.

Chevalier, Louis. *Classes laborieuses et classes dangereuses à Paris pendant la première moitié du XIXe siècle.* Paris: Plon, 1958. Of methodological interest for its use of literary sources and of general interest for its portrait of the social problem to which engineers' education was never addressed.

Daumard, Adeline. *La bourgeoisie parisienne de 1815 à 1848.* Paris: SEVPEN, 1963. Daumard's major *thèse*, indispensable for all students of the history of social stratification in this period.

———. "L'évolution des structures sociales en France à l'époque de l'industrialisation." *Revue historique* (April–June 1972), 325–326. An extremely important article, rich in provocative generalizations, which emphasizes the continuity and persistence of patterns of social stratification during French industrialization from 1815 to 1914. "The division of wealth and the economic hierarchy remain almost constant."

————— et. al. *Les fortunes françaises au XIXe siècle.* Paris: Mouton, 1973. A study of the distribution and composition of wealth in Paris, Lyons, Lille, Bordeaux, and Toulouse, based on evidence from the *déclarations de succession.*

Dunham, Arthur L. *The Industrial Revolution in France, 1815–1848.* New York: Exposition Press, 1955. One may argue over whether or not Dunham shows that there was really a revolution during this time, but the book is undeniably a major accomplishment. The Société d'Encouragement pour l'Industrie Nationale receives little attention, however. It is a subject that bears further investigation.

Dupeux, Georges. *La société française, 1789–1970.* Paris: Armand Colin, 1972. Another indispensable work for the historical study of French social stratification.

Duroselle, Jean-Baptiste. *Les débuts du catholicisme social en France (1822–1870).* Paris: P.U.F., 1951. Despite the prominence of Le Play, it is difficult to know whether or not engineers were very prominent in the development of social Catholicism in France. This book does not help much to answer the question. He briefly mentions a certain Alcan, a *représentant du peuple* in 1848, as active in certain social Catholic circles, which may have been Michel Alcan of the Ecole Centrale, but none of the other biographical sources on Alcan mentions this.

Faure-Soulet, J.-F. *De Malthus à Marx: L'histoire aux mains des logiciens.* Paris: Gauthier-Villars, 1970. A sophisticated study of the French political economists.

Ferkiss, Victor C. *Technological Man: The Myth and the Reality.* New York: NAL, 1970. Searching for Technological Man in the late 1960s, Ferkiss concludes that such a being does not exist: "There is no new man emerging to replace the economic man of industrial society or the liberal democratic man of the bourgeois political order. The new technology has not produced a new human type, provided with a technological world view adequate to give cultural meaning to the existential revolution. Bourgeois man continues dominant just as his social order persists, while his political and cultural orders disintegrate." He then enumerates the qualities that Technological Man should possess. By phrasing his conclusion this way, Ferkiss seems to view technical change not as a creation of human society and will but as a peculiarly autonomous force capable of independently creating a "new human type." The present study has posed a more modest hypothesis: I ask whether an engineer's education, itself a social and cultural activity, produced a group of men (not an entire society) who may still have been bourgeois, but whose philosophical outlook and social actions somehow distinguished them from others in that status group.

Gerschenkron, Alexander. *Continuity in History and Other Essays.* Cambridge, Mass.: Harvard Univ. Press, 1968.

—————. *Economic Backwardness in Historical Perspective.* Cambridge, Mass.: Harvard Univ. Press, 1966.

Bibliography

Gide, Charles, and Charles Rist. *Histoire des doctrines économiques.* Paris: Sirey, 1942, and later editions. The standard work.

Henderson, W. O. *Britain and Industrial Europe, 1750–1870.* Leicester: Leicester Univ. Press, 1972.

Hoffmann, Stanley et al. *In Search of France.* Cambridge, Mass.: Harvard Univ. Press, 1963. The article by Pitts develops the idea of the "delinquent community" in the schools.

Huber, Michel, Henri Bunle, and Fernand Boverat. *La population de la France: Son évolution et ses perspectives.* Paris: Hachette, 1963.

Iggers, Georg. *The Cult of Authority.* The Hague: Mouton, 1958. Iggers' first study of the Saint-Simonians. Some of the points are superseded by his introduction to his edition of the Doctrine of Saint-Simon (1972).

Landes, David S. *The Unbound Prometheus: Technological Change and Industrial Development in Western Europe from 1750 to the Present.* Cambridge, England: Cambridge Univ. Press, 1969.

Langer, William L. *Political and Social Upheaval, 1832–1852.* New York: Harper and Row, 1969.

Leuillot, Paul. *L'Alsace au début du XIXe siècle: Essais d'histoire politique, économique, et religieuse (1815–1830).* 3 Vols. Paris: SEVPEN, 1956. No systematic consideration of the engineers' role, but many scattered details, especially about the activities of the *ingénieurs des mines.*

✓ Lévy-Leboyer, Maurice. "La décélération de l'économie française dans la seconde moitié du XIXe siècle." *Rev. d'hist. écon. et sociale* 4(1971), 485–507.

Lhomme, Jean. *La grande bourgeoisie au pouvoir.* Paris: P.U.F., 1960.

Manuel, Frank. *The Prophets of Paris.* Cambridge, Mass: Harvard Univ. Press, 1962.

Markovitch, Tihomir. *L'industrie française de 1789 à 1964. Conclusions générales.* vol. 7 of *Histoire quantitative de l'économie française,* Paris: ISEA, 1966.

Martin, Marie-Madeleine. *Les doctrines sociales en France.* Paris: Editions du Conquistador, 1963.

McClelland, David. *The Achieving Society.* New York: The Free Press, 1967.

McKay, Donald C. *The National Workshops: A Study in the French Revolution of 1848.* Cambridge, Mass: Harvard Univ. Press, 1933. A model monograph.

Mornet, Daniel. *Les origines intellectuelles de la Révolution française.* Paris: A. Colin, 1933.

Moss, Bernard H. *The Origins of the French Labor Movement: The Socialism of Skilled Workers, 1830–1914.* Berkeley: Univ. of Cal. Press, 1976. Emphasizes ideological determinants of the fact that French skilled workers did not produce very many Michel Alcans or Denis Poulots.

Mousnier, Roland. *Les institutions de la France sous la monarchie absolue.* Vol I. *Société et état.* Paris: P.U.F., 1974.

Palmade, Guy P. *French Capitalism in the Nineteenth Century.* Newton Abbott: David and Charles, 1972. A study of capitalists, i.e., bankers and financiers of enterprise, rather than of capitalism as a whole. Graeme Holmes states in an introductory essay that reviews the scholarship produced in the decade since the work first appeared, "No sociologist or social historian has made a study of recruitment or career patterns in French entrepreneurship." The more recent works of Maurice Lévy-Leboyer and Louis Bergeron have begun to meet this challenge, however.

Ponteil, Felix. *Les classes bourgeoises et l'avènement de la démocratie.* Paris: Albin Michel, 1968.

Ponteil, Felix, *Les institutions de la France de 1814 à 1870.* Paris: P.U.F., 1966.

Reader, W. J. *Professional Men: The Rise of the Professional Classes in Nineteenth-Century England.* London: Weidenfeld and Nicolson, 1966.

Remond, René. *La droite en France de 1815 à nos jours.* Paris: Aubier, 1954.

Serman, William. *Les origines des officers français 1848–1870.* Paris: Publications de la Sorbonne, 1979.

Shorter, Edward, and Charles Tilly. *Strikes in France, 1830–1968.* Cambridge, England: Cambridge Univ. Press, 1974.

Toutain, J.-C. *La population de la France de 1700 à 1959.* Paris: ISEA, 1953.

Viennet, Odette. *Napoléon et l'industrie française.* Paris: Plon, 1947. Well-researched, but distinctly uncritical.

Zeldin, Theodore. *France, 1848–1945.* Vol. I. *Ambition, Love, and Politics.* Emphasizes the fragmentation of the bourgeoisie. Some minor mistakes concerning technical education, but, in general, a rich and delightful work.

C. Works of Methodological Value

Barber, Bernard, and Walter Hirsch. *The Sociology of Science.* New York: Free Press, 1962.

Baudelot, Christian, Roger Establet, and Jacques Malemort. *La petite bourgeoisie en France.* Paris: Maspero, 1974.

Bourdieu, Pierre, and Jean-Claude Passeron. *Les héritiers: Les étudiants et la culture.* Paris: Minuit, 1964. An influential work.

———. *La réproduction: Eléments pour une théorie du systéme d'enseignement.* Paris: Minuit, 1970.

Clark, Terry N. "Institutionalization of Innovations in Higher Education: Four Models." *Administrative Science Qtrly,* 13:1 (June 1968), 1–25. The four models: differentiation, organic growth, diffusion, and (Clark's own contribution) "combined-process." Based upon Clark's work on the history of French sociology, but applicable to innovations in engineering education or to the history of the engineering profession.

Clifford-Vaughan, Michalina. "Some French Concepts of Elites." *British J. of Sociology.* 11 (December 1960), 319–332. Argues that French thinkers spent much time prescribing agendas for their favorite elites but little time studying what they actually did.

Cole, Jonathan R., and Stephen Cole. *Social Stratification in Science.* Chicago: Univ. of Chicago Press, 1973.

Craft, Maurice, ed. *Family, Class, and Education,* London: Longman, 1970. Useful articles on the subculture of school and on social class factors in education achievement.

Crozier, Maurice. *The Bureaucratic Phenomenon.* Chicago: Univ. of Chicago Press, 1964. Still the major work in the field. Crozier's comments on the educational system (accepting uncritically the historical judgments of writers such as Taine) seem a bit overdrawn, but the attempt to connect the style of classroom relations and the content of the curriculum is most suggestive. The major methodological weakness (shared by the equally influential Hoffmann study) is that it is not a comparative study.

Daumard, Adéline, and François Furet. "Problèmes de méthode en histoire sociale: Réflexions sur une critique." *Revue d'hist. moderne et contemp.* XI (October–December 1964), 241–248.

Daumard, Adéline. "Une référence pour l'étude des sociétés urbaines en France aux XVIIIe et XIXe siècles: Project de code socio-professionnel." *Revue d'hist. moderne et contemp.* X (July 1963), 185–210. The social-professional code used in the present study is a slightly modified version of the code presented here.

Drake, Michael, ed. *Applied Historical Studies: An Introductory Reader.* London: Methuen, 1973. Important articles by E. G. West and J. S. Hurt on the question of the returns to educational investment in early nineteenth-century England and a useful caveat by Stuart Blumin on the historical study of vertical mobility.

Duncan, Otis D., and Robert W. Hodge. "Education and Occupational Mobility: A Regression Analysis." *Am. J. of Sociol.* LXVIII:6 (May 1963), 629–644.

Febvre, Lucien. *Pour une histoire à part entière.* Paris: SEVPEN, 1962. Contains a chapter on the sources of prestige of the village blacksmith that shows the importance of the local context in status rankings. Much of the *forgeron's* prestige came from his "marginal" activities, such as veterinary services.

Girard, Alain. *La réussite sociale en France.* Paris: P.U.F., 1961. One of the best studies of the subject available, even if not always reliable in its statistics.

Goblot, Edmond. *La barrière et le niveau.* Paris: Félix Alcan, 1925.

Halsey, A. H., Jean Floud, and C. Arnold Anderson. *Education, Economy, and Society,* New York: Free Press, 1961. A selection of the best articles from the preceding decade, including Anderson's important study of intergenerational mobility through education.

Hans, Nicholas. *Comparative Education.* London: Routledge and Kegan

Bibliography

Paul, 1967. A comparative survey of the history of education, too broad to be of much use.

Hohenberg, Paul M. *Chemicals in Western Europe, 1850–1914: An Economic Study of Technical Change.* Chicago: Rand McNally, 1966. Hohenberg's study is an attempt to explain why Germany took the lead in chemical production over France and Switzerland after the middle of the century. He claims that in comparison to mathematics, mechanics, and physics, chemistry was a "low-prestige field" and that by the end of the century, "an almost complete divorce had occurred between science, especially academic science, and industry." He seems to be unaware of the first-rate chemical engineering program at the Ecole Centrale, claiming that "until 1872 there were no schools of industrial chemistry, and those founded after that time did not compare in prestige with the Grandes Ecoles, which for their part gave no emphasis at all to chemical subjects." He also talks about the great gap between the university, "where applications and technology had no place, and the low-grade practical schools, insufficient as these may have been." The record of La Martinière, the *écoles d'arts et métiers*, and especially the Ecole Centrale Lyonnaise, founded in 1857 in imitation of the Ecole Centrale des Arts et Manufactures, suggests that France in fact had schools that could turn out good chemical engineers. In any case, Hohenberg concludes with observations that tend to belie the heavy emphasis upon the connection between technical education and economic growth found in the statements of educators or men like Lyon Playfair: "It is possible, of course, that French chemists were poor, and therefore worth less to the firm just because of their weaker training. Yet there was no rush to hire foreign chemists, or to send young Frenchmen abroad. Even this might be in keeping with the political temper of the time, but the same held true in England. Finally, it is not clear that French chemists received poor training, since a considerable amount of scientific research continued to flow from France. . . . On balance, it is probably true that French technical education supplied an inadequate number of graduates and trainees, and did not train those few effectively for industrial research and operations. But this inadequacy is with respect to the size of France's chemical industry and the opportunity for it, not with respect to the demand for technical manpower actually obtaining. In terms of job opportunities, and of the duties chemists were called on to perform in industry, not to mention promotion to leading positions, the chemists were too many and often overtrained."

Jackson, John A. *Social Stratification.* Cambridge: Cambridge Univ. Press, 1968.

Labrousse, Ernest. "Voies nouvelles vers une histoire de la bourgeoisie occidentale aux XVIIIe et XIXe siècles." *Relazioni del X Congresso Internazionale di Scienze Storiche*, Vol. IV. Florence: Sansoni, 1955, 367–396.

Lefebvre, Georges. "Recherches sur les structures sociales aux XVIIe et XIXe siècles." *Bulletin d'histoire mod. et contemp.* I (1956), 53–61.

Lenski, Gerhard E. "Status Crystallization: A Non-vertical Dimension of Social Status." *Am. Sociological Rev.* 19:4 (August 1954), 405–413.

Le Yaouanq, Jean. "La mobilité sociale dans le milieu boutiquier parisien au XIXe siecle." *Le mouvement social.* 108 (July 1979), 89–112.

Merton, Robert K. *The Sociology of Science: Theoretical and Empirical Investigations.* Chicago: Univ. of Chicago Press, 1973.

Mousnier, Roland. "Le concept de classe sociale et l'histoire." *Revue d'hist écon. et sociale* XVIII:4 (1970), 449–454.

Musgrave, P. W. *Technical Change, the Labour Force, and Education.* London: Pergamon, 1967. A study of the roles of technically trained and generally educated manpower in the British and German iron and steel industries, 1860–1914. Argues that the important differences in education between the two countries lay not in the training of technical specialists to respond to changes such as the Bessemer or open hearth processes, but in general education that produced men who were of use in the new economic organizations (especially changes in their size) produced by gradual general economic changes such as the growth of railroads. What Britain needs is "a broad general education including science at all levels" equal to, if not greater than, the "need for particular scientific instruction." Uses no German sources.

Papy, Michel. "Professions et mobilité à Oloron sous la monarchie censitaire." *Revue d'hist. écon. et sociale* XIX:2 (1971), 225–264.

Reiss, Albert J. *Occupations and Social Status.* Gencoe: Free Press, 1961.

Rogoff, Natalie. "Social Stratification in France and the United States." *Am. J. of Sociol.* LVIII:4 (January 1953), 347–357. A rare attempt at a comparison between the two structures based upon interviews in the two countries. Shows the greater extent to which the French bourgeoisie's self-perception is based upon education and life-style rather than occupation and income.

Smelser, Neil J., and Seymour Martin Lipset, eds. *Social Structure and Mobility in Economic Development.* Chicago: Aldine, 1966.

Thernstrom, Stephan. "Notes on the Historical Study of Social Mobility." *Comp. Studs. in Soc. and Hist.* 10 (1967–1968), 162–172.

Tilly, Charles. *The Vendée.* Cambridge, Mass: Harvard Univ. Press, 1964. Used for the construction of the social-professional code as well as for his findings concerning separate rank orders in rural and urban societies.

Tirat, Jean-Yves. "Problèmes de méthode en histoire sociale." *Revue d'hist. moderne et contemp.* (July 1963), 211–218.

Weber, Max. "Class, Status, and Party." in H. H. Gerth and C. Wright Mills, *From Max Weber: Essays in Sociology.* Oxford: Oxford Univ. Press, 1964.

Wilensky, Harold, and Jack Ladinsky. "From Religious Community to Occupational Group: Structural Assimilation Among Professors, Lawyers, and Engineers." *Am. Sociological Rev.* 32:4 (August 1967), 541–561. The major finding—that a "deviant occupational specialty recruits marginal men (e.g., ethnic-religious or racial minorities) who then form a community based on common minority status and occupation" that continues that pattern of recruitment for some time before occupation alone becomes the basis for community—might be tested against the case of the French *ingénieurs civils,* if more information about their religious backgrounds or other forms of "deviance" could be found. Colladon and Perdonnet were both Swiss Protestants, and many of the Saint-Simonians came from minority backgrounds.

D. Biographies and Studies in the History of Science, Technology, and Education

Adam, Jean-Paul. *L'instauration de la politique des chemins de fer en France.* Paris: P.U.F., 1972. An examination of the debate over State vs. private development of railroads during the period 1837–1840. A few brief passages describe the role of *Ponts* engineers such as Becquey, Cavenne, and Kermaingant in the Chamber of Deputies and the administration. Having *Ponts* such as Becquey in the Chamber gave the proponents of State control an important advantage.

Ahlström Göran. "Higher Technical Education and the Engineering Profession in France and Germany during the 19th Century." *Economy and History* XXI:2, 51–88.

Alglave, Emile. "L'Ecole La Martinière." *Revue scientifique* 7 (15 August 1874), 146–155. Part of Alglave's series on scientific institutions in Lyons. Contains the best short description of Tabareau's special method of teaching mathematics, as well as some new material on his methods of student organization and discipline.

Allain, Ernest. *L'oeuvre scolaire de la Révolution.* Paris: Delalain, 1891.

Allemagne, Rene d'. *Les Saint-Simoniens.* Paris: Grund, 1930. An archivist's presentation of texts and engravings. Adds little to Charlety.

Ambrose, Stephen E. *Duty, Honor, Country: A History of West Point.* Baltimore: Johns Hopkins Univ. Press, 1966. The best history to date.

Anderson, Philip A. *Case Studies in the Acceptance of Dalton's Atomic Theory.* Thesis: Harvard University, 1970.

Anderson, Robert. *Education in France, 1848–1870.* Oxford: Clarendon Press, 1975. An excellent study paying special attention to secondary education and to local developments.

———. "Secondary Education in Mid-Nineteenth-Century France: Some Social Aspects." *Past and Present* 53 (November 1971), 121–146. A study of the social origins of secondary students based on an 1864 *enquête* by the Ministère de l'Instruction Publique. Suggests that from the standpoint of economic growth secondary education was *too* accessible to the economic sector of the bourgeoisie: "If the French system had been less democratic, in fact, a more dynamic entrepreneurial class might have emerged. Excluded from the existing schools, the new class might have created new ones embodying their own values, as happened to some extent in England." Suggests that the crucial developments occurred during the Second Empire.

Archer, R. L. *Secondary Education in the Nineteenth Century.* London:

Frank Cass, 1966. The chapters on Huxley and scientific education have still not been superseded.

Argles, Michael. *South Kensington to Robbins: An Account of English Technical and Scientific Education since 1851.* London: Longman, 1964.

Armstrong, John A. *The European Administrative Elite.* Princeton: Princeton Univ. Press, 1973. Excellent comparative, historically informed study of bureaucratic elites.

Armytage, W. H. G. *A Social History of Engineering.* Cambridge, Mass: MIT Press, 1961. As badly organized as it is unreliable. Claims that Eiffel graduated from the Ecole Polytechnique: one among many inaccuracies.

Artz, Frederick B. *The Development of Technical Education in France, 1500–1850.* Cambridge, Mass.: MIT Press, 1966. Standard institutional history, but extremely useful for chapter 1.

Ashby, Eric. *Technology and the Academics.* New York: St. Martin's Press 1963. A rather superficial treatment of the growth of science in nineteenth-century British universities. Largely superseded by Sanderson's works.

Association Polytechnique. *Histoire de l'Association Polytechnique et du developpement de l'instruction populaire en France.* Paris: Chaix, 1880. Much of the history of the Association Polytechnique has been obscured by the fire which destroyed its records at the Halle-aux-Draps before Georges Dumont and Victor Hudelo were given the task of assembling the documents for the writing of this history. Reprints a number of speeches, mostly from the post-1848 period.

Astier, Paul. "L'enseignement technique et l'éducation générale." *Revue politique et parlementaire* (December 1913), 431–464. Astier's law of 1919 made him the Guizot of elementary technical education.

Aubry, Paul V. *Monge: Le savant ami de Napoléon Bonaparte.* Paris: Gauthier-Villars, 1954. The best general biography.

Aulard, Alphonse. *Napoleon Ier et le monopole universitaire.* Paris: Armand Colin, 1911.

Ballot, Charles. *L'introduction du machinisme dans l'industrie française.* Paris: Rieder, 1923. A pioneering work by a young scholar who died before it was published. Still useful on many subjects.

Barker, Richard J. "The Conseil Général des Manufactures under Napoleon." *French Historical Studies.* VI:2 (Fall 1969), 185–214.

Barnard, Henry. *Science and Art: Systems, Institutions, and Statistics of Scientific Instruction Applied to National Industries in Different Countries.* New York: Steiger, 1872. One of the best of the several useful compendia produced by Barnard and his staff.

Barral, Georges. *Le panthéon scientifique de la Tour Eiffel.* Paris: Nouvelle Librairie Parisienne, 1892. Short biographies of seventy-two scientists and engineers whose work contributed to building the tower.

Ben-David, Joseph. *Fundamental Research and the Universities: Some Comments on International Differences.* Paris: OECD, 1968. Pluralism and competition in institutional structures seems a more productive solution than the centralization in France.

————. "The Rise and Decline of France as a Scientific Centre." *Minerva* VIII:2 (April 1970), 160–179.

————. *The Scientist's Role in Society.* Englewood Cliffs, N.J.: Prentice-Hall, 1971. An ambitious institutional history of science since the rise of scientific societies by one of the most influential sociologists of science. Ben-David is prone to judge his sources rather uncritically, however, and is somewhat obsessed by the question of which nation wins the gold medal of scientific primacy in a particular period.

Bendix, Reinhard. *Work and Authority in Industry: Ideologies of Management in the Course of Industrialization.* New York: Harper, 1963. Still useful essay.

Bernal, John D. *Science and Industry in the Nineteenth Century.* Bloomington: Univ. of Indiana Press, 1953. Somewhat selective arguments by the leading Marxist historian of science. Should be read in conjunction with something like Gillispie or Cardwell.

Berthelot, Marcelin. "La crise de l'enseignement secondaire." *Revue des deux mondes* (15 September 1891), 336–374. Berthelot had considerable influence on scientific education during the Third Republic. Here he repeats many of Dumas' arguments. *Plus ça change.*

Bien, David D. "Military Education in Eighteenth-Century France: Technical and Non-technical Determinants." in Monte D. Wright and Lawrence J. Paszek, eds., *Science, Technology, and Warfare: Proceedings of the Third Military History Symposium.* Washington: USAF HQ, 1959, 51–59. The emphasis of Bien and Shy upon the nontechnical determinants coincide with my own interpretive propensities. The eighteenth-century case nevertheless deserves more extensive investigation.

Bigot, Charles. *Questions d'enseignement secondaire.* Paris: Hachette, 1886. The arrogance of a *Normalien* humanist emergences in Bigot's attacks on technological education.

Biucchi, Basilio. "Ecole et révolution industrielle en Suisse." *Annales cisalpines d'histoire sociale* I:2 (1971), 105–123. Emphasizes the importance of the intellectual and institutional preconditions for the relatively thorough and rapid industrialization of Switzerland: the activities of eighteenth-century educational reformers and the widespread network of rural schools.

Blanchard, A. "Les ci-devant ingénieurs du roi." *Revue internationale d'histoire militaire.* 30 (1970), 97–108. Important. One of the first publications resulting from Mlle. Blanchard's extensive research for a collective biography of eighteenth-century military engineers.

Bode, Carl. *The American Lyceum: Town Meeting of the Mind.* Carbondale: Southern Illinois Univ. Press, 1968. A study of the principal American institution for the diffusion of useful knowledge in the first half of the nineteenth century. For Thomas Greene, one of the key figures in the lyceum movement, the advantages were as much moral and fraternal as they were scientific and intellectual: "From all the divisions, ranks, and classes of society, we are to meet . . . to instruct and to be instructed. While we mingle together in these pursuits . . . we shall remove many of the prejudices which ignorance or partial acquaintance with each other fostered." A comparative study of the diffusion of useful knowledge seems a promising topic for future research.

Bois, Benjamin. *La vie scolaire et les créations intellectuelles en Anjou pendant la Révolution.* Paris: Alcan, 1929. Probably the best local study of education during the Revolution, especially valuable because it does not confine itself to full-time schools.

Boissier, Gaston. "Les methodes dans l'enseignement secondaire." *Revue des deux mondes* (15 June 1868), 683–696. A call to classicists to adopt more "scientific" methods in the teaching of the classics.

Bolgar, R. R. "Victor Cousin and Nineteenth-Century Education." *The Cambridge Journal* V:2 (March 1949), 357–368. A favorable assessment of Cousin's achievements in institutional reform, especially his strengthening of the Ecole Normale and improvement of examination standards in the *facultés.*

Boltanski, Luc. *Prime éducation et morale de classe.* Paris: Mouton, 1969. An examination of the shift from appeals to the authority of religion and civic morality to appeals to the authority of science in child-rearing manuals throughout the nineteenth century. Engineers were among the first to adopt this latter approach in their lectures on *hygiène.*

Booker, P. J. "Gaspard Monge and His Effect on Engineering Drawing and Technical Education." *Transactions of the Newcomen Society* 34 (1961–1962), 15–36. Deals with both France and England. A full study of Monge's influence has not yet appeared.

Bouglé, Célestin. "Le bilan du saint-simonisme." *Annales de l'Université de Paris* 5 (September–October 1931), 446–463; 6 (November–December 1931), 539–556. Useful review of the work on Saint-Simon and his school which appeared in the 1920s, when such ideas were much in vogue.

———. *The French Conception of Culture Générale and Its Influence Upon Instruction.* New York: Columbia Univ. Press, 1938. Lectures given at Columbia University Teachers' College. A leading French sociologist explains *culture générale* to an American audience, but with no historical comments whatsoever. The whole subject badly needs a historian.

Bourgin, Georges, and Hubert Bourgin. *Le régime de l'industrie en France de 1814 à 1830.* 3 Vols. Paris: Alphonse Picard, 1912. An extensive collection of important documents, but little commentary and a poor index.

Bourlet, Carlo. "La chaire de géométrie descriptive du Conservatoire des Arts et Métiers." *Revue scientifique* 25 (22 December 1906), 769–773. A brief historical review, but principally a scientific lecture on the nature of geometry. And yet this was an opening lecture. The days when Dupin used his opening lectures to give wide-ranging lectures on the social importance of education seem far away.

Bouillé, Michel. *Enseignement technique et idéologique au XIXe siècle.* Doctoral thesis. Paris: 1972.

Boutmy, Emile. *Le baccalauréat et l'enseignement secondaire.* Paris: Armand Colin, 1899. A critical examination of the *baccalauréat* system that testifies eloquently to its influence upon French social stratification. Very little material on the history of the degree.

Bramwell, Sir Frederic. "L'art de l'ingénieur et la science." *Revue scientifique.* 42:11 (15 September 1888), 321–329.

Brooke, Michael Z. *Le Play: Engineer and Social Scientist.* London: Longmans, 1970. The first full-length study of Le Play, the "technologist with a social conscience." Lacks a systematic examination of his influence, however, and contains some puzzling passages about technical education, such as the comment that "a university education for engineers was provided free of charge at the Ecole Normale."

Brunel, Isambard. *Life of Isambard Kingdom Brunel: Civil Engineer.* London: Longmans, 1870.

Buisson, Ferdinand, ed. *Dictionnaire de pédagogie et d'instruction primaire.* Paris: Hachette, 1882. Intended as a kind of professional encyclopedia for the new generation of Republican primary school teachers. A useful reference work as well.

Calhoun, Daniel H. *The American Civil Engineer: Origins and Conflict.* Cambridge, Mass.: MIT Press, 1960. Based principally upon the study of the professional careers of the most prominent civil engineers, but includes valuable material on the relative importance of West Point and of projects such as the Erie Canal in giving engineers their education.

Callot, Jean-Pierre. *Histoire de l'Ecole Polytechnique.* Paris: Les Presses Modernes, 1958. A coffee-table book intended, apparently, for alumni. Little information about the earlier period not already dealt with by Fourcy and Pinet.

Calvert, Monte A. *The Mechanical Engineer in America, 1830–1910: Professional Cultures in Conflict.* Baltimore: Johns Hopkins Univ. Press, 1967. Emphasizes the debate between the proponents of on-the-job training and advocates of education in full-time schools. In this regard America provides a middle case between France, where school training won the debate very early, and Britain, where a suspicion of classroom education held sway throughout the profession until after World War I.

Cardwell, D. S. L. *From Watt to Clausius: The Rise of Thermodynamics in the Early Industrial Age.* Ithaca: Cornell Univ. Press, 1971.

———. *Turning Points in Western Technology.* New York: Science History Publications, 1972.

Carpenter, Kenneth E. "European Industrial Exhibitions before 1851 and Their Publications." *Technology and Culture* 13:3 (July 1972), 466–476.

Carré, Gustave. *L'enseignement secondaire à Troyes.* Paris: Hachette, 1888.

Chabot, C., and Sébastien Charléty. *Histoire de l'enseignement secondaire dans le Rhône de 1789 à 1900.* Lyons: Rey, 1901. In view of Charléty's eminence, a disappointing study.

Caullery, Maurice. *La science française depuis le XVIIe siècle.* Paris: Armand Colin, 1933. A French biologist presents the history of French science as the history of a small number of isolated individuals heroically producing works of brilliance.

Le cent-cinquantième anniversaire de la Société d'encouragement pour l'industrie nationale et les problèmes actuels de l'économie française, 1801–1951. Paris: Société d'Encouragement pour l'Industrie Nationale, 1951.

Chaline, Jean-Pierre. "Les contrats de mariage à Rouen au XIXe siècle." *Revue d'hist. économique et sociale* 2 (1970), 238–275.

Charléty, Sébastien. *Histoire du saint-simonisme (1825–1864).* Paris: Paul Hartmann, 1931. The best narrative history of a movement that, as a movement, rather than a *courant d'idées*, still has not been the subject of systematic analysis.

Chartier, Roger. "Un recrutement scolaire au XVIIIe siècle: l'Ecole royale du Génie de Mézières." *Revue d'hist. moderne et contemp.* XX (July–September, 1973), 353–375.

Château, Léon. *Notice biographique sur Pierre-Philibert Pompée.* Paris: Boyer, 1875. Life of the founder of the Ecole Turgot by his son-in-law.

Chatelein, A. "La formation de l'enseignement technique populaire à Paris au XIXe siècle." *Bulletin de la Société d'etudes de la region parisienne* (October 1954), 1–10.

Chérest, Aimé. "La vie et les oeuvres de A.-T. Marie: Avocat, membre du Gouvernement provisoire, etc." *Bulletin de la Société des sciences, arts, et belles-lettres de l'Yonne* (1873), 2–386. Other than the accounts by Thomas and Lalanne and the study by McKay, probably the most important source on the fate of the National Workshops.

Chauvin, Victor. *Histoire des lycées et collèges de Paris.* Paris: Hachette 1866.

Cipolla, Carlo. *Literacy and Development in the West.* Baltimore: Penguin, 1969. A useful introduction to the question, if superseded in parts by the work of Sanderson and Stone. Shows how industrialization had effects upon education that depended to a large extent upon previously existing patterns of State support for schooling.

Cobban, Alfred. "The 'Middle Class' in France, 1815–1848." *French Historical Studies* V:1 (Spring 1967), 41–52. Not surprisingly, Cobban criticizes the idea that the "middle class" was a homogeneous social group during this period, despite contemporary propaganda to the contrary.

Collignon, Edouard. "La science et l'art de l'ingénieur." *Revue scientifique* L:12 (17 September 1892), 354–358. Refers to Lamé's designation of engineering as an "incomplete science." I have not yet been able to locate this reference in Lamé's own writings.

Collins, Randall. *The Credential Society: An Historical Sociology of Education and Stratification.* New York: Academic Press, 1979. An extremely rich and provocative synthesis.

Comberousse, Charles De. *Histoire de l'Ecole Centrale des Arts et Manufactures.* Paris: Gauthier-Villars, 1879. The first history of the school by a graduate and long-time member of the faculty. In five places it reprints sections from the *Rapport à présenter* without attribution, which suggests that De Comberousse may have helped Dumas write it. On the other hand, De Comberousse is inaccurate on a number of details, especially about the early years, as Pothier demonstrated in his own later history. One of the most useful parts of the book is the set of *notices nécrologiques* reprinted at the end.

Compayré, Gabriel. *Histoire critique des doctrines de l'éducation en France depuis le seizième siècle.* 2 Vols. Paris: Hachette, 1880. Perhaps still the best survey.

Contamine, Henry. "La révolution de 1830 à Metz." *Revue d'hist. moderne* 6 (1931), 115–123. Notes that Metz was a town with strong Liberal, anti-Bourbon sentiments, but makes no mention of the role of the movement for popular education.

Coolidge, Julian L. *A History of Geometrical Methods.* Oxford: Oxford Univ. Press, 1940.

Costabel, Pierre. "L'Oratoire de France et ses collèges." in René Taton, ed. *Enseignement et diffusion des sciences en France au XVIIIe siècle.* Paris: Hermann, 1964. Revises the previously favorable view of the Oratorians' scientific education.

Crosland, Maurice. *The Society of Arcueil: A View of French Science at the Time of Napoleon.* Cambridge, Mass: Harvard Univ. Press. 1967. Essentially a collection of juxtaposed scientific biographies rather than an analysis of the institutional structure of French science.

Çrouzet, François. "Essor, déclin, et renaissance de l'industrie française des locomotives, 1838–1914. *Revue d'hist. écon. et soc.* 55 (1977).

Daumard, Adeline. "Les élèves de l'Ecole Polytechnique de 1815 à 1848." *Revue d'hist. moderne et contemp.* 3 (1958), 226–234.

Daumas, Maurice. "L'école des chimistes français vers 1840." *Chymia* 1(1948), 55–65.

Davenport, William H., and Daniel Rosenthal, eds. *Engineering: Its Role and Function in Human Society.* New York: Pergamon, 1967.

Day, C. Rod. "Education, Technology, and Social Change in France: The Short, Unhappy Life of the Cluny School, 1866–1891." *French Historical Studies* VIII:3 (Spring 1974), 427–444. As his papers show, J. B. Dumas was the major scientific patron of this school, which recruited its students in large part from the *classes populaires*, especially those attending upper parts of the primary system. Its eager, industrious students manufactured excellent scientific educations for themselves until the school fell victim to bureaucratic intrigue and the prejudices of the Ecole Normale Supérieure.

———. "The Development of Higher Primary and Intermediate Technical Education in France, 1800 to 1870." *Historical Reflections* III:2 (Winter 1976), 49–68.

———. "The Making of Mechanical Engineers in France: the Ecoles d'Arts et Métiers, 1803–1914." *French Hist. Studs.* X:3 (Spring 1978), 439–460.

———. "Technical and Professional Education in France: the Rise and Fall of Enseignement Secondaire Spécial." *Journal of Social History* (Winter 1972–1973), 177–201.

DeDainville, François. "Effectifs des collèges et scolarité aux XVIIe et

XVIIIe siècles dans le nord-est de la France." *Population* 10 (1955) 455–488.

―――. "L'enseignement des mathématiques dans les collèges jésuites de France du XVIe au XVIIIe siècle." *Revue d'histoire des sciences et de leurs applications* 7 (1954), 12–31.

―――. "L'enseignement scientifique dans les collèges des jésuites." in René Taton, *Enseignement et diffusion des sciences en France au XVIIIe siècle*. Paris: Hermann, 1964.

De Camp, L. Sprague. *The Ancient Engineers*. Cambridge, Mass: MIT Press, 1970.

Derry, T. K., and Trevor I. Williams. *A Short History of Technology*. Oxford: Oxford Univ. Press, 1961. Intended as a condensation of the five-volume history of technology published under the editorship of these authors and three others.

Domenghino, Michael C. "Events Leading to the Establishment of the Federal Polytechnic School in Zurich." *Annales cisalpines d'histoire sociale* I:2 (1971), 125–137. Around 1800, Swiss thinkers had wished to model their school on the Ecole Polytechnique. By 1855, however, the teaching of a single curriculum to all students seemed impossible, given the growth of knowledge. Thus when the Swiss institution was founded it included separate schools of architecture, civil engineering, mechanical engineering, applied chemistry and pharmacy, forestry, and "philosophical and economic studies."

Donkin, S. B. "The Society of Civil Engineers (Smeatonians)." *Transactions of the Newcomen Society* XVII (1936–7), 51–60.

Donnard, Jean-Hervé. *La vie économique et les classes sociales dans l'oeuvre de Balzac*. Paris: Armand Colin, 1961.

Donnay, Maurice. *Centrale*. Paris: Nouvelle Société d'Edition, 1930. A mixture of personal reminiscences, speculations on the current role of the engineer, and history of the school (which adds little to other accounts).

Dreyfus-Brisac, Edmond. "Les réformes de l'enseignement secondaire en France." *Revue internationale de l'enseignement* (15 January 1881), 1–24.

Dupont-Ferrier, Gustave. *Du Collège de Clermont au Lycée Louis-le-Grand*. 2 Vols. Paris: Boccard, 1925. The most detailed history of a single French secondary school. Still useful for the periods not covered in Palmer's documentary history.

Dupuy, R. Ernest. *Sylvanus Thayer: Father of Technology in the United States*. West Point: US Military Academy, 1958. Even if technology be taken as the application of science to the arts, many others can claim the title of being its father in the United States.

Durkheim, Emile. *L'évolution pédagogique en France*. Paris: Alcan, 1938. One of Durkheim's few ventures into history, but based in large part on secondary sources—and on the *Essais* of Lacroix.

―――. *Socialism*. New York: Collier Books, 1962. The surviving text of Durkheim's lectures on Saint-Simon.

Duroselle, Jean-Baptiste. "Michel Chevalier, saint-simonien." *Revue historique* LXXXII:215 (1956), 233–266. Argues that Chevalier's brief period of Saint-Simonian discipleship was far more influential upon his later thought than Chevalier himself ever admitted.

Duruy, Albert. "L'instruction publique et la Révolution." *Revue des deux mondes* 45 (1881), 851–859.

Duveau, Georges. *La pensée ouvrière sur l'éducation pendant la Seconde République et le Second Empire.* Paris: Domat-Montchrestien, 1948. Of fundamental importance to the present study.

Eckalbar, John C. "The Saint-Simonians in Industry and Economic Development." *Am. Journal of Econ. and Sociology* 38:1 (1979) 83–96.

Ecole Centrale des Arts et Manufactures. *Compte rendu des fêtes du centenaire.* Paris: de Brunoff, 1929.

Ecole Polytechnique. *Livre du centenaire, 1794–1894.* 3 Vols. Paris: Ecole Polytechnique, 1894. A history of the school as a society, the teaching of the various subjects (although this is the weakest part), biographical sketches of prominent graduates, and a history of each of the specialized technical services in which *Polytechniciens* served.

Edmonson, James. *From Mécanicien to Engineer: Technical Education and the Machine-Building Industry in Nineteenth-Century France* (PhD Dissertation, Univ. of Delaware, 1981). An original and important work nearing completion as this book was being drafted.

Eiffel, Gustave. *Discours prononcé à la distribution des prix de Sainte Barbe.* Paris: Goupy and Jourdan, 1886. The role of the engineer in economy and society has vastly increased, especially in the second half of the century. We now need personnel with special training, not just narrow functionaries. We need men "disengaged from all administrative ties," just as are both the (private) Collège Sainte-Barbe and the (State-supported but relatively autonomous) Ecole Centrale. The graduates of these schools are "thrown into life with no other support than their personal worth, with no State to protect them or assure them an honored career." Eiffel thus emphasizes one of the principal themes in the portrait of the Centrale engineer presented throughout the school's history, a theme largely dictated by the rivalry with the *Ponts* and *Mines* engineers: the *ingénieur civil*'s professional autonomy and ability to prove himself without the cushion of bureaucratic protection.

Emmerson, George S. *Engineering Education: A Social History.* New York: Crane, Russak and Co., 1973. As wide-ranging as Armytage but better written, better organized, and generally more useful.

Euvrard, F. *Historique de l'Ecole Nationale des Arts et Métiers de Châlons-sur-Marne.* Châlons-sur-Marne: Union Républicaine, 1895. A full-length, if amateur, history that adds a number of details to Guettier's study, especially on curriculum and teaching methods.

Evan, William M. "The Engineering Profession: A Cross-Cultural Analysis." In Robert Perrucci and Joel E. Gerstl, eds., *The Engineers and the Social System.* New York: John Wiley and Sons, 1969, 99–137. Compares national engineering professions along most of the usual dimensions: size of profession, membership in professional associations, migration to the United States, etc. But in some of the key tables, such as prestige ranking of the profession and number of engineering students enrolled, French data are missing. The article also foregoes an excellent opportunity to apply statistical analysis to such questions as the relation between R & D investment, patents issued, and number of engineers.

Farrell, Albert. *The Jesuit Code of Liberal Education.* Milwaukee: Denier Press, 1938. The standard presentation.

Falcucci, Clément. *L'humanisme dans l'enseignement secondaire en France.* Toulouse: Privat, 1939. A very valuable work although relatively weak on the period 1800–1848.

Farrington, Frederic E. *French Secondary Schools.* London: Longmans, 1910. One of the best short historical surveys in any language coupled with a detailed investigation of the life of the schools in the decade before World War I.

Fayet, Joseph. *La Révolution française et la science, 1789–1795.* Paris: Marcel Rivière, 1960. Argues that the Revolution did not profoundly affect the march of science because the "philosophy that inspired the revolutionary movement" had penetrated scientific institutions and "the majority of scientists" long before 1789. The matter is more complex, however, as is shown by works of Gillispie, Hahn, and Guerlac. Fayet is nevertheless a mine of information on a variety of subjects, from the Conservatoire to Marat.

Ferdinand-Dreyfus, Camille. *Un philanthrope d'autrefois: La Rochefoucauld-Liancourt.* Paris: Plon, 1903. Essential work by a skilled historian.

Ferguson, Eugene S. *Kinematics of Mechanisms from the time of Watt.* Washington, DC: US National Museum Bulletin, 1962.

Finch, James Kip. *The Story of Engineering.* Garden City, NY: Doubleday, 1960. The best brief survey.

Fisher, Marvin. *Workshops in the Wilderness: The European Response to American Industrialization, 1830–1860.* Oxford: Oxford Univ. Press, 1967. America was the testing and shaping ground for many Europeans' hopes and fears concerning the impact of industrialization. Compares the (ambivalent but generally favorable) reactions of the most prominent French engineer who visited America during this period, Michel Chevalier, with those of other visitors, but does not deal with the question of how American models were used in debates in Europe itself, a topic that bears further investigation although René Rémond's research has laid an excellent foundation for the study of the French case.

Fohlen, Claude. "France, 1700–1914." In Carlo Cipolla, ed., *The Emergence of Industrial Societies.* Part 1. London: Collins/Fontana, 1973, 7–75. A good introduction, although the attempt to adjudicate the conflict in views among Rostow, Marczewski, Crouzet, and Lévy-Leboyer is a bit too brief.

Fox, Robert W. "Scientific Enterprise and the Patronage of Research in France, 1800–1870." *Minerva* XI:4 (October 1973), 442–473. An important article that argues that State support for scientific research was a far less important factor in its advance than the personal motivations of individual scientists. The leading scientists might have put far more pressure on the government during the July Monarchy if they themselves had not been so easily diverted from research by the temptations of popular lecturing and politics, then so much in vogue. Fox also suggests that "at the very least, the classical bias of the *lycées* probably contributed to a diversion of able students from science and hence to a weakening of the scientific community."

―――. "Education for a New Age: The Conservatoire des Arts et Métiers, 1815–1830," in D. S. L. Cardwell, ed., *Artisan to Graduate.* Manchester: Manchester Univ. Press, 1974. An excellent brief study.

Geiger, Reed G. *The Anzin Coal Company, 1800–1833.* Newark, Del.: Univ. of Delaware Press, 1974. No systematic treatment of engineers as a managerial group, but some useful discussion of the role of the State mining engineers. Apparently the *porions* (overseers) were drawn from "some fifteen or twenty interrelated local families" who "claimed the same nearly hereditary right to fill Anzin's managerial ranks as did the leading *chefs de famille* to positions on the board." They probably also provided a resistance to the introduction of school-trained beginning engineers from outside their ranks. The question, then (to which Geiger does not address himself), is whether or not they sought to acquire a technical education for their sons in order to protect their position.

Genette, Gérard. "Enseignement et rhétorique au XIXe siècle." *Annales: E.S.C.* (March–April 1966), 292–305.

Gerbod, Paul. *La condition universitaire en France au XIXe siècle.* Paris: P.U.F., 1965. A study of the teachers who served in *enseignement secondaire* between 1840 and 1880, emphasizing the impact of political changes. Different categories are used for the analysis of the social origins of the teachers at the beginning of the period and at the end. A gold mine of information.

―――. *Paul-François Dubois: Universitaire, journaliste, et homme politique.* Paris: Klincksieck, 1967. A sympathetic biography that mentions neither Dubois' failure as a Polytechnique lecturer nor his role in the founding of the Ecole Centrale.

―――. *La vie quotidienne dans les lycées et collèges au XIXe siècle.* Paris: Hachette, 1968. The topical rather than chronological organization makes this a difficult work to use. Most of the material comes from the second half of the century.

―――. "La vie universitaire à Paris sous la Restauration de 1820 à 1830." *Revue d'hist. moderne et contemp.* 13 (1966), 5–48.

Gille, Bertrand. "Instruction et développement économique en France au XIXe siècle." *Annales cisalpines d'histoire sociale* I:2 (1971), 95–104. Raises a number of questions that the present study attempts to answer, such as the importance of the movement for the diffusion of useful knowledge and the social effects of the configuration of institutions of technical education. Concludes that "there would certainly be a very interesting study to be done on 'the engineer.'"

―――. *Recherches sur la formation de la grande entreprise capitaliste (1815–1848).* Paris: SEVPEN, 1959.

Bibliography

————. *The Renaissance Engineers*. London: Lund Humphries, 1966.

Gillispie, Charles C. *The Edge of Objectivity: An Essay in the History of Scientific Ideas*. Princeton: Princeton Univ. Press, 1960. Gillispie's seductively elegant style occasionally lends a certain opacity to his presentation, as a comparison of his discussion of Sadi Carnot with that of Cardwell reveals.

————. *Lazare Carnot: Savant*. Princeton: Princeton Univ. Press, 1971. Reinhard made no serious attempt to consider Carnot's contribution to science, but Gillispie fills the gap admirably.

————. "Science and Technology." *New Cambridge Modern History*, IX. Cambridge, England: Cambridge Univ. Press, 1957.

Gillmor, C. Stewart. *Coulomb and the Evolution of Physics and Engineering in Eighteenth-Century France*. Princeton: Princeton Univ. Press, 1971. A model study that will not be replaced for a long time.

Gilpin, Robert. *France in the Age of the Scientific State*. Princeton: Princeton Univ. Press, 1968. Not always reliable in its historical details.

Gobron, Louis. *Législation et jurisprudence de l'enseignement public et de l'enseignement privé en France*. Paris: Larose, 1896.

Gontard, Maurice. *L'enseignement primaire en France de la Révolution à la loi Guizot*. Paris: Les Belles Lettres, 1959. The basic work.

Goode, G. Brown. "The Origin of the National Scientific and Educational Institutions of the United States." *Papers of the American Historical Association*. IV. New York: G. P. Putnam's, 1890, 5–112. Still a valuable source.

Gouhier, Henri. *La jeunesse d'Auguste Comte et la formation du positivisme*. 3 Vols. Paris: J. Vrin, 1933–1941. A monumental study and a gold mine of information on both Saint-Simon and Comte.

Les grandes industries modernes et les Centraux. Paris: Brunoff, 1929. A centenary volume listing the achievements of Centrale graduates in various sectors.

Graves, Norman J. *Education for Industry and Commerce in French Public Elementary and Secondary Schools During the Nineteenth Century*. PhD Thesis: Univ. of London, 1964.

————. "The 'Grandes Ecoles' in France." *The Vocational Aspect* XVII:36 (Spring 1965), 41–49.

————. "Technical Education in France in the Nineteenth Century." *The Vocational Aspect* XVI:34 (1964), 148–159. Argues that the *écoles d'arts et métiers* were slow in responding to the needs created by French industrialization.

Guerlac, Henry. "Science and French National Strength." In Edward M. Earle, *Modern France: Problems of the Third and Fourth Republics*. New York: Russell and Russell, 1964, 81–105. A good short introduction to the history of French science in the nineteenth century that emphasizes the individualistic character of French scientists and the power of the pure-science ideal within the culture. Should be read in conjunction with Robert Fox's more recent article.

————. "Vauban: The Impact of Science on War." In Edward M. Earle, *Makers of Modern Strategy*. Princeton: Princeton Univ. Press, 1943, 26–48.

Guettier, A. *Histoire des écoles nationales d'arts et métiers*. Paris: Dejey, 1880. A centenary volume that is still one of the most useful studies of the schools. Prints 200 pages of documents selected from the schools' archives.

Guillermin, V., and M. Gillot. *L'Ecole Nationale des Mines de Saint-Etienne*. Saint-Etienne: Société des Anciens Elèves, 1921. Brief sketch by an alumnus, but still the most accessible source on the history of the school. Began teaching about railroads in 1836.

Guillet, Léon. *Cent ans de la vie de l'Ecole Centrale des Arts et Manufactures*. Paris: De Brunoff, 1929. Centenary volume by the director of the school, a distinguished metallurgist. Some of the details on the earlier period are not mentioned in Pothier, but the latter work still stands as the single best source on the early years of the school.

Guinot, Jean-Pierre. *Formation professionnelle et travailleurs qualifiés depuis 1789*. Paris: Domat, N.D. Most of the material is covered more thoroughly by Duveau.

Harant, Henri. *Enseignement populaire d'après les documents contenus dans la classe 90 de l'exposition universelle de 1867*. Paris: Eug. Lacroix, 1867. Primary education recruits from "the children of the *classe peu aisée de la nation*," which includes workers in the cities, workers in the country side, artisans, merchants, *petits industriels* [!!], and lower employees. Mentions Dupin as the founder of adult education.

Hayek, Friedrich A. *The Counterrevolution of Science*. New York: The Free Press, 1955. Hayek argues that the authoritarianism of the Saint-Simonians and the followers of Comte was derived from their application of a model of science to social thought that stressed the authoritative determination of a single set of goals. Social action then consisted of organizing men according to the one best way to achieve this goal. The engineer's control of all parts of the process was essential. Hayek opposes to this a market model in which the best results are obtained by the free exchange of goods and services among agents in the process. Whether engineers in the first half of the century held to an authoritarian, plan-rational model as opposed to a more liberal market-rational model is a question not yet adequately answered, but men like Dumas and the later Chevalier certainly allowed for the play of market forces in their social philosophies. In explaining the authoritarianism of the Saint-Simonians, moreover, Iggers points to the influence of De Maistre and the early Lamennais rather than scientific models.

Hazen, William B. *The School and the Army in Germany and France*. New York: Harper, 1872.

Herivel, J. W. "Aspects of French Theoretical Physics in the Nineteenth Century." *British Journal for the History of Science* 3:10 (1966), 109–132. Emphasizes the deleterious effects of scientists' increasing involvement in politics and public affairs.

Bibliography

Higonnet, Patrick. "La composition de la Chambre des Députés de 1827 à 1831." *Revue historique* CCXXIX:2, 351–378.

Hinton, Lord, of Bankside. *Engineers and Engineering.* Oxford: Oxford Univ. Press, 1970.

Hughes, Thomas P. "Commentary." In Monte Wright and Lawrence Paszek, eds., *Science, Technology, and Warfare.* Washington: USAF HQ, 1969, 69–74.

————, ed. *Lives of the Engineers: Selections from Samuel Smiles.* Cambridge, Mass: MIT Press, 1966. Excellent introductory essay.

Institut National de la Statistique et des Etudes Economiques. "L'enquête par sondage sur l'emploi." *Bulletin mensuel de statistique.* N.S. Supplément (April–June 1953). Of methodological value.

Isambert-Jamati, Viviane. "L'autorité dans l'éducation française." *European J. of Sociol.* 6:2 (1965), 149–166.

————. *Crises de la société, crises de l'enseignement.* Paris: P.U.F., 1970.

————. "La rigidité d'une institution: Structure scolaire et systèmes de valeurs." *Revue française de sociologie* VII (1966), 306–347.

Janet, Paul. *Victor Cousin et son oeuvre.* Paris: Calmann Lévy, 1885. Stresses the importance of Cousin's lectures in the period 1817–1820 for the development of his later thought. Some of the key ingredients in the ideology of *culture générale* were thus being formulated at the same time that the *baccalauréat* was entrenching itself. The whole subject bears further investigation.

Jefferson, Carter. "Worker Education in England and France, 1800–1914." *Comp. Studs. Society and History* 6:3 (April 1964), 345–366.

Jewkes, John, David Sawers, and Richard Stillerman. *The Sources of Invention.* London: Macmillan, 1969. Detailed case studies of some of the most important technological innovations in the twentieth century. Few deal with France, however. Comparable studies for the nineteenth century are sorely lacking.

Joachim, Jules. *L'Ecole centrale du Haut-Rhin à Colmar.* Colmar: Paul Hartmann, 1935.

Julia, Dominique, et Paul Pressly "La population scolaire en 1789: Les extravagances du Ministère Villemain." *Annales: E.S.C.* (November–December 1975), 1516–1561.

Karady, Victor. "L'expansion universitaire et l'évolution des inégalités devant la carrière d'enseignant au début de la IIIe République." *Revue française de sociologie* XIV (1973), 443–470.

————. "Normaliens et autres enseignants à la Belle Epoque." *Revue française de sociologie* XIII (1972), 35–58.

Kemper, John D. *The Engineer and His Profession.* New York: Holt, Rinehart, and Winston, 1967.

Kindleberger, Charles. "Technical Education and the French Entrepreneur." In Edward C. Carter II, Robert Forster, and Joseph N. Moody, eds., *Enterprise and Entrepreneurs in Nineteenth- and Twentieth-Century France.* Baltimore: Johns Hopkins Univ. Press, 1976, 3–40. The best discussion to date.

Klemm, Friedrich. *A History of Western Technology*. Cambridge, Mass: MIT Press, 1964. Largely superseded by the Derry and Williams volume.

Kline, Morris. *Mathematics in Western Culture*. Oxford: Oxford Univ. Press, 1953. Excellent chapters on mathematics as an influence on science, art, religion, and literature, but nothing on mathematics as a social cachet.

Kuhn, Thomas S. "Energy Conservation as an Example of Simultaneous Discovery." In Marshall Clagett, ed., *Critical Problems in the History of Science*. Madison: Univ. of Wisconsin Press, 1959, 321–356.

Laurent, Charles. *J.-L.-C. Dauban: Notice biographique*. Saint-Nicolas: Trenel, 1868. Dauban was a director of the *école d'arts et métiers* at Angers during its early years. His father had helped Lafayette organize the National Guard during the Revolution of 1789. At the age of ten he became an orphan. He was sent to the *école d'arts et métiers* at Compiègne in 1802, staying on after graduation as a teacher in the workshops. In the final years of the Empire he organized a *compagnie franche* of Bonapartists that clashed with the Bourbon sympathizers in the area of the school. "Too compromised by his patriotism," he lost his job under the Bourbons. In 1817, however, he obtained a post as a tutor in the Collège Henri IV. The July Revolution brought to power friends from his days at Henri IV, and he was appointed director of the Angers school in 1830.

Lavollée, C. "L'Ecole Centrale des Arts et Manufactures." *Revue des deux mondes*. (15 May 1872), 415–438. The identity of the author remains a mystery. Lavallee did not die until 1873, so it is possible that this was either a transparent pseudonym or a double typographical error. It is a completely favorable article, but the information in it would have been easily available to anyone through published documents.

Layton, Edwin T., Jr. *The Revolt of the Engineers: Social Responsibility and the American Engineering Profession*. Cleveland: Case Western Reserve Press, 1971. The early chapters contain an especially useful discussion of the ideology of engineering, especially its assumption of the equivalence of moral laws and the laws of nature.

———. "Millwrights and Engineers: Science, Social Roles, and the Evolution of the Turbine in America." In Wolfgang Krohn, Edwin T. Layton, Jr., and Peter Weingart, eds., *The Dynamics of Science and Technology: Social Values. Technical Norms and Scientific Criteria in the Development of Knowledge*. Boston: D. Reidel, 1978.

———. "Mirror-Image Twins: The Communities of Science and Technology in 19th-Century America." *Technology and Culture* 12:4 (October 1971), 562–580. Of fundamental importance for the argument of chapter 4.

———. "Science, Business, and the American Engineer." In Robert Perrucci and Joel E. Gerstl, eds., *The Engineers and the Social System*. New York: John Wiley and Sons, 1969, 51–72. The best short statement of the place of the engineer in the American social structure.

———. "Technology as Knowledge." *Technology and Culture*. 15:1 (January 1974), 31–41.

Bibliography

Leblanc, René. *L'enseignement professionel en France.* Paris: Cornely, 1905.

Lechner, Alfred. *Geschichte der technischen Hochschule in Wien, 1815–1840.* Vienna: Technische Hochschule in Wien, 1942.

Léon, Antoine. "Promesses et ambiguités de l'oeuvre d'enseignement technique en France, de 1800 à 1815." *Revue d'hist. moderne et contemp.* XVII (July–September 1970), 846–859.

———. *La Révolution française et l'éducation technique.* Paris: Société d'Etudes Robespierristes, 1968. An important comparative institutional history that updates Artz in many areas. Emphasizes the persistence of preindustrial goals for technical education, such as the provision of livelihoods to the impecunious sons of military veterans and the maintenance of artisanal specialties and apprenticeship patterns. Léon does not deal directly, however, with the schools' contribution to the development of the role of *ingénieur.*

Lemaistre, Alexis. *Potaches et bachots.* Paris: Firmin-Didot, 1893. The history of the secondary schools as folklore.

Lévy-Leboyer, Maurice. "La croissance économique en France au XIXe siècle." *Annales: E.S.C.* XXIII (1968), 788–807.

———. "La décélération de l'economie française dans la seconde moitiée du XIXe siècle." *Revue d'hist. économique et sociale* 4 (1971), 485–507.

———. "Le patronat français a-t-il été malthusien?" *Le Mouvement social* 88 (July–September 1974).

Lexis, Wilhem. *Die Technischen Hochschulen im Deutschen Reich.* Berlin: A. Asher, 1904. The fourth volume of his survey of the history of German education. Still the most valuable single source for the history of the *Technischen Hochschulen* in the nineteenth century.

Liard, Louis. *L'enseignement supérieur en France, 1789–1893.* Paris: Armand Colin, 1894. The most valuable single work.

Locke, Robert R. "Industrialisierung und Erziehungssystem in Frankreich und Deutschland vor dem l. Weltkrieg." *Historische Zeitschrift* 225 (1977), 265–296.

Lundgreen Peter. *Bildung und Wirtschaftswachstum im Industrialisierungsprozess des 19. Jahrhunderts.* Berlin: Colloquium, 1973.

———. "Technicians and the Labour Market in Prussia, 1810–1850." *Annales cisalpines d'histoire sociale.* I:2 (1971), 9–29.

Lundgreen, Peter, and Wolfram Fischer. "The Recruitment and Training of Administrative and Technical Personnel." In Charles Tilly, ed., *The Formation of National States in Europe.* Princeton: Princeton Univ. Press, 1975. A useful comparative treatment marred by syntactic errors and errors in diction that make certain passages all but incomprehensible. If, as I suspect, neither of the authors is a native speaker of English, they should at least have been given good editors.

Magendie, Maurice. *La politesse mondaine et les théories de l'honnêteté en France au XVIIe siècle.* Paris: P.U.F., 1925.

Manegold, Karl-Heinz. "Eine Ecole Polytechnique in Berlin." *Technikgeschichte* 33:2 (1966), 182–6.

———. "Das Verhältnis von Naturwissenschaft and Technik im 19. Jahrhundert im Spiegel der Wissenschaftsorganisation." *Technikgeschichte in Einzeldarstellungen* 11, 141–187.

Maneuvrier, Edouard. *L'éducation de la bourgeoisie.* Paris: Cerf, 1888.

Manuel, Frank E. *The Prophets of Paris,* Cambridge, Mass.: Harvard Univ. Press, 1962.

———. *The New World of Henri Saint-Simon.* Cambridge, Mass.: Harvard Univ. Press, 1964. The starting point for all studies of the Saint-Simonian phenomenon.

Mason, E. S. "Saint-Simonism and the Rationalisation of Industry." *Quarterly Journal of Economics* XLV (1930–1931), 640–683.

Mayer, Jean. "Science." In Julian Park, ed., *The Culture of France in Our Time.* Ithaca: Cornell Univ. Press, 1954. Discussion of importance of mathematics in French culture.

Mazoyer, Louis. "La bourgeoisie du Gard et l'instruction au début de la monarchie de Juillet." *Annales d'hist. économique et sociale* (1964), 20–39. Argues that although the Protestant bourgeoisie led the way, by the end of the July Monarchy both Protestant and Catholic bourgeoisies had launched a joint effort to bring enlightenment—hence social peace—to the common people.

McKie, Douglas. "Science and Technology." In *New Cambridge Modern History* XII. Cambridge, England: Cambridge Univ. Press, 1957, 87–111.

Melon, Paul. *L'enseignement supérieur et l'enseignement technique en France.* Paris: Armand Colin, 1893. Contains probably the best survey of nineteenth-century French research facilities.

Mendelsohn, Everett. "The Emergence of Science as a Profession in Nineteenth-Century Europe." In Karl Hill, ed., *The Management of Scientists.* Boston: Beacon Press, 1964.

Merz, John Theodore. *A History of European Thought in the Nineteenth Century.* 4 Vols. New York: Dover Publications, 1965. Originally published in 1911. One of the great works of wide-ranging nineteenth-century erudition. A masterful survey. And yet at the time he wrote it, Merz was a full-time practicing chemical engineer living in Scotland.

Meuriot, Paul. *Le baccalauréat.* Nancy: Drion, 1919. A brief survey of the history of the degree.

Ministère du Commerce, de l'Industrie, des Postes, et des Télégraphes. *L'Ecole Centrale des Arts et Manufactures.* Paris: Librairie des Arts et Manufactures, 1903. An *annuaire* that contains a statistical breakdown of the various fields in which *Centraux* were active, but not much else of interest.

———. *L'enseignement technique en France.* Paris: Imprimerie Nationale, 1900. This volume, edited by Paul Buquet, contains several valuable articles, especially one by Buquet on the Conservatoire des Arts et Métiers.

Monmartin, Antonin. *Précis sur l'Ecole La Martinière.* Lyons: Perrin, 1862. Extremely brief discussion of the school's history, but valuable for containing a full set of the school's regulations.

Bibliography

Monod, Gabriel. "Les idées de Michelet sur les réformes de l'enseignement en 1845, d'après des notes inédites." *Revue universitaire* (1904), 315–321.

Moody, Joseph N. *French Education Since Napoleon* Syracuse: Syracuse Univ. Press, 1978. The most recent survey of France alone. Should be read in conjunction with the relevant chapters in Ringer's comparative study.

Mouchelet, E. *Notice historique sur l'Ecole Centrale des Arts et Manufactures.* Paris: Dunod, 1913. The only work on the school to appear in the forty-three years between Pothier and Guillet. The slightest of the histories. Yet there are a few details not reported in the others.

Muret, M. G. "Antoine de Chézy, histoire d'une formule d'hydraulique." *Annales des Ponts et Chaussées* 2 (1921), 165–269.

Musson, A. E., and Eric Robinson. *Science and Technology in the Industrial Revolution.* Toronto: Univ. of Toronto Press, 1969.

Nash, Leonard K. *The Atomic-Molecular Theory.* Cambridge, Mass.: Harvard Univ. Press, 1973.

Naville, Pierre. "Technical Elites and Social Elites." *Sociology of Education* 371 (Fall 1963), 27–29.

Neuschwander, Claude. *L'Ecole Centrale des Arts et Manufactures.* Casablanca: Lacour, 1960. The most recent history, very much of the coffee-table-book-for-alumni variety. Adds little to previous studies.

Newman, Edgar. "The Blouse and the Frock Coat: The Alliance of the Common People of Paris with the Liberal Leadership and the Middle Class during the Last Years of the Bourbon Restoration." *Journal of Modern History* 46 (March 1974), 26–59. Claims that the role of students in the Revolution has been exaggerated and that the feelings of solidarity between workers and bourgeoisie were quite genuine, despite the fact that the bourgeoisie "offered nothing" to the workers. Perhaps, in fact, the workers saw them as teachers?

Oberlé, Raymond. *L'enseignement à Mulhouse de 1798 à 1870.* Paris: Les Belles Lettres, 1961. Indispensable.

O'Boyle, Lenore. "Klassische Bildung and soziale Struktur in Deutschland zwischen 1800 und 1848." *Historische Zeitschrift* 207 (1968), 584–608.

———. "The Problem of an Excess of Educated Men in Western Europe, 1800–1850." *Journal of Modern History* 42:4 (December 1970), 471–495. Opens up an interesting question without really resolving it.

Ocagne, Mortimer d'. *Les grandes écoles de France.* Paris: Hetzel, 1873. One of the most popular guides, which includes a short history of each school. Reissued in 1887.

Outram, Dorinda. "The Language of Natural Power: The *Eloges* of Georges Cuvier and the Public Language of Nineteenth-Century Science." *History of Science* 16 (September 1978), 153–178.

Padberg, John W. *Colleges in Controversy: The Jesuit Schools in France from Revival to Suppression, 1815–1880.* Cambridge, Mass.: Harvard Univ. Press, 1969.

Pacquier, J.-B. *L'enseignement professionel en France: Son histoire, ses différentes formes, ses résultats.* Paris: Hetzel, 1908. A convenient survey of the various schools with occasional information on parents' occupations and the career choices of students.

Paul, Harry W. "The Issue of Decline in Nineteenth-Century French Science." *French Historical Studies* VII:3 (Spring 1972), 416–450. The first shot in a full-scale revision of the previous view of the trajectory of French science in the nineteenth century. Should be read as an antidote to Ben-David and Gilpin.

————. *The Sorcerer's Apprentice: The French Scientists' Image of German Science, 1840–1919.* Gainesville, Fla.: Univ. of Florida Press, 1972.

Palmer, Robert R. "Free Secondary Education in France before and after the Revolution." *History of Education Quarterly* (1974), 437–452.

Perreux, Gabriel. *Au temps des sociétés secrètes: La propagande républicaine au début de la Monarchie de Juillet.* Paris: Hachette, 1931.

Perrot, Marguerite. *La mode de vie des familles bourgeoises.* Paris: Armand Colin, 1961. Very useful for the history of social stratification.

Petot, Jean. *Histoire de l'administration des ponts et chaussées.* Paris: Riviere, 1958. An excellent study, well balanced and well documented, which ends just as the story gets really interesting: after the fall of the First Empire.

Picavet, François. *Les Idéologues.* Paris: Alcan, 1891. Fundamental.

Pierre, Victor. *L'école sous la Revolution.* Paris: Douniol, 1881.

Pigeire, Jean. *La vie et l'oeuvre de Chaptal.* Paris: Domat-Montchrestien, 1931. Uncritical, innocent of interpretive themes, but still the best work on Chaptal available.

Pinet, Gaston. "La fondation de l'Ecole Polytechnique." *Revue scientifique* 11 (15 March 1884), 332–337.

————. *Histoire de l'Ecole Polytechnique.* Paris: Baudry, 1887. By far the most useful history for the period of the July Monarchy. Includes material from a large number of interviews with *Polytechniciens* who attended the school during those years.

Piobetta, Jean Baptiste. *Le baccalauréat de l'enseignement secondaire.* Paris: Baillière, 1937. The most useful history to date, but hardly the last word. Chiefly an administrative history of the degree. Publishes 600 pages of documents.

Poincaré, Henri. *Science and Method.* London: Dover, 1957.

Pollard, Sidney. *The Genesis of Modern Management.* Baltimore: Penguin Books, 1968. Good chapters on technical education and recruitment of managers.

Poncetton, François. *Eiffel: Le magicien du fer.* Paris: Editions de la Tournelle, 1939. The most complete biography to date, but uncritical and undocumented. It is clear, however, that Poncetton had access to the Eiffel family records.

Poteil, Felix. *Histoire de l'enseignement, 1789–1965.* Paris: Sirey, 1966.

More conventional in its approach than Prost, but more extensive bibliographies.

Porter, Whitworth. *History of the Corps of Royal Engineers*. 2 Vols. London: Longmans, 1889.

Pothier, Francis. *Histoire de l'Ecole Centrale des Arts et Manufactures*. Paris: Delamotte, 1887. Clearly the most important single source for the history of the Ecole Centrale. Pothier, Lavallée's son-in-law, had studied under many of the teachers he describes. He also had access to two sets of documents not available to the present writer: Olivier's private register and the records of the *Conseil d'ordre*. Apart from an understandable circumspection in his criticisms of individuals, Pothier demonstrates no more than the usual old boy's partiality for his school. The work is not really a connected history, however, but a chronicle, reviewing events year by year in no particular order. Like De Comberousse, Pothier publishes a large number of relevant documents, but in his case these were taken from the school's archives.

Rae, John B. "Engineers Are People." *Technology and Culture* 16:2 (July 1975), 404–418.

Ratcliffe, Barrie M. "The Economic Influence of the Saint-Simonians: Myth or Reality?" *Proc. Annual Meeting Western Soc. French Hist.* 5 (1977: publ. 1978), 252–262.

Ringer, Fritz K. *The Decline of the German Mandarins*. Cambridge, Mass.: Harvard Univ. Press, 1969.

Ringer, Fritz K. *Education and Society in Modern Europe*. Bloomington, Ind.: Indiana Univ. Press, 1979. First full-length comparative work on the social history of education in France, Germany, Britain, and the United States. Indispensable.

Roderick, Gordon W., and Michael D. Stephens. *Scientific and Technical Education in 19th Century England*. New York: Barnes and Noble, 1972. A series of narrowly conceived local studies.

Roy, Joseph-Antoine. *Histoire de la famille Schneider et du Creusot*. Paris: Marcel Rivière, 1962. Some material about the Schneiders' activities in promoting popular education, but not very much about their own schooling, or that of their managers.

Royle, Edward. "Mechanics' Institutes and the Working Classes, 1840–1860." *The Historical Journal* XIV:2 (1971), 305–321. Argues against the previous view that the Mechanics' Institutes failed and that they benefited only the middle classes. An important counterbalance to the work of Brian Simon, E. P. Thompson, and Engels.

Ruhlmann, Georges. *Cinq siècles au Collège Sainte-Barbe: 1460–1960*. Paris: Association Amicale des Anciens Barbistes, 1960. Not really a history but a collection of biographies of prominent alumni. No significant history of the school has been written since Quicherat.

Sadi Carnot et l'essor de la thermodyamique. Paris: CNRS, 1976. Important collection of papers from a CNRS colloquium on the 150th anniversary of the publication of the *Réflexions sur la puissance motrice du feu*. Five papers deal with aspects of engineering education in Carnot's time.

Saint-Martin, Monique de. *Les fonctions sociales de l' enseignement scientifique*. Paris: Mouton. Scientific education, as currently organized, serves to perpetuate *la classe dirigeante* in contemporary France. And elsewhere?

Sanderson, Michael. "Literacy and Social Mobility in the Industrial Revolution in England." *Past and Present* 56 (August 1972), 75–103.

———. *The Universities and British Industry, 1850–1970*. London: Routledge and Kegan Paul, 1972.

Schmidt, Charles. *La réforme de l'Université impériale en 1811*. Paris: Armand Colin, 1906.

Scholer, Walter. *Geschichte des naturwissenschaftlichen Unterrichts*. Berlin: Walter de Gruyter, 1970.

Schwickerath, Joseph. *Jesuit Education: Its History and Principles*. St. Louis: Herder, 1903.

Shapin, Steven, and Barry Barnes. "Science, Nature, and Control: Interpreting Mechanics' Institutes." *Social Studies of Science* 7 (1977), 31–74. Compares the British pattern briefly with America. Further research on the French diffusion of useful knowledge is badly needed.

Shinn, Terry. "Des Corps de l'Etat au secteur industriel: Génèse de la profession d'ingénieur, 1750–1920." *Rev. Franc. de social.* XIX (1978), 39–71. Reprinted in the work listed next.

Shinn, Terry. *L'Ecole Polytechnique 1794–1914: Savoir scientifique et pouvoir social*. Paris: Presses la fondation nationale des sciences politiques, 1980. Of major importance for the "social recruitment" comparisons in chapter 3 of this study.

Shy, John. "Western Military Education, 1700–1850." In Monte D. Wright and Lawrence J. Paszek, eds., *Science, Technology, and Warfare*. Washington, D.C.: USAF HQ, 1969, 60–68.

Sicard, Augustin. *Les études classiques avant la Revolution*. Paris: Pettin, 1887.

Simon, Brian. *Studies in the History of Education, 1780–1870*. London: Lawrence and Wishart, 1969. Marxist essays on the interaction between education and class structure in England.

Sinclair, Bruce. *Philadelphia's Philosopher Mechanics: A History of the Franklin Institute, 1824–1865*. Baltimore: Johns Hopkins Univ. Press, 1974.

Smiles, Samuel. *Lives of the Engineers*. 3 Vols. London: John Murray, 1862.

Snyders, Georges. *La pédagogie en France aux XVIIe et XVIIIe siècles*. Paris: P.U.F., 1965. Psychologically informed essay of major importance.

Société des Ingénieurs Civils. *Annuaire de 1911*. Paris: Hotel de la Société, 1911. Contains important historical essays by Benoit-Duportail and Mallet.

———. *L'enseignement technique superieur devant la Société des Ingénieurs Civils*. Paris: Hotel de la Société des Ingénieurs Civils, 1917. Reports the long debates over engineering education during the period from November 1916 to July 1917 held at the Société. The importance of *enseignement secondaire* was stressed by numerous speakers, as was the necessity for avoiding specialization.

Sociéte des Ingénieurs Civils. *125 ans de progrès technique: 1848–1973.* A few details on the founding of the Société in the opening pages.

Soubeiran, Max. *Etudes sur les ecoles pratiques de commerce et de l'industrie.* Paris: Nony, 1900.

Stephens, Michael D., and Gordon Roderick. "Science, the Working Classes, and the Mechanics' Institutes." *Annals of Science* 29:4 (December 1972), 349–360.

Straub, Hans. *A History of Civil Engineering.* London: Leonard Hill, 1952.

Suleiman, Ezra N. *Elites in French Society: The Politics of Survival.* Princeton: Princeton Univ. Press, 1978.

Sutter, Jean. "L'évolution de la taille des Polytechniciens." *Population* 13:3 (July–September 1958), 373–406. In the course of an investigation of the changes in *Polytechniciens'* height over the course of the nineteenth century, Sutter tabulated their social origins.

Talbott, John E. *The Politics of Educational Reform in France, 1918–1940.* Princeton: Princeton Univ. Press, 1969. The first chapter is still one of the best short introductions to the social role of *enseignement secondaire* in the nineteenth century.

Taton, René. "Algebra and Geometry." In Taton, ed., *Science in the Nineteenth Century.* New York: Basic Books, 1965.

Taton, René. *L'histoire de la géométrie descriptive.* Paris: Université de Paris, 1954.

———. "L'Ecole Polytechnique et le renouveau de la géométrie analytique." In *L'aventure de la science: Mélanges Alexandre Koyré.* Paris: Hermann, 1964, 552–564.

———. "Laplace et Sylvestre-François Lacroix." *Revue d'hist. des sciences* VI:4 (October–December 1953), 273–285.

———. *L'oeuvre scientifique de Monge.* Paris: P.U.F., 1951.

Trénard, Louis. "L'enseignement secondaire sous la Monarchie de Juillet: Les réformes de Salvandy." *Revue d'hist. moderne et contemp.* 12 (April–June 1965), 81–133. Better on the substance of the curriculum debates than on the relation between education and social stratification.

Timoshenko, Stephen P. "The Development of Engineering Education in Russia." *The Russian Review* 15:1 (January 1956), 173–185. Discusses the role of French engineers such as Lamé and Clapeyron in establishing the Russian equivalent of the Ecole des Ponts et Chaussées.

———. *A History of the Strength of Materials.* New York: Columbia Univ. Press, 1953.

Troux, Albert. *L'école centrale du Doubs à Besancon.* Paris: Felix Alcan, 1926. One of the best of the *école centrale* monographs.

Tudesq, Andre. *Les grands notables en France (1840–1849),* 2 Vols. Paris: P.U.F., 1964.

Usher, Abbott Payson. *A History of Mechanical Inventions.* Cambridge, Mass.; Harvard Univ. Press, 1970.

Vallery-Radot, René. *La vie de Pasteur.* Paris: Hachette, 1918. The "official" biography.

Van Duzer, Charles H. *The Contribution of the Ideologues to French Revolutionary Thought.* Baltimore: Johns Hopkins Univ. Press, 1935.

Vaughan, Michalina. "The 'Grandes Ecoles.'" In Rupert Wilkinson, ed., *Governing Elites.* Oxford: Oxford Univ. Press, 1969.

Vaughan, Michalina, and Margaret S. Archer. *Social Conflict and Educational Change in England and France, 1789–1848.* Cambridge: Cambridge Univ. Press, 1971. An exercise in rather muscular retrospective sociology that focuses on the struggle for ideological domination of French society through control of religious education in the schools.

Velluz, Léon. *Histoire brève de la chimie.* Paris: Maloine, 1966.

Vial, Francisque, *Trois siècles d'histoire de l'enseignement secondaire.* Paris: Delagrave, 1936. A testimony to Lacroix's influence and a defense of the *cours.*

Vignery, Robert J. *The French Revolution and the Schools: Educational Policies of the Mountain, 1792–1794.* Madison: State Historical Society of Wisconsin, 1965.

Webb, R. K. *The British Working-Class Reader, 1790–1848.* London: George Allen and Unwin, 1955. Good passages on how the movement for the diffusion of useful knowledge attempted to suppress "politics" while justifying technological change.

Weill, Georges. *L'école saint-simonienne.* Paris: Alcan, 1896. A lively narrative history of the movement.

———. *Histoire de l'enseignement secondaire en France.* Paris: Payot, 1921. Still the best short overview.

———. "Les théories saint-simoniennes sur l'éducation." *Revue internationale d'enseignement* XXXI (1897), 237–246.

Wilkinson, Rupert. *Gentlemanly Power: The Public Schools and the Victorian Tradition.* New York: Barnes and Noble, 1962. A provocative portrait of the formation of the British political elite that emphasizes the "totalitarian" tendency of sports and the prefect system to "internalize" conformity to the traditions of the group. Comparisons with Jesuit schools and with the education of Chinese mandarins.

Williams, L. Pearce. *Michael Faraday.* New York: Basic Books, 1964. Sympathetic and technically informed portrait of a great scientist who also took seriously his mission to diffuse useful knowledge.

———. "Science, Education, and Napoleon I." In Leonard M. Marsak, ed., *The Rise of Science in Relation to Society.* New York: Macmillan, 1964, 80–90.

———. "Science, Education, and the French Revolution." *Isis* 44 (December 1953), 311–330.

Wolf, A. *A History of Science, Technology, and Philosophy in the 18th Century.* New York: Harper and Row, 1961.

Wylie, Laurence W. *Saint-Marc Girardin, Bourgeois.* Syracuse: Syracuse Univ. Press, 1947.

Zeldin, Theodore. "Higher Education in France, 1848–1940." *Journal of Contemporary History* 2:3 (July 1967), 53–80.

Ziman, John. *Public Knowledge: The Social Dimension of Science.* Cambridge, England: Cambridge Univ. Press, 1968.

Index